The Wealth of Forests

Edited by Chris Tollefson

The Wealth of Forests:
Markets, Regulation, and
Sustainable Forestry

UBCPress / Vancouver

Printed in Canada on acid-free paper ∞

ISBN 0-7748-0682-6

Canadian Cataloguing in Publication Data

Main entry under title:

The wealth of forests

  Includes bibliographical references and index.
  ISBN 0-7748-0682-6

  1. Sustainable forestry – British Columbia. 2. Forest policy – British Columbia.
  3. Forestry law and legislation — British Columbia. I. Tollefson, Chris, 1959-
  SD146.B7W42 1998   333.75′09711   C98-910800-7

UBC Press gratefully acknowledges the ongoing support to its publishing program from the Canada Council for the Arts, the British Columbia Arts Council, and the Multiculturalism Program of the Department of Canadian Heritage.

UBC Press
University of British Columbia
6344 Memorial Road
Vancouver, BC V6T 1Z2
(604) 822-5959
Fax: 1-800-668-0821
E-mail: orders@ubcpress.ubc.ca
http://www.ubcpress.ubc.ca

Contents

Preface

These are turbulent, unpredictable, and yet opportune times for Canadian forestry. Never before have competing demands on our forest resources been so great. Simultaneously, we are finally being forced to confront and reckon with the sustainable limits of these resources. Due to these and various other forces, the improbable has happened: government, industry, First Nations, and the NGO sector now appear to be part of an emerging consensus that the current pattern of industrial forestry in Canada must change. This book is about defining and responding to this challenge.

The Wealth of Forests evolved out of a workshop sponsored by Environment Canada that was held at the Faculty of Law at the University of Victoria in June 1995. The goal of the workshop was to identify innovative policy options for the protection of forest resources. The range of opinion at the workshop was diverse. There was a spirited discussion of the BC Forest Practices Code of British Columbia, which had come into force earlier that month. Critical debate also focussed on a variety of other groundbreaking BC government-sponsored forest policy initiatives, including the Commission on Resources and Environment (CORE) land use zoning process, stumpage reform, the creation of Forest Renewal BC (FRBC), and pending changes to the allowable annual cut determination process.

As editor, I have tried to corral, without unduly curtailing, the lively ideological and cross-disciplinary debates that emerged at this workshop and that have continued as this collection has taken shape. In this collection, readers will discern that many significant differences in perspective, analysis, and prescription remain. Notably, however, many shared conclusions have emerged about where we are going, what challenges lie ahead, and what must be done.[1]

Precisely where the growing pressures to reform Canadian forestry will lead is no more certain today than when we embarked on this project. Over the past year, the BC government, in response to a softening market for BC forest products and the Asian crisis, has retreated from many of the ambitious sustainable forestry initiatives it introduced in the mid-1990s – by, among other things, rolling back forest practice regulation and decreasing stumpage rates.

Another pressure for change has been the 1997 decision of the Supreme Court of Canada in *Delgamuukw*. In this decision, for the first time the country's highest court held that First Nations are entitled to make land claims against untreatied, unalienated Crown land, which in BC represents over 90 percent of the land base. This decision has sparked renewed debate on the question of tenure reform. For different reasons, tenure reform has emerged as an issue in Ontario largely at the behest of a forest industry eager to capitalize on the deregulatory inclinations of the Mike Harris government. It is expected that this push for the privatization of Crown forest lands will begin in earnest once Ontario's newly launched land zonation process, known as "Lands for Life," concludes in early 1999.

Given the complexities of, and competing pressures for, changing the way we do forestry, can we be confident about government's ability to provide leadership? Intriguingly, leadership may be emerging from new quarters. Earlier this month, MacMillan Bloedel announced that, in response to changing consumer preferences and market conditions, it will phase out clearcut logging and embark on a new stewardship strategy of old-growth and habitat conservation. Three years ago, when we commenced this project, the prospect of BC's largest forest company making a public commitment of this kind could scarcely have been imagined.

As both the momentum and need for change builds, let us offer the hope that a spirited and constructive debate over the future of sustainable forestry in Canada – as reflected in this collection – can and will continue.

Chris Tollefson
Victoria, BC
23 June 1998

Note
1 See my conclusion at 374 et seq.

Acknowledgments

Support and encouragement for the project that has evolved into *The Wealth of Forests* has come from a variety of personal and institutional sources.

I would like first to recognize the authors of this collection, all of whom have contributed original, previously unpublished work. I would also like to express my gratitude for the fine research and editorial assistance I have received from students at the University of Victoria, including Candice Alderson, Victoria Bartulabac, Zita Botelho, Tracey Cook, Steve Ferance, Elyssa Lockhart, Matt Pollard, Chris Rhone, and Mark Underhill. As always, I am also indebted to my secretary at the Faculty of Law, Rosemary Garton.

Permission to use the original artwork that appears on the cover of this collection was generously donated by the artist, Will Gorlitz, of Wellington, Ontario.

This book would not have been possible without the support from Environment Canada that funded the original workshop out of which this collection emerged, and from Forest Renewal BC that provided research and publication funding. Special thanks are due to Dr. Roger McNeill (Environment Canada) and Dr. Louise Rees (FRBC-Science Council of BC). I would also like to acknowledge ongoing institutional support for this project provided by the University of Victoria through the offices of the Dean of Law (David Cohen) and the President (Dr. David Strong). I have also benefited greatly from the ongoing encouragement and advice of Dr. Peter Pearse.

The Wealth of Forests project has received critical and continuing support from the Eco-Research Chair of Environmental Law and Policy at the University of Victoria. Special thanks are owed to its chair, Dr. Michael M'Gonigle.

I am also grateful to UBC Press for its support through the review and publication process. I would particularly like to thank Peter Milroy, Jean Wilson, and Randy Schmidt.

Finally, I would like to recognize the "in-kind" contribution of my children, Hannah and Rory, whose support has truly been a sustaining force throughout this project.

Part 1:
Introduction

1
Introduction

Chris Tollefson

Industrial forestry in North America is at a crossroads. Technological change, driven in part by increased international competition, has led to extensive restructuring within the forest industry, dramatically reducing forest sector employment, particularly in the Pacific Northwest.[1] At the same time, industry and government regulators alike have come under harsh and continuing criticism over logging and silviculture practices and their perceived failure to protect other forest values linked to biodiversity, tourism, and non-timber forest products. There is a pervasive and lingering sense that much of the wealth of our forests has been squandered.

A broad consensus is emerging that the practice *and* theory of industrial forestry must change. In large measure, this consensus centres on the need to achieve "sustainable forestry." The meaning of this ubiquitous and elusive concept is a matter of ongoing controversy. Few would now argue that sustainable forestry should be equated with sustaining timber production or sustaining timber-dependent communities. Nor, for many, is it adequate to conceive of the concept as implying a responsibility for sustaining an optimal mix of human-valued resource benefits, even if those benefits are interpreted as including biodiversity, aesthetics, and other passive uses.[2] Increasingly, it is argued that sustainable forestry can only occur where the overriding goal is to sustain forest ecosystems.[3]

Debate over the term sustainable forestry occurs within a broader, ongoing public dialogue focused on the need to achieve sustainability. In this context, the sustainability imperative is typically conceived as applying equally to the economy, the environment, and society as a whole. For example, the BC Commission on Resources and Environment, in a landmark discussion paper advocating the enactment of a sustainability law, characterized sustainability as requiring the maintenance of

- a prosperous and diverse economy that maximizes the jobs created through resource use without compromising the ability of the land to replenish natural resources;

- the integrity of the natural environment, including biological diversity; quality of soil, air and water; and special natural features;
- the quality of life, for both individuals and communities throughout British Columbia.[4]

Although there is broad agreement on the need to achieve sustainable forestry, there is considerable disagreement over which policy instruments should be deployed to promote this goal. Many economists trained in the neoclassical tradition contend that, particularly in the areas of resource management and environmental protection, governments have relied too heavily on "command and control" regulation which defines, in specific terms, permissible and prohibited conduct and contemplates prosecution for non-compliance. Often, as Stanbury and Vertinsky elaborate elsewhere in this book, the same policy objectives can be achieved more efficiently and effectively, and less coercively, by employing "market" or "economic" instruments that influence private behaviour using positive or negative incentives.[5] Illustrations in the environmental policy context include pollution taxes, subsidies for green technologies, and, of increasing interest lately, marketable pollution permit regimes.

In the United States, this "free market environmentalist" critique has, over the last decade or so, gathered considerable force both in academe and in key policy-making circles.[6] Marketable pollution permit regimes, considered one of the more radical aspects of the free market prescription, have been entrenched in the federal Clean Air Act since the early 1980s.[7] More recently, free market environmentalism has played a central role in the Republican campaign in Congress to overhaul federal natural resource and environmental law.[8] In Canada, market instruments have not as yet been championed to the same extent, but the free market perspective in Canadian policy circles is growing, as manifested in initiatives such as the federal government's proposed Regulatory Efficiency Act,[9] and in criticisms levelled against the BC government's newly enacted Forest Practices Code. Here, these criticisms are examined by Stanbury and Vertinsky in Chapter 3 and by Cook in Chapter 9.[10]

But while the principle of harnessing the market for sustainability has a strong political constituency, there are many unanswered questions relating to whether, and to what extent, this approach could serve as a basis for reforming Canadian forestry. At present, as Haley and Luckert point out in Chapter 6, within the complex web of rights and regulation that characterizes our forest industry, market instruments tend to be used for timber-utilization purposes (including conferring and taxing timber rights), while command and control instruments have generally been used to advance non-timber values (prescribing forest practices and silviculture).[11] Given this pattern, various questions concerning the use of market instruments for sustainability arise. For example, how effectively do the market instruments currently used in the forest policy area – including stumpage and forest tenures – serve the objective of sustainable forestry? Can the role of market instruments be expanded to sustain non-timber values? In this quest for sustainability, what is the role and potential of emerging instruments such as land use zoning and eco-certification? These are some of the questions this collection seeks to address.

Forest Policy in Transition

The practice of forestry ... is currently undergoing the most profound and rapid change since its establishment a century ago. The evolution from sustained yield management of a relatively small number of commercial tree species to the protection and sustainable management of forest ecosystems is changing some of the fundamental premises of forest management.[12]

Forestry scholars tell us that industrial forestry typically evolves through a series of fairly predictable stages.[13] In the initial phase, forests are typically exploited in a virtually unregulated fashion. To promote rapid development of the resource, forest rents are kept low and companies are provided with long-term rights to fibre supply as consideration for establishing and maintaining processing operations in the jurisdiction.

Later, with increasing recognition of the need to manage the transition from first to second growth, a sustained yield paradigm takes hold. In this stage, the state seeks to maximize the long-term production of timber by achieving an equilibrium that balances timber growth and harvest rates. Among other things, establishing this equilibrium requires the state to undertake inventories of timber supply and, by projecting growth and harvest rates over time, identify an allowable annual cut (AAC) that is theoretically sustainable in perpetuity.

Inevitably, as Dellert recounts in Chapter 11 on the history of cut regulation in British Columbia, the dominance of economic and timber values within the sustained yield paradigm is called into question.[14] As pressure builds for a more ecological approach that takes non-timber values into account, the paradigm is modified to reflect a recognition of the need to manage the forest for "multiple uses" such as recreation, wilderness protection, and aesthetic values. While the goal of maximizing timber production in a sustained manner does not change, timber-management planning processes become operationally more complex as forest managers are given the additional mandate of sustaining non-timber values.

The uneasy compromise between timber and non-timber values reflected in this "multiple use-sustained yield" (MUSY) paradigm characteristically proves to be an unstable one. Where the goal of multiple use is pursued within a single defined forest area, non-timber uses are often undermined by timber uses. Often multiple use planning becomes an exercise in "adjacency" through zoning some forest lands for industrial uses and others for all other uses.[15]

Most proponents of an evolutionary analysis of industrial forestry claim that the paradigm that most accurately describes contemporary forest policy in Canada is the same one that has dominated since the end of the Second World War: namely, multiple use-sustained yield.[16] As such, that maximization of timber production should be the overriding goal of forest policy has, until very recently, gone largely unquestioned. Increasingly, however, policy debate is focusing on means *and* ends. A similar pattern has evolved in the United States.[17] There, the MUSY approach to public forest lands management has been enshrined in federal law since 1960. Currently, it is manifested in an exceedingly complex regime of interrelated forest and natural resource laws.

The experience of the last thirty-five years has persuaded many commentators to conclude that MUSY has failed, that the goal of sustained yield maximization has consistently been allowed to take precedence over non-timber uses.[18] As a result, both in Canada and in the United States, there has been a strong sense that forestry has been governed by the sustained yield approach for too long and that a new paradigm – one informed by a broad notion of forest sustainability – is urgently needed.

Increasingly, there are calls – exemplified here in contributions by M'Gonigle, Dellert, and Gale – for a new paradigm which proceeds from an explicit recognition of the forest as a complex and interdependent ecosystem, and which sets, as its overriding priority, the goal of maintaining that system.[19] As yet, this ecosystem paradigm remains largely inchoate; its implications for industrial forestry, as we now know it, uncertain.

Forest Policy in Canada

For most of this century, Canadian forest policy in many critical respects has been a picture of continuity not change. During this period, our forests have remained primarily (in BC, almost 95 percent) in public hands. But while most provinces have retained substantial ownership of the forests, control over timber resources has largely passed into the hands of a relatively small number of large forest companies. The system of property rights in timber resources currently in place – forest tenures – has also remained essentially the same as that which has prevailed for most of this century. Under these tenure systems, companies lease long-term property rights in trees. Most of these licences confer on the company exclusive harvesting rights for a geographical area. In some provinces (notably, BC), non-exclusive volume-based licences also constitute a substantial proportion of the allowable annual cut.[20]

In the early days, allowing companies to secure (at low or no cost) long-term timber harvesting rights without requiring them to make the investment necessary to purchase the land itself was regarded as a legitimate and effective means of promoting the fledgling Canadian forest industry.[21] Indeed, the tenure structure was just one of several ways in which the expansion and development of industrial forestry in Canada has been publicly subsidized. For much of this century, the economic rent the industry has been required to pay (primarily in the form of stumpage, payable upon harvesting) has been kept relatively low, both by North American and even by global standards.[22] In addition, provincial governments have traditionally underwritten the cost of reforestation. This pattern continues today in all jurisdictions except for BC and Alberta, where, since the late 1980s, there have been attempts to pass these costs on to licensees.[23]

A key justification for maintaining these industry-friendly policies has been the argument that not only are they necessary to foster a favourable investment climate, but that they create and maintain forest sector jobs and, in the long run, stable and prosperous forest communities. To this latter end, many provincial forest tenure systems impose upon forest companies, as a condition of obtaining or renewing a licence, an obligation to operate a processing facility that uses some or all of the timber supply generated from the licence. Maintaining community stability has also been a reason for governments to impose restrictions on the transfer and divisibility of licences.

Policy Failures

Over the last two decades or so, a broad consensus has emerged that, in fundamental ways, Canadian forest policy is not working. The dimensions of this policy failure are threefold. First, there is a strong sense that forest policies have failed to promote efficient production and utilization of the resource itself. The stumpage system has been criticized as encouraging companies to use timber inefficiently.[24] It has also been argued that because forest tenures are of shorter duration than the time it normally takes for trees to grow to a harvestable size, forest companies have little incentive to engage in reforestation or, more generally, in advanced silviculture.[25]

A more commonly perceived failure of existing forest policy is its ineffectiveness in protecting other values and attributes of our forests. Under current policies, forest companies have no economic stake in non-timber values such as protecting streams, wildlife, and recreational opportunities, to say nothing of less tangible values such as carbon sequestration and forest biodiversity. The underlying precept of sustained yield – the notion that all old growth should ultimately be eliminated and replaced by a "normal," even-aged forest – has also come under continued attack. So, too, has the authority of professional foresters and resource economists to make, on behalf of the public, decisions about whether, when, where, and how to log.

Finally, there is a broad perception that provincial forest policies have not delivered in economic terms either. During the 1980s, employment in the forest sector slumped dramatically with predictable and disastrous implications for many small, forestry-based communities. The immediate cause of the slump was mechanization, which in turn was necessitated by the need to be internationally competitive in the low-priced, minimally processed commodities – manufactured lumber and pulp and paper – which are Canada's main forest exports.

In retrospect, it was argued that one of the principal reasons Canada has become so heavily reliant on these types of exports, as opposed to more value-added products, are provincial forest policies that subsidize a volume-oriented approach to forestry.[26] To avoid further job loss and maintain international competitiveness, M'Gonigle, Gale, and others therefore argue for policies that facilitate a transition from a volume-based to value-based approach to managing the forest resource.[27] Corporate concentration within the forest sector has been another continuing concern. As of 1997, 44.4 percent of the AAC available to forest licencees operating in BC was held by just five companies. The top ten companies held rights to 68.1 percent of the forest licencee portion of the AAC. The remainder, less than 32 percent, was shared by over thirty other companies.[28] Moreover, because much of the timber cut by these large companies is committed to processing facilities which they own or control – often as a term of their licence – many independent processors experience chronic timber shortages. In short, existing forest policies not only have failed to promote prosperity and community stability but arguably have actually thwarted these goals.

Pressures for Change: 1975 to 1990
The need to reform how public forest lands are managed in Canada has been widely recognized for at least two decades. In the early 1970s, this recognition prompted the BC government to appoint a Royal Commission on Forest Resources chaired by forest economist Peter Pearse (see Chapter 2).[29] The Pearse Report (1975) advocated broadening the notion of sustained yield to embrace multiple forest uses. To this end, the report urged that forest-planning processes be opened up to provide for public participation and to incorporate non-timber values. The report did not perceive, however, that these reforms would necessarily have an adverse impact upon timber production. This expectation was in part due to an assumption that forest productivity could be greatly enhanced by more intensive, government-funded silviculture. The other assumption was that, through careful planning, a multiple use approach could accommodate several potentially inconsistent resource values – including logging – within the same area. In so doing, the report contended, dedication of timber-rich forest for economically inefficient "single uses" such as recreation or conservation could be avoided.

In the wake of the Pearse Report and the growing popularity of the MUSY notion within forest policy circles, many provinces began to manage for non-timber values by requiring licensees to submit detailed short- and long-term plans for the management of their timber allocation.

However, these plans and planning processes continued to be regarded by forest regulators and licensees alike as being contractual matters that were most appropriately handled with little or no public participation.

Despite these and other reforms designed to implement the MUSY approach, Canadian forest policy and practice changed very little in substance from the late 1970s to the late 1980s. Fundamentally, forest policy continued to be driven by the perceived imperative of maximizing timber production.[30] Publicly funded, intensive silviculture programs, of the type urged by Pearse, did not materialize, falling victim to recessionary pressures and government restraint. Nor did government finances permit forest regulators to monitor compliance with the management plans which had become the principal vehicle for protecting non-timber values. Meanwhile, the rate of cut during this period, far from stabilizing or declining, actually went up from 40 percent to 50 percent.[31]

While the forest industry and forest regulators carried on business as usual, external pressures for change were building. By the late 1980s, as discussed, extensive forest sector job-loss resulting from industry restructuring and mechanization, triggered a debate over the whether the public interest was still being served by forest policies originally intended to subsidize the expansion and profitability of a young forest sector. As Scarfe's chapter on timber pricing policy describes, further fuel for this debate was added by trade complaints brought by American forest interests claiming that low provincial stumpage rates amounted to an unfair trade subsidy to Canadian-based producers.[32]

Meanwhile, in Canada, and indeed globally, the environmental movement began to focus its attention squarely on the forests issue. In BC and elsewhere in Canada, environmentalists were engaged in "valley-by-valley" battles aimed at protecting old growth and other ecologically sensitive areas from logging. Moreover, dissatisfaction with forest policies extended well beyond the environmentalist community. Within the public at large, which was becoming increasingly sensitized to and aware of forest issues, strong support was emerging for policies which would compel forest companies to undertake and underwrite silviculture and which would penalize those that engaged in harmful forest practices. Support was also strong for increasing protection of old growth and other ecologically sensitive areas, and for broadening opportunities for public participation in land use planning decisions.

Forest Policy Reforms

Particularly in British Columbia, and to a lesser extent in other provinces, these pressures led to some significant changes in the management of forest resources.

One significant area of reform was the reallocation of responsibilities between the industry and government with respect to forest management. Several provinces have now adopted measures that make forest companies responsible for not only carrying out basic silviculture but also paying for the full cost of this work.[33]

There has been a concerted attempt to capture a greater rent from the forest resource through stumpage charges, a policy shift discussed in detail by Scarfe.[34] To this end, the province of BC raised stumpage rates by 15 percent in 1987, and implemented another increase in 1994. Notably, the revenues generated by the latter increase were statutorily targeted to support the work of a Crown corporation known as Forest Renewal BC. This new agency has been given responsibility for funding local and regional silviculture, environmental remediation initiatives, value-added manufacturing, and community development.

There have also been significant changes in how land use decisions, affecting the use of public forest lands, are made. The most dramatic changes have occurred in British Columbia. There, over the last several years, community-based planning processes have tried, with mixed success, to delegate to affected interests the task of reaching a consensus on how provincial Crown lands should be allocated among competing potential uses.[35]

In response to continuing criticism that it was condoning environmentally harmful logging activity by the forest industry, British Columbia recently became the first Canadian province to have legally enforceable forest practices legislation. Under the Forest Practices Code, enacted in summer 1995, companies engaged in logging on public lands must comply with prescribed rules and standards relating to a wide variety of activities – including roadbuilding, site preparation, and silviculture.

More recently, there are signs that the BC government's commitment to protecting the forest resource, particularly non-timber values, is weakening. In 1997, the environmental protection provisions of the controversial Forest Practices Code were softened and the government entered into a historic agreement – known as the Jobs and Timber Accord – with the province's major forest companies, under which companies will be granted additional cutting rights and be eligible for increased harvesting cost subsidies if they meet specified job creation targets.[36]

Instruments for Sustainable Forestry

These policy initiatives, particularly the Forest Practices Code, have triggered an ongoing and sometimes heated debate about how best to achieve the goal of sustainable forestry. For economists committed to a market-based approach to resource management, the Code is a quintessential illustration of the much-maligned command and control style of regulation which not only fails to protect non-timber values but severely affects the ability of companies to produce competitively priced timber for world markets.

The latter is best achieved, they contend, by adopting measures that strengthen and broaden the private property rights conferred on forest companies under the forest tenure system. In the context of timber values, greater security of tenure would, it is claimed, lead to significantly more private investment in silviculture than is currently expended. Using a similar reasoning, in his contribution to this collection, Pearse advocates reforming the forest tenure system to confer legal rights to non-timber values – water, wildlife, and recreation – so as to provide these new rights-holders with the incentive to ensure that these values are not adversely affected by timber operations.[37]

Reforming the tenure system is also a key priority for those who contend sustainable forestry cannot be achieved unless we replace the volume-oriented, capital-intensive system of timber rights currently in place with another system that is value-added, labour intensive, and locally controlled. To bring this transition about, M'Gonigle and Gale argue, property rights to the timber supply – currently heavily concentrated among a handful of major corporations – must be redistributed to provide a larger share for local communities, private woodlot owners, and First Nations.[38] Out of this reallocation of rights would emerge, it is claimed, a more sustainable forestry in terms of both human benefits and ecological considerations.[39]

There has also been considerable optimism concerning the potential for using new, voluntarist-style instruments to promote sustainable forestry. One such instrument, analyzed in Chapter 12 by Gale and Burda, is "eco-certification," which is intended to identify "sustainably produced" wood products with a view to harnessing consumer preferences for sustainable forestry practices.[40] Another alternative, considered by Rayner in Chapter 10, is land use zoning that aims to "zone" the forest resource by use, with the goal of "producing" timber and non-timber values more efficiently.[41]

The Wealth of Forests

While much has been written on the theory and practice of sustainable forestry and on the relative merits of regulatory versus market approaches to environmental protection, these literatures have not, as yet, been bridged. *The Wealth of Forests* seeks to build that bridge by analyzing – in relation to this critical renewable natural resource – the potential and limits of market and other policy instruments as means of achieving sustainability.

This book does not purport to offer a conclusive definition of "sustainable forestry" and work backwards to derive the implications for forestry policy in British Columbia or Canada. Instead, the book is squarely situated in the here and now, focusing on the extent to which existing and emerging policy alternatives are bringing us closer to the elusive goal of sustainable forestry. In some measure, the decision to focus on *means* without seeking to resolve the ongoing debate over *ends* is driven by practical considerations, including the diverse disciplinary perspectives of the project collaborators and the inherently open-ended nature of the concept of sustainable forestry. At the same time, we are confident that this approach has yielded a collection that is grounded in sustainability theory but retains a practical orientation, ensuring its relevance well beyond the walls of academe.

In addressing the concept of sustainable forestry, the collection focuses on the experience of British Columbia as an illustrative case study. Several factors justify this focus: the significance of BC's timber-producing role in domestic and global terms; the scope of forest policy reform undertaken by the BC government, particularly from 1992 to 1996; the intensity of the public debate over the merits and future of these reforms; and the continuing high levels of international interest in BC forests *and* forest practices dating back to the Clayoquot protests of 1993. For these and other reasons that will be elaborated further, BC arguably finds itself more deeply immersed in the debate over sustainable forestry and how to get there than any other jurisdiction in North America.

The Wealth of Forests is an interdisciplinary endeavour to which contributions were invited from leading scholars with a variety of backgrounds including forestry, economics, public administration, political science, and law.

It is in five parts. Part 1 provides an introduction to sustainable forestry and to the book. Part 2 provides a context for the two intersecting debates that provide the framework for the book. It includes four chapters. The first, authored by Peter Pearse, is an overview of the potential for sustainably managing forest resources – including timber, water, fish, wildlife, and recreation – by means of various market instruments.[42] A second chapter, by W.T. Stanbury and Ilan Vertinsky, identifies various technical and political barriers to deploying market instruments for sustainable forestry as part of a general review of the relative benefits and drawbacks of market and regulatory policy instruments.[43] In the third chapter, Rod Dobell critically assesses, from an ecological economics perspective, the claims made by proponents of market instruments. Ultimately, he argues, given the complexity and uncertainty associated with the quest for sustainability, we must adopt an adaptive approach to institution design. A key implication of this approach, he observes, is that "in a changing, complex world, it is not possible to aim at certainty in property rights in natural capital or resource systems themselves any more than it is in other claims on jobs, markets, or other income streams in a rapidly changing economy."[44] The final chapter in Part 2, by Michael M'Gonigle, also cautions against relying on market instruments and traditional notions of property rights to achieve sustainability. Indeed, he contends that a focus on instrument choice and deployment confounds the more basic requirement for sustainable forestry: sustainable ecosystem-based governance structures.[45]

Part 3 is a series of chapters focusing on the operational implications of a sustainability approach in key selected areas of existing or emergent forest policy. Befitting its central role in a transition to sustainable forestry, forest tenure is addressed in two chapters, one jointly authored by David Haley and Martin Luckert and a second authored by Michael M'Gonigle.[46] A chapter considering the relationship between timber pricing and sustainable forestry, including an analysis of the ongoing Canada-US softwood lumber dispute, is provided by Brian Scarfe.[47] This discussion is followed by an evaluation of BC's controversial Forest Practices Code by Tracey Cook; an assessment of the potential of "priority-use" zoning as a "sustainable solution" by Jeremy Rayner; an analysis of sustained yield, cut regulation policies by Lois Dellert; and, finally, an examination of forest eco-certification as a market incentive for sustainable forestry by Fred Gale and Cheri Burda.[48]

Part 4 examines the question of how, and to what extent, domestic and international legal regimes might impede the adoption of sustainable forest policies identified elsewhere in *The Wealth of Forests*. In their co-authored chapter, David Cohen and Brian Radnoff consider the realm of domestic law. In particular, they consider the question of whether, and to what extent, governments should be required to compensate a private party whose interest in a forest resource has been devalued due to environmental regulation.[49] In the other contribution to Part 4, Fred Gale explores the manner and extent to which the ability of Canadian provinces to pursue sustainable forest policies, as a component of what he terms an "eco-conversion" strategy, is constrained by obligations under international trade and investment laws including the North American Free Trade Agreement (NAFTA), the General Agreement on Tariffs and Trade (GATT), and the World Trade Organization (WTO).[50]

In Part 5, I conclude with a chapter that reviews and synthesizes the themes and arguments that emerge in this collection.

Notes

1 M.P. Marchak, *Logging the Globe* (Montreal and Kingston: McGill-Queen's Press 1995). See especially Ch.1 to Ch.3.
2 A useful discussion of competing paradigms of sustainable forestry is found in R. Gale and S. Cordray, "What should the forests sustain? Eight answers," *Journal of Forestry* 89 (1991): 33.
3 Gale and Cordray, ibid., 34-36, distinguish three variants of the ecosystem approach: "ecosystem type" sustainability, "ecosystem insurance" sustainability, and "ecosystem-centred" sustainability, each of which is premised on a different rationale and has different policy implications.
4 Commission on Resources and Environment, *A Sustainability Act for British Columbia* (Victoria: Queen's Printer 1994), 14.
5 The terms "market instrument" and "economic instrument" are used synonymously in this book. See W.T. Stanbury and I.B. Vertinsky, "Governing Instruments for Forest Policy in British Columbia: A Positive and Normative Analysis," infra, Ch.3.
6 See, for example, R. Stroup and J. Baden, *Natural Resources: Bureaucratic Myths and Environmental Management* (San Francisco: Pacific Institute for Public Policy Research 1983); T. Anderson and D. Leal, *Free Market Environmentalism* (San Francisco: Pacific Institute for Public Policy Research 1991); "Special Issue: Symposium – Free Market Environmentalism," *Harvard Journal of Law and Public Policy* (1992): 15.
7 See R. Hahn and R. Stavins, "Incentive-based environmental regulation: A new era from

an old idea," *Ecology Law Quarterly* 18 (1991): 15; T. Tietenberg, *Environmental Economics and Policy* (New York: HarperCollins 1994), 240-5.

8 See, for example. J. Cushman, "Republican campaign to undo major environmental programs reaches full force," *New York Times* (17 July 1995): A-1.

9 See Bill C-62, *Regulatory Efficiency Act*, as tabled 6 December 1994; Standing Joint Committee for the Scrutiny of Regulations, *Report on Bill C-62* (16 February 1995). In this regard, see generally R. Hirshhorn and J.-F. Gautrin, "Competitiveness and Regulation," in *Towards Efficient Regulation* (Kingston, ON: Queen's School of Policy Studies 1993).

10 T.L. Cook, "Sustainable Practices? An Analysis of BC's *Forest Practices Code*," infra, Ch.9; Stanbury and Vertinsky, supra, note 5.

11 D. Haley and M.K. Luckert, "Tenures as Economic Instruments for Achieving Objectives of Public Forest Policy in British Columbia," infra, Ch.6.

12 V. Sample, N. Johnson, G. Aplet, and J. Olson, "Introduction: Defining Sustainable Forestry," in *Defining Sustainable Forestry*, ed. Sample et al.(Washington, DC: Island Press 1993), 4.

13 H. Kimmins, "Sustainable Forestry: Can We Use and Sustain Our Forests?" *Forest Industry Lecture No. 27* (1991): 1 (positing a three-stage evolution from "exploitive, pre-forestry" to "administrative forestry" to "social forestry"). See also M. Howlett and J. Rayner, "The Framework of Forest Policy in Canada," in *Forest Management in Canada*, ed. M. Ross (Calgary: CIRL 1995), 43 at 63-5; and L. Dellert, *Sustained Yield Forestry in British Columbia: The Making and Breaking of a Policy: 1900-1993* (MA thesis, York University 1994).

14 L. Dellert, "Sustained Yield: Why has it Failed to Achieve Sustainability?" infra, Ch.11.

15 R.W. Behan, "Multi-resource forest management: A paradigm challenge to professional forestry," *Journal of Forestry* 88 (1990): 13.

16 See Kimmins, supra, note 13; Dellert, supra, note 13 at 133; Howlett and Rayner, supra, note 13 at 25-28 and 67-72.

17 Canada-US comparisons in terms of management of public forest lands are complicated by differences in the forms of land tenure. In the United States, a substantial portion of the working forest is privately owned, in most states well over half. The *Multiple Use Sustained Yield Act* applies only to federal forest lands, which in most "forestry states" comprise only about 25 to 30 percent of the forest land base. For further comparative analysis, see G. Hoberg, *Regulating Forestry: A Comparison of British Columbia and the U.S. Pacific Northwest* (Kingston: Queen's University School of Policy Studies 1994); M. Haddock, *Forests on the Line: Comparing the Rules for Logging in British Columbia and Washington State* (Vancouver: Sierra Legal Defence Fund/Natural Resources Defense Council 1995).

18 See Behan, supra, note 15; M. Blumm, "Public choice theory and the public lands: Why 'multiple use' failed," *Harvard Law Review* 10 (1994): 405; S. Hardt, "Federal land management in the twenty-first Century: From wise-use to wise stewardship," *Harvard Law Review* 10 (1994): 345.

19 Behan, ibid. 16. See also M. M'Gonigle, "Structural Instruments and Sustainable Forests: A Political Ecology Approach," infra, Ch.5; F. Gale, "Ecoforestry Bound: How International Trade Agreements Constrain the Adoption of an Ecosystem-based Approach to British Columbia's Forests," infra, Ch.14; and Dellert, supra, note 14.

20 In BC, the two primary forms of forest tenure are Tree Farm Licences (TFLs) and Forest Licences, which together accounted for 81 percent of the AAC as of 1995. TFLs are a form of area-based tenure while Forest Licences are a volume-based tenure. Of the 81 percent total, 56 percent of the AAC is assigned to Forest Licences and 25 percent is assigned to TFLs. See M. Ross, "Volume Based Tenures," in *Forest Management in Canada*, ed. Ross, 190-1.

21 Dellert, supra, note 13 at 128.

22 Marchak, supra, note 1 at 88.

23 Ross, supra, note 20.

24 P. Pearse, "Economic Instruments for Promoting Sustainable Forestry: Opportunities and Constraints," infra, Ch.2 at 31.

25 Ibid. at 30.

26 R.M. M'Gonigle and B. Parfitt, *Forestopia: A Practical Guide to the New Forest Economy* (Madeira Park, BC: Harbour Publishing 1994), Ch.4.

27 Ibid. See also M'Gonigle, supra, note 19, and Gale, supra, note 19.
28 BC Ministry of Forests (Resource Tenures and Engineering Branch), "Apportionment System – Licencees' Annual Commitment in Cubic Metre" (MOF, 22 December 1997).
29 Pearse, supra, note 24.
30 Dellert, supra, note 13 at 115-16.
31 Ibid. at 116.
32 B. Scarfe, "Timber Pricing Policies and Sustainable Forestry," infra, Ch.8; Hoberg, supra, note 17 at 62-63.
33 Ross, supra, note 20.
34 Scarfe, supra, note 32.
35 Much of the pioneering work in this area was facilitated by the Commission on Resources and the Environment (CORE), an independent agency created under special legislation in 1990 with a mandate to develop a provincial land use strategy. In 1996 CORE was superceded by the Land and Resource Management Plan (LRMP) process. Like CORE, LRMP is a stakeholder process. The focus of LRMP, however, is subregional implementation issues. It is overseen by the BC government's Land Use Coordination Office (LUCO).
36 For further discussion of the Accord and references, see the editor's conclusion at 373-4.
37 See Pearse, supra, note 24 at 32-3. The notion that existing forest tenures should be made more secure and/or broadened to encompass new values is itself highly controversial. While some non-priced forest values could be subjected to the market and provided through a system of property rights (i.e., recreational use of forest lands), there is considerable debate within the economics literature about whether all currently non-priced forest values can be provided in this fashion. It has been argued, for example, that forest resources which provide passive use values – values enjoyed for their mere existence – are incapable of market valuation given their broad diffusion throughout society as a whole (see Haley and Luckert, supra, note 11 at 137-8).
38 See M'Gonigle, "Living Communities in a Living Forest: Towards an Ecosystem-Based Structure of Local Tenure and Management," infra, Ch.7; and Gale, supra, note 19. See also M'Gonigle and Parfitt, supra, note 26 at Ch.9.
39 Ibid. See also the discussion of the implications of replacing industrial tenures with community forests in Haley and Luckert, supra, note 11 at 145.
40 F. Gale and C. Burda, "The Pitfalls and Potential of Eco-Certification as a Market Incentive for Sustainable Forest Management," infra, Ch.12. See also (1993),"Certifying sustainable forest products: A roundtable discussion," *Journal of Forestry* 91: 33.
41 J. Rayner, "Priority-Use Zoning: Sustainable Solution or Symbolic Politics?" infra, Ch.10 at 223. This line of argument is elaborated by Haley and Luckert, supra, note 11 at 148 (fn 26), who indicate that initial economic analysis of intensive timber zoning in the Revelstoke Forest District suggests the possibility of significant productivity increases.
42 Pearse, supra, note 24.
43 Stanbury and Vertinsky, supra, note 5.
44 R. Dobell, "Compliance and Constraint: Economic Instruments in Context," infra, Ch.4.
45 M'Gonigle, supra, note 19.
46 Haley and Luckert, supra, note 11; M'Gonigle, supra, note 38.
47 Scarfe, supra, note 32.
48 See Cook, supra, note 10; Rayner, supra, note 41; Dellert, supra, note 14; Gale and Burda, supra, note 40.
49 D. Cohen and B. Radnoff, "Regulation, Takings, Compensation, and the Environment: An Economic Perspective," infra, Ch.13.
50 Gale, supra, note 19.

Part 2:
Perspectives on Sustainable Forestry

Rhetorical agreement on the goal of sustainable forestry often tends to conceal substantial analytic and prescriptive differences. A central purpose of this Part is to shed light on these differences by offering several distinct and, in some ways, competing perspectives on the public policy challenges posed by the concept of sustainable forestry.

From the perspective of resource economics, the challenge of sustainable forestry is to manage more sustainably the various timber and non-timber resources that constitute our forest wealth. Employing this approach, Pearse concludes in Chapter 2 that there is broad scope for enhancing sustainable forestry through better-developed property rights, markets, and pricing mechanisms in various forest resources.

Other commentators are concerned with understanding why, in pursuit of the stated policy goal of sustainable forestry, governments come to rely on certain policy instruments and eschew others. For Stanbury and Vertinsky, the answer lies in a public choice analysis that focuses on the incentives and disincentives associated with deploying the various governing instruments that are available to advance public policy objectives. In Chapter 3, they contend that this mode of analysis helps to explain why, in their view, governments have tended to respond to the challenge of sustainable forestry by using regulatory, rather than market-based, instruments.

For some analysts, however, meeting the challenge of sustainable forestry requires a rethinking of the traditional categories and precepts of liberal economics. Sustainability, Dobell argues in Chapter 4, requires us to elaborate an ecological economics that transcends the traditional "hard distinction" regulatory and market instruments and that takes account of social equity and the need to conserve and sustain social and natural capital.

Others would go further. Like Dobell, M'Gonigle is highly critical of the dominant tendency to portray the challenge of sustainable forestry as involving a more effective deployment of market and/or regulatory instruments. For M'Gonigle, however, it is not enough to broaden or redefine the categories of liberal economic analysis. Achieving sustainability, he argues in Chapter 5, requires a radical new approach to law and policy-making, a "political ecology" that self-consciously proceeds from the needs of the ecosystem itself.

2
Economic Instruments for Promoting Sustainable Forestry: Opportunities and Constraints

Peter H. Pearse

Introduction

Today, nearly a decade after the Brundtland Report promoted the notion of sustainable development, its popularity continues to grow (Brundtland 1987). Notwithstanding ambiguities about its meaning in operational terms, sustainability is embraced, in principle, not only by environmental organizations but also by business groups and governments. Like the idea of conservation earlier in this century, about which W.H. Taft is reported to have said, "There are a great many people in favour of conservation no matter what it means," sustainability has become an undisputed, and seemingly indisputable, precept of progressive economic and social policy.

The growing interest in sustainability has converged with two other trends in public attitude that raise the importance of the subject of this book. One of these is the growing anxiety about how natural resources are being used and managed. In Canada, much of this concern is focused on forests. It includes concern about the management and harvesting of industrial timber, but even greater apprehension surrounds the non-commercial, or environmental, products and services of forests, such as water, wildlife, recreation, aesthetic values, and atmospheric balance.

The other trend is the waning confidence in governmental regulation, both as a means of organizing economic activity (most conspicuously in the former communist states of eastern Europe) and as a means of controlling private economic activities (in the western industrialized countries). Correspondingly, confidence in markets and economic incentives as means of directing producers and consumers has been gaining favour.

The convergence of these trends has drawn attention to possibilities for engaging additional market forces to guide private behaviour in directions consistent with the public interest, as an alternative to increasing governmental command and control regulation. Both federal and provincial governments have shown growing interest in economic instruments as adjuncts of environmental policy (Government of Canada 1992; Cassils 1991; Manitoba Environment n.d.).

There is now considerable literature on the general subject of economic instruments, the various forms they can take, their incentive effects, and their administrative implications. There is also some documentation of experiments with new economic instruments, notably new forms of rights for commercial fisheries, tradable emissions permits for managing air pollution, and tax-rebate schemes to promote recycling (OECD 1992; Government of Canada 1992). But so far, little attention has been given to the potential application of these ideas to the pervasive problems of managing forests in Canada. This chapter explores the potential scope for economic instruments, their implications, and related institutional changes to support the integrated sustainability of forests in British Columbia, consistent with declared objectives of public policy.

The first part of this chapter draws attention to the link between economic behaviour and the framework of law and institutions within which economic activity takes place. A sketch of the economic instruments relevant to forest management follows. The next part explores the potential applicability of particular instruments to particular forest values – timber and a wide variety of other products and services. (Other chapters in this book examine some of these instruments in more detail.) The final part of the chapter deals with means of integrating multiple demands on forest land.

Economic Behaviour and Economic Organization
Economics textbooks give much attention to "market failures" and how they result in misallocated resources and inefficiencies. The theory of the market system predicts that a perfect market economy will create incentives that drive individual producers and consumers to behave in ways tending to maximize aggregate public welfare, given the resources and technology available to them. But real economies are always fraught with imperfections. The resulting inefficiencies in resource allocation, waste, and inequities provide the rationale for governmental regulation of economic activity to forestall or offset these unwanted effects.

Thus the environmental problem, and the misuse of natural resources, can be explained in terms of market failures, and the extensive regulatory activity of governments in these areas can be seen as interventions to "correct" market failures (Stager 1988). But the rapid expansion of governmental regulatory activity in recent decades has not won the battle against environmental deterioration. Moreover, concern has grown about the economic distortions that result from regulations and about the inefficiencies and costs that those regulations impose. Consequently, policymakers have begun to search for alternative means of inducing producers and consumers to adopt environmentally friendly behaviour.

Growing concern about degradation of the environment and natural resources has also led economists to explore the causes of these market failures and ways of preventing them. The result is a growing literature about deficiencies in property rights, economic structures, and laws affecting economic behaviour. For example, the resource depletion and excess harvesting capacity characteristic of commercial fisheries is explained by the distorted incentives of fishers under the traditional common property regime for managing ocean fish. Riparian law, by preventing the transfer of water from one use to another as circumstances change, results in its inefficient use. Unpriced access to air and water to dispose of waste leads to pollution – and so on.

These systems of laws, property rights, and organizational structures, which govern economic behaviour, are here referred to generally as *economic institutions*. Deficiencies in the institutional framework lead to market failures, which provide scope for governmental intervention to prevent the unwanted results of market failures.

The recent emphasis on institutions has served to focus attention on the fact that institutional arrangements are the product of public choice; they can be created and modified through collective decisions. Once decided upon, the institutional framework sets the exogenous conditions – the framework – for the decision-making of producers and consumers which in turn determines the allocation of scarce resources. Our ability to create and modify the institutional arrangements that guide economic behaviour offers scope for policy development, and so warrants the attention of economic analysts (Bromley 1989).

Economic Instruments Relevant to Forest Management

Command and Control versus Incentives

Most governmental interventions in the natural resources field are of the command and control type, such as restrictions on users, regulations on the way that resources are used, performance standards, and required technologies. These regulatory arrangements are aimed at controlling the behaviour of users by penalizing those failing to comply with governmental rules.

The alternative is to control behaviour and improve economic performance by means of incentives. Appropriate pricing of resources, property rights, taxes, and subsidies can be invoked to induce people to use resources more efficiently and, when desired, more conservatively. The defining characteristic of these *economic instruments* is that they depend on altering economic incentives to change behaviour. This approach often involves changing the institutions that guide the actions of producers and consumers. Much of the recent literature on this subject suggests considerable scope for developing economic instruments to forestall or correct market failures of the kind that lead to degradation of natural resources and the environment (Portney 1988; Opschoor and Vos 1989).

In comparison with control by fiat, economic instruments are often claimed to have important advantages (Tietenberg 1995; Government of Canada 1992). The most important of these are:

- They are less costly for governments to administer, because they are less demanding of technical information, monitoring, and data. They are also less costly to enforce, because they do not entail policing users' adherence to rules of behaviour that conflict with their private self-interest.
- The costs of compliance incurred by producers and consumers is also lower, because these instruments leave wider choice and greater flexibility than administrative regulations.
- Insofar as they ensure that resource costs are reflected in the prices paid by users, they encourage more efficient allocation and use of all resources.
- They tend to ensure that users bear the cost of the resources they use.
- They respond more effectively, and with less uncertainty, to changing economic circumstances.
- They encourage research and development on resource-saving technologies.
- They allow greater flexibility in managing resources among users and under diverse circumstances, thus allowing environmental goals to be achieved at lower cost.

Whether particular economic instruments will realize these advantages depends, of course, on their design and on the circumstances in which they are invoked. Thus, it is important to examine carefully the feasibility and likely effects of developing economic instruments to improve the management of specific resources in specific social and economic circumstances.

The remainder of this chapter builds on the considerable literature about economic instruments in general by examining their applicability and promise as means of improving the management of forest resources in British Columbia. It is intended to be a first step in exploring opportunities for advancing forest and environmental policies through institutional changes and related development of economic instruments.

Economic Instruments for Forest Management

If economic instruments are defined widely, the number and variety that can potentially be invoked in managing natural resources and the environment is considerable. Environment Canada's 1990 Green Plan Consultations produced a list of over 40 such instruments (Canada's Green Plan 1990). However, for present purposes the array can be grouped into five general categories: property rights, pricing, taxes and subsidies, deposit-refund systems, and compliance incentives.

Property Rights

Markets can be expected to operate efficiently only if producers have control over their inputs. This calls for a system of property rights for resources, to enable users to acquire control over them and organize their use without interference from others (Posner 1977). Thus, a sole owner of the right to hunt game or trap furbearing animals in a particular area has an incentive to manage and conserve the resources to generate maximum value over time. In contrast, someone who shares these rights with an unlimited number of others lacks such an incentive, so this person and the other users are likely to deplete the resources unless they are restrained from doing so.

Property has several dimensions that are embodied in varying degrees in different forms of property (Pearse 1990). The most important dimensions of property for present purposes are:

Exclusiveness. The extent to which the holder of the property can exclude others. The degree of exclusiveness ranges in natural resource rights, from the completely exclusive rights of private owners of land and certain minerals, to the common property rights of fishers and recreationists.

Duration. The time over which the rights will endure. Again, the range in natural resource rights is wide, from a year or less in the case of permits for hunters and fishers, to perpetuity in the case of freehold forest landowners.

Transferability. The ease with which the holder can transfer rights to someone else. Transferability varies from the readily divisible and transferable rights of private landowners to non-transferable rights to water and waste discharges.

Comprehensiveness. The extent to which the holder's rights extend to all the attributes of the resource. Thus, the rights of a fee simple landowner include timber and any agricultural or developmental values and sometimes water and the sub-surface minerals as well, whereas the holder of a timber licence can claim only the rights to the timber.

Benefits Conferred. The holder's right to the economic benefits that the resources can generate. This right is often truncated in varying degrees by obligations to pay governmental royalties, stumpage fees, rentals, taxes, and other charges that share the benefits with the Crown.

Quality of Title. The security of the holder's rights. Quality of title refers to the ability of the holders to enforce their rights, to protect them from encroachment by others (including governments), and to secure their entitlement (Scott and Coustalin 1996).

These are only the most fundamental dimensions of property, and because they all vary in degree, the variety of rights to resources is very broad.

These characteristics of property have important implications for the way that holders of rights to resources treat those resources, and for the value the resources generate. Exclusiveness, duration, and rights to the economic benefits obviously weigh heavily in the holder's incentives to conserve and manage the resources over time, and to invest in their continuing production and enhancement. Comprehensiveness determines users' inclinations to take account of the impact of their actions on other resource values, and to search for the best combination of uses where multiple uses are feasible. And transferability enables resources to be continuously allocated and reallocated to those uses and users that can make the best use of them. Holders of property are much influenced by the quality and security of their rights, which depend on all these characteristics as well as the legal form that the rights take.

Theoretically, "complete" property gives its holder each of these characteristics in the maximum degree – exclusive rights, forever, with unrestricted transferability, comprehensive entitlement, and a claim on all the economic benefits, all of which are secure and enforceable against others. However, all rights to forest resources in British Columbia today are more or less truncated or "incomplete." The earliest Crown grants of land, notably on Vancouver Island, come closest to the model of complete property: they give the owner exclusive, perpetual, transferable rights to sub-surface resources and surface water as well as to the forest and land itself. However, some attributes (such as wildlife) have been severed by legislation and are reserved to the Crown, the owner's use of the land is restricted by zoning, and taxes force owners to share the economic benefits with the Crown.

Most resources are used under much weaker forms of rights. Almost all users of timber, rangelands, and minerals hold only *usufructory* rights, which are rights to the temporary use of resources belonging to someone else – in this case, usually the Crown. These rights, issued by the provincial government, have limited terms, convey rights only to a specific resource or use, are transferable only under certain conditions, and require sharing of resource rents with the Crown through royalties or other charges. Many of these licences and permits do not provide exclusive use – rights to water, the waste-assimilative capacity of water, fish, and wildlife afford access to resources only in common with other holders of rights. Some, such as hunting and fishing rights, have terms as short as one season. Many are non-transferable and all require payments to the Crown.

The general pattern of rights across the spectrum of forest resources and values is broadly consistent with theories about the development of property (Coase 1960; Demsetz 1967). The essential point of this theory is that where demands on a resource are low relative to the resource availability, resource value is low. Furthermore, the system users' rights are likely to be crude, and appropriately so. But as the value of a resource rises, so too does the potential gain from more efficient allocation arrangements. As a result, more sophisticated forms of property rights can be expected to emerge.

The history of natural resource development in British Columbia fits this model well. The first Europeans found such an abundance of fish, timber, water, and other resources that there was no scarcity in the economic sense, and no allocation problem, and therefore no need for individual property rights. But gradually, one natural resource after another became scarce – furbearing animals with the development of the fur trade, minerals with the gold rush, agricultural land with settlement, then timber, water, game, and fish. As pressures on each resource developed, its value increased, and some system was needed to allocate it among competing users. Rights in the form of grants, leases, licences, and permits were issued over resources as they became scarce and valuable. The process continues, with new forms of rights being developed for users of fish, the waste-assimilative capacity of water and air, and outdoor recreational resources (Pearse 1988a).

This trend suggests that the system of property rights over resources should be viewed in an evolutionary context, gradually responding and adapting to increasing demands and pressures. In a modern, rapidly changing economy, governments can play a useful, if not essential, role in encouraging and assisting in this development.

The important point here is that, as a rule, the more complete the property rights held by resource users, the more effective market forces will be in ensuring that resources are allocated efficiently, and consequently the less will be the need for governmental intervention to correct failures. Thus, we rely on freehold owners of farmland or mineral deposits to use their resources efficiently without much governmental control of their operations. In contrast, much less confidence can be placed on commercial fishers to self-regulate because they compete with each other in common property fisheries. Close regulatory control of their vessels, gear, fishing time, and location, and other details of their fishing activity is therefore required.

Because more complete property rights allow holders to realize greater value from resources, the property rights themselves are more valuable. For example, a right to timberland is more valuable the longer its term, and the more exclusive, comprehensive, transferable, and secure it is. The character of property rights thus governs the extent to which the potential economic rent in the resources will be realized in the value of the property rights themselves. These rents can, of course, be redistributed from the holder of the property rights to governments or others through various forms of taxes and charges.

Later parts of this chapter suggest a variety of ways in which property rights to forest resources in British Columbia might be strengthened (i.e., made more complete) so that market forces will align the incentives of users more closely to the public interest in sustainability.

Pricing

Prices are the critical signals for producers and consumers in a market system, so it is important that they accurately reflect relative costs and values. (The general term "pricing" is meant, here, to include "user fees," "resource charges," and similar levies, as well as conventional market prices. The essential characteristic of all these terms is that they require users to incur a cost, in addition to their production costs, for each unit of a resource they use.) To provide the appropriate guidance, the price of each resource product or service should correspond to its marginal cost, which in turn should reflect its opportunity cost in its next-best use and utilization in its next-best point in time. Such prices will be generated only in perfectly competitive markets for well-defined resources or resource rights.

The exacting conditions for such pricing of forest-related resources rarely exist in British Columbia. Title to most attributes of forest land is retained by the provincial Crown, which seldom allocates rights by competitive processes. The value of forest land is therefore not reliably reflected in market prices.

Restrictions on the transferability and divisibility of resource rights are major obstacles to the emergence of effective prices. Access to some resources is provided without charge as a matter of policy (e.g., outdoor recreation), and the fees charged for most other resources are fixed administratively and arbitrarily rather than by competitive processes. Nevertheless, as discussed below, the opportunities for developing markets and prices for forest resources are considerable.

This chapter is concerned primarily with the effect of economic instruments on the behaviour of resource users. This is important, because charges for natural resources are often advocated or justified on other grounds, such as the equity of "user pay," the desirability of "cost-recovery" in government programs, the right of public owners to resource rents, or the need for public revenue in general. All these considerations influence policymakers. But the focus here is on the capacity of economic instruments to alter economic behaviour in ways consistent with sustainability, so we leave aside these other fiscal objectives (though it is worth noting that pricing is likely to advance them as well).

Taxes and Subsidies

Taxes add to the price of goods and services, which tends to dampen demand for them. Subsidies have the opposite effect, and so can be employed to encourage desired forms of consumption or production behaviour.

A broad array of taxes and subsidies can be used to support resource management. Subsidies in the form of grants, soft loans, or offsets against taxes are sometimes used to assist producers in complying with pollution control standards. Subsidies are common when new standards requiring new technologies are imposed. Tax credits, or elimination or reduction of tax, are often designed to encourage environmental investments.

Conversely, tax measures can be used as disincentives, to discourage undesirable behaviour. Environmental charges on products penalize the use of products that impose a burden on the environment. Effluent charges encourage polluters to abate their wastes. Revenues from such levies are sometimes earmarked for funds for environmental improvement. Examples of all of these measures are found in British Columbia.

Deposit-Refund Systems
A fourth category of economic instruments for environmental protection is deposit-refund systems. These systems involve adding a charge to the price of a product that may pollute if discarded, and refunding the charge when the product is returned. In British Columbia, deposit-refund systems are associated with beverage containers and automobile batteries, but not with natural resources. However, performance bonds, a common feature of resource licences, have a similar influence in encouraging licensees to meet performance requirements.

Compliance Incentives
The final class of economic instruments is compliance incentives. The primary mechanism of this kind is non-compliance fees, charged as penalties for failure to meet regulations. Such fees are ultimately enforced through the courts.

All of the above five categories of economic instruments have some present or potential application to forest resource management in British Columbia. However, some will not be considered further in this chapter. Deposit-refund schemes appear to have limited scope in natural resource-based activities, and so will be ignored. Compliance incentives, or penalties for non-compliance, are more properly viewed as adjuncts to regulatory controls rather than as economic instruments in themselves. And taxes and subsidies affect economic behaviour through their effects on costs of production and prices of goods and services, and so can be considered under the general heading of pricing. Accordingly, the remainder of this chapter focuses on property rights and pricing as economic instruments for promoting sustainable forest management.

Potential Economic Instruments to Support the Sustainability of Forest Resources
What are the prospects for wider application of these economic instruments to specific forest resources and values? In the following discussion, property rights and pricing mechanisms are considered in terms of their

potential application to the major forest values in British Columbia – timber, water, fish, wildlife, forage, and recreation.

As stated, there appears to be considerable scope for development of economic instruments to improve resource management and use, and economic instruments have significant advantages over command and control regulation. But those points must not be overstated. While the following discussion suggests that market mechanisms might be beneficially invoked for a range of forest values for which they are now lacking or deficient, they cannot eliminate altogether the need for regulation. Some values, such as the aesthetic appeal of songbirds and landscapes, or the cultural and scientific value of wilderness, do not lend themselves well to economic instruments for management. As Rod Dobell explains in Chapter 4, governments cannot assign everything to markets; governments have an essential role both in setting social goals in resource management and in ensuring that the activities of producers and consumers are consistent with achieving those goals. Thus, the best way of ensuring socially desirable activity will sometimes involve regulation.

This point is important to emphasize here, because advocates of markets and advocates of governmental intervention are often identified with right-wing and left-wing ideology respectively, and each resists the opposing view. However, most commentators agree that market forces are powerful influences on economic behaviour and are capable of assisting in the protection of environmental values. But markets will never be perfect, so some regulation will always be useful. The task is to strike the right balance between reliance on markets and reliance on regulation.

The following discussion suggests that the objective of sustainability of forest resources can be advanced by engaging market forces more than at present, thus reducing, but not eliminating altogether, the need for regulatory controls. The opportunities for developing property rights and pricing as means of promoting sustainable forest use are examined, first for timber and then for other forest use values.

Timber

Historically, timber production has been the dominant use of forests in British Columbia. Timber is now managed and used under a tenure system that embraces a complex array of property rights. A small fraction of the forest land in British Columbia, in the order of 4 percent, is privately owned. Almost all the rest is the property of the provincial Crown, and rights to the timber on it are conveyed to private users through various forms of usufructory rights. With few exceptions, these rights to timber on Crown land are allocated without competition. Their terms range from one to twenty-five years, usually with provisions for renewal or replacement. Holders' rights extend only to industrial timber, and require the holder to pay a stumpage fee to the Crown as the timber is harvested.

Most forms of tenure give their holders the exclusive right to harvest timber within a defined forest area. However, the major form – the Forest Licence – entitles the holder to harvest a specified volume of timber each year within a general administrative area in which others also hold cutting rights. The specific timber stands to be harvested by the licensee are identified by supplementary authorizations from time to time during the term of the licence. Under all forms of licence, the Ministry of Forests closely regulates activities, notably the rate and pattern of harvesting, utilization of the timber cut, roadbuilding, environmental protection, and silviculture.

Thus, for the most part, timber is managed and used under highly truncated forms of property rights. As policy instruments for promoting sustainability of forests, these tenure arrangements have a number of shortcomings that obstruct efficient use and blunt incentives to invest in silviculture and forest enhancement. These weaknesses can be identified with reference to the dimensions of property noted earlier:

- Licences that do not convey exclusive, long-term rights over a defined area of forest undermine licensees' incentives to invest in future forest production because the benefits will accrue to others (Pearse 1985).
- The terms of all licences are much shorter than the time it normally takes to grow forest crops, and renewals are always conditional. This fact deters licensees from investing in potentially profitable timber production (Haley and Luckert 1990).
- Rights that extend only to timber discourage holders from considering the impact of their logging and other operations on water, wildlife, and other forest resources.
- Restrictions on transferability and divisibility of licences impede competition, promote industrial concentration, and prevent advantageous reallocation of resources among potential users as conditions change. The absence of competition in the initial allocations of licences aggravates these tendencies (Pearse 1992a).
- Recurrent legislative interventions by the provincial government have curtailed the rights conveyed under licences and regulated the activities of licensees. This intervention has weakened the perceived security of property rights.

These and other weaknesses of the tenure system suggest that strengthening the property rights of users would advance the sustainability of timber resources.

The most pervasive impediment to users' incentives to manage industrial forests and enhance timber production is the prevailing uncertainty of their rights. The limited, temporary, and closely controlled usufructory rights that characterize the present tenure system, coupled with uncertainties in contemporary British Columbia arising from Native land claims, environmental pressures, and a history of legislative intervention to cancel or alter the rights held by timber companies, leave present licensees too uncertain about the security of their rights to voluntarily invest in long-term forest management (Luckert and Haley 1990).

Accordingly, policy aimed at promoting sustainability of timber resources should address means of improving the quality, security, and value of users' rights (CORE 1994). Several avenues warrant consideration:

- Strengthening the proprietary interest of holders of timber rights. Possibilities include the sale or transfer of Crown lands to private owners: the conversion of licences to leases or other, stronger, forms of property that give holders a registered interest enforceable against third parties.
- Providing assurance of compensation in the event of cancellation or curtailment of rights by the government.
- Incorporating less restrictive terms and conditions in usufructory rights. This approach might include longer terms and more certain provisions for renewal, relaxation of restrictions on the divisibility and transferability of rights, abolition of the prohibition on sales of logs outside the province, and simplification and streamlining of burdensome controls on harvesting and utilization.
- Defining more clearly the resources to which the rights apply. For example, licences that do not identify the timber that the licensee is authorized to cut might be altered to apply to defined geographical areas, within which licensees are assigned exclusive rights to the timber for the duration of their licences.

Strengthening the tenure system in such ways would stimulate the licensees' proprietary interest and encourage investment in the long-term productivity of the forests they use. The magnitude of the impact of such changes is difficult to estimate, because few empirical studies have been made. However, the available evidence suggests it may be substantial. Two recent studies indicate that timber companies in British Columbia spend several times as much, per hectare, on voluntary silviculture on their freehold lands as they do on similar Crown lands held under licence (Zhang and Pearse 1996; Haley and Luckert 1990). On the reasonable assumption that they invest on their own lands when the benefits exceed the costs, the investment shortfall on licensed Crown lands provides an indication of the potential benefits in forest enhancement currently foregone as a result of the weaknesses of the present tenure system.

The weaknesses of property rights to timber extend beyond their insecurity. Licensing arrangements for Crown forests provide licensees with rights to the timber only, leaving them no incentive to protect or cultivate other forest values such as water, wildlife, and recreational opportunities, which are often adversely affected by timber operations. Moreover, the users of these other forest resources typically hold much weaker rights, which they cannot assert against detrimental timber operations, thus biasing forest management against non-timber forest values. A later part of this chapter considers economic instruments for promoting the integrated management of forests for the full range of benefits they can yield.

Present arrangements also suggest considerable scope for improving markets for timber, and promoting sustainability by ensuring that prices reflect the full value of resources used. The absence of competition in the allocation of most licences precludes market valuation of timber rights. Until recently at least, substantially higher prices were paid in the few competitive timber sales, suggesting that the administered prices paid under the usual non-competitive licences significantly undervalues Crown timber (BC Ministry of Forests 1993a). Provincial legislation prohibits the sale of raw logs to buyers outside the province. As a result, the value of some timber has been reduced to as little as one-fifth of the value it would have in the absence of this restriction (Pearse 1993a). Furthermore, requirements on licensees to manufacture their timber in an appurtenant mill depresses its potential value – in one case, the loss was estimated at $60 per cubic metre, several times the stumpage price (Binkley 1995). These restrictions on the way timber can be used and marketed inevitably depress its value, and the available evidence suggests the impact is substantial.

The value of timber is also depressed by regulatory requirements that increase the cost of harvesting timber. These requirements include utilization standards, cut controls, scaling rules, and a wide range of regulations on logging. The controversial Forest Practices Code, only one of several regulatory initiatives in recent years, has been conservatively estimated to cost the province $2.1 billion per year in reduced timber availability, increased information requirements, compliance, training, administration, and enforcement (Haley 1996). The environmental and other benefits have not been quantified, but professional foresters complain that this command and control regulation is inefficient and often inappropriate. They contend that the results could be achieved at a much lower cost through requirements to achieve specified results rather than the present detailed specification of operational methods and practices (PFLA 1995).

Stumpage charges for Crown timber distort users' incentives to use resources efficiently. These charges are levied as a fixed amount per cubic metre on all the timber recovered under a licence, regardless of its varying quality and value. This practice encourages "high-grading" and waste, and in turn demands more regulation in the form of utilization standards.

These impediments to realization of the full potential value of timber and timber rights, and to the reflection of these values in market prices, weaken the incentives of forest enterprises to use resources efficiently and to invest in silviculture and forest development. Several policy changes would strengthen the role of prices in guiding timber users towards improved forest management:

- Expansion of provisions for competitive bidding as the means of allocating timber rights.
- Promotion of markets for intermediate wood products, such as logs and chips, by removing existing restrictions, including export restrictions and requirements to manufacture timber in designated mills.
- A shift from command and control regulation of forest practices to results-based regulatory requirements.
- Revision of the manner in which stumpage charges are collected, from a fixed amount per cubic metre of wood recovered, to a lump-sum or annual payment for the timber the licensee is authorized to harvest.

Available evidence suggests that such measures would substantially increase the value and price of timber and improve its utilization. Resulting higher values for timber and timber rights would strengthen incentives to utilize resources for maximum value, and to invest in continuing timber production.

Forest Resources Other Than Timber

Forests provide many benefits in addition to commercial timber, such as water, livestock grazing, and recreation. A wide variety of these other goods and services are produced under varying forest conditions, and they are produced, or consumed, in varying combinations.

Forest resources other than timber are allocated and used in British Columbia under even more rudimentary management regimes than that for timber. Users' rights are usually not exclusive, and they are often highly restricted, short term, and non-transferable. Most of these values are isolated from markets and market prices altogether.

Not coincidentally, these non-commercial values, used under weak forms of property rights, give rise to some of the most serious market failures in forest resource use. This issue is reflected in the environmental problems associated particularly with water, fish, wildlife, and recreation. It explains, as well, the almost complete reliance on government to manage and develop these resource values. Forest values, other than timber, thus offer wide scope for development of economic instruments.

Table 2.1 lists some of the most important forms of rights to resources on Crown forest land in British Columbia. The list is not exhaustive: it excludes rights on private lands, rights to sub-surface minerals, and rights to a wide range of minor resources. Moreover, there are many variants of the rights listed. The list includes only the rights to activities bearing most heavily on the way forests are managed and used, and is intended only to illustrate the general pattern of rights on Crown forest lands.

Scope for Strengthening Rights to Forest Resources Other than Timber
The rights governing the use of various forest resources in British Columbia,

Table 2.1

Major forms of rights held by users of resources on Crown forest land in BC

Resource	Subject of property rights	Form of users' property rights
timber	timber harvesting and management	tree farm licences forest licences
water	water flow waste-absorbing capacity of watercourses	water licence waste discharge permit
fish	recreational fishing aboriginal fishing	sport fishing licence varies (see text)
wildlife	recreational game hunting aboriginal hunting trapping of furbearing animals commercial guiding	hunting licence varies (see text) trapline licence guiding territory licence
forage	grazing	grazing lease grazing licence grazing permit
recreational resources	non-consumptive recreation	general right of public access

their limitations, and ways in which they might be strengthened to promote sustainability have been examined in detail elsewhere (Pearse and Jessee 1994). For present purposes, it is sufficient to note certain general features of the rights to forest resources other than timber. Most take the form of licences or permits, which convey only non-exclusive and highly restrictive usufructory rights that are not enforceable against third parties. An exception is the leases used to convey some grazing rights; another is the aboriginal rights to fish and hunt, which are rooted in the Constitution, and sometimes set out in treaties and agreements. But under most of these forms of tenure, the quality of title and the security of the holder's rights are notably weak.

Other limitations of these rights can be seen in terms of the dimensions of property noted earlier. Most are not *exclusive*, insofar as more than one party usually holds rights to the same resources (the exceptions are grazing leases and trapline licences). Nor, in most cases, is the number of licensed users limited. Aboriginal rights are communal, and the number of eligible users depends on the size of the aboriginal communities. In the case of non-consumptive recreation, access is not usually regulated at all.

The *duration* of these rights is also limited, with the exception of aboriginal rights and water licences. Their terms vary, as do their provisions for renewal. Some, such as hunting and fishing rights, have terms of only one year or less.

None of these rights is freely *transferable*. Some, such as hunting and fishing licences and aboriginal rights, are not transferable at all, while most other licences may be transferred only with the consent of the responsible Minister. None is divisible.

The *comprehensiveness* of all of these rights is highly restricted insofar as they convey rights to only one forest resource, and most restrict the holders to a particular form of use, such as recreational or commercial use.

The *benefits conferred* under the rights are constrained in two general ways. One constraint is through levies and fees charged for the rights themselves; these imply that the holders of the rights must share some of the economic benefits with the Crown. However, not all of the rights listed in Table 2.1 involve governmental charges and, for those that do, the charges are nominal and bear little relation to the values conferred. The other way in which benefits are constrained is through regulations that restrict the holders in the way they may exercise their rights, thereby preventing them from generating the highest possible returns from the resources they use.

These limitations suggest several general ways in which rights to forest resources other than timber could be changed to make them more effective economic instruments for achieving sustainable forest development:

Exclusivity. Rights that are now held in common could be made more exclusive. For example, grazing licences and permits that convey non-exclusive rights to forage could be converted to exclusive leases or licences, thus giving their holders incentives to manage the resources for long-term sustainability. Guiding licences, which now convey exclusive rights to guide non-resident hunters only, could be extended to resident hunting also, thereby giving the holders exclusive rights to hunting and strengthened incentives to manage the game. Resort owners could be assigned rights to manage and regulate fishing on individual lakes and streams, as in some fly-in lakes in Manitoba (Pearse 1988a). And fishing and hunting rights could be granted to private parties, as in Europe, or to local organizations of users along the lines of Quebec's system of *zones d'exploitation contrôlée* (Pearse and Wilson forthcoming).

Duration. All rights with fixed terms can be strengthened by extending their duration and provisions for renewal. The longer the terms, the more secure and valuable will be the rights and the longer the time perspective of their holders in making their management decisions.

Transferability. Restrictions on transferability and divisibility, which now encumber most licences and permits, could be reduced or eliminated, thereby enabling their acquisition by users who can make the most beneficial use of them (OECD 1994).

Comprehensiveness. Rights that restrict the form in which a resource may be used (such as non-commercial use hunting licences) or to the specific purpose for which rights may be exercised (such as water licences issued specifically for agriculture or mining) could be broadened to permit reallocation among uses as circumstances change.

Benefits Conferred. The economic benefits that accrue to holders of resource rights could be increased in two ways: by allowing the holders more freedom to use the resources to best advantage, and by reducing the required payments to the Crown. The first of these changes might apply to the regulations requiring the use of particular technologies (such as those that usually attach to waste discharge permits), and regulations restricting the way rights may be exercised (as in hunting and fishing rights). The second applies to licence fees and related charges, especially those that blunt licensees' incentives to enhance the resources.

Scope for Pricing
Pricing now has almost no role in the allocation and use of forest resources other than timber in British Columbia, so present policies leave a great deal of scope for pricing mechanisms as instruments for promoting sustainable resource development.

Policy reform in this area should aim at providing the private holders of rights to resources with access to competitive markets for their products, so they will be able to realize the economic benefits that can be generated through careful protection, management, and enhancement of the resources.

Governments can develop markets either at the level of the final consumer or at the level of the private producer. Examples of the former include the present arrangements through which provincial agencies sell licences to individual sport fishers and hunters and users of campgrounds (although the prices charged are not determined by market forces). The second involves allocation of rights to resources to producing enterprises, which in turn retail their product to individual consumers. Examples include municipal utilities that obtain water under provincial water licences and distribute it to urban households and businesses, and private operators of ski resorts and campsites. In these latter cases, there are two markets to consider: the market through which the government allocates resource rights to enterprises, and the market through which enterprises sell services to consumers.

Governments traditionally find it difficult to act as competitive suppliers of resources to individual consumers. They encounter heavy resistance to pricing, especially prices that differ among locations in response to supply and demand. Furthermore, the provincial government is typically in a monopoly position, making it difficult for it to determine proxies for competitive prices. Non-governmental bodies, especially private enterprises, find it less difficult to engage in competitive market pricing; indeed, they are expected to do so.

The way the government allocates rights to enterprises has less lasting consequence for efficient resource use if the rights are divisible and transferable. However, if an objective is to capture the resource rent for the Crown, it is important to do so in a way that minimizes distortions. As with timber, the least distortive means of capturing resource rents are initial or lump-sum payments, or fixed annual rental charges for rights. The practicality of these sorts of levies is obviously related to the forms of tenure employed, discussed earlier.

In short, pricing can be developed in several general ways to promote the sustainable use of non-timber resources:

- Where governments provide access to these resources directly to consumers, by adopting pricing policies that encourage efficient resource use – generally, prices that reflect the long-run marginal cost of supply, including the value (the opportunity cost) of the resource itself.
- Where governments allocate rights to enterprises that in turn supply final consumers, by ensuring that any charges for those rights do not distort patterns of use and resource allocation.
- Where consumers are supplied by enterprises, by minimizing restrictions on price rationing and competitive pricing.

Development of pricing, as a means of encouraging efficient and sustainable use of resources, would almost certainly be facilitated by institutional changes to enable non-governmental organizations to acquire resource rights and serve the consumers of these other forest values.

Integration of Forest Uses
One of the precepts of sustainability is that all resource and environmental values must be appropriately recognized and weighed in management decisions. If users are not confronted with all the benefits and costs associated with their production of a particular forest value, they will lack the economic signals and incentives to use and to develop the value efficiently. For example, if a resource such as water or livestock forage is unpriced or underpriced, users will tend to use it wastefully and excessively, and their incentives to protect and invest in sustainable or enhanced production of the resource will be weakened.

Inefficiencies can also result from bias in the combination of resource values produced on a given tract of forest. Usually, the most beneficial management regime is one that accommodates, simultaneously or intermittently, production of two or more types of goods or services. Thus, "multiple use" is a long-established principle of forest management. However, if different resource values are treated differently, so that the incentives of users to use and develop them are biased in varying degrees, the resulting combination of uses will be distorted. Thus, users can be expected to favour commercial products that yield revenue over forest values that are unpriced and unmarketable. Indeed, much of the environmental concern about forest management arises from the economic incentives favouring industrial values, such as timber, at the expense of unpriced values, such as recreation, wildlife, and aesthetic benefits.

The problem of producing the optimum combination of values on each tract of forest is complicated by their interdependence in production. Production of timber affects wildlife, recreation, and aesthetic values, and the effect may be adverse or beneficial in different circumstances. Moreover, the degree to which one form of production affects the other depends on the intensity of its production.

If a private owner's property rights to a forest were fully comprehensive (i.e., the owner holds complete property in all attributes of the forest, receiving all the benefits and bearing all the costs of producing the various goods and services), the owner could be expected to search for the combination of uses that would generate the greatest net return in the aggregate. The same result could be achieved by separate owners of each forest attribute who were able to bargain with each other, so that one form of production could expand at the expense of another whenever the former was more valuable.

The first of these models is approximated in the traditional freehold title to land, which provides owners with rights to all the resources on and under their land. The second can be found in some European countries where ownership of the resources on a tract of land is typically stratified; separate owners hold rights to the surface of the land, sub-surface minerals, water, fisheries. and other attributes, and they bargain among themselves to accommodate beneficial changes in the pattern of resource use.

Neither model is found in British Columbia, however. Tenure arrangements provide forest users with rights to only one attribute at a time. Even owners of private forest land lack rights to the water, sub-surface minerals, fish, and wildlife on it, these attributes having been reserved (with minor exceptions) to the Crown by legislation. Rights to use different attributes are allocated separately, to different parties, through varying forms of tenure. But the holders of rights to different resources on the same forest tract cannot bargain among themselves and benefit by shifting the pattern of uses towards the most valuable combination. As a result, the extent to which each value is compromised in the interest of others is determined, not by rent-maximizing efforts of users, but by regulation and administrative discretion, without systematic reference to the values involved.

The goal of integrated resource use calls for weighing conflicting forest values and uses against each other to determine the use or combination of uses of each forest tract that will generate the greatest benefit. The present imbalance in the rights available to users of different forest attributes militates against optimum patterns of resource use. These distortions could be eliminated or reduced by:

- Strengthening the rights available to users of non-timber forest resources, putting them on a comparable legal footing with holders of timber rights (as discussed).
- Expanding markets for forest products and services now available free or at nominal charge, to enable holders of rights to them to benefit from efficient management and use.
- Providing for divisibility and transferability of rights to all forest resources, thus enabling competing forest users to bargain among themselves to determine the most beneficial compromises in resource values.

These changes would provide the framework for developing the second (stratified) model for integrated resource use referred to earlier.

Conclusion

This chapter has provided an overview of opportunities for the development of economic instruments to promote sustainability of forest resources in British Columbia. The opportunities appear to be numerous and substantial, resulting in large part from the prevalence of public ownership of land and resources and a long-standing governmental policy of providing access to forest resources without charge, or at least without recourse to market mechanisms.

Though arrangements vary widely across the array of timber and non-commercial values, most forest products and services are accessible to users only under crude forms of property rights. These truncated rights blunt incentives to use land efficiently and to invest in sustainable resource use and development. Moreover, many forest goods and services such as wildlife, water, and recreation are insulated from markets. This prevents their value from being registered in prices, and obscures decisions about the most beneficial trade-offs among conflicting uses.

Thus, there appears to be broad scope for improving resource management through development of property rights, markets, and pricing as instruments for promoting sustainability of forest resources in British Columbia. As pressures on forest resources grow and increase in variety, and as reliance on nature's bounty must increasingly be replaced by management effort and investment, the benefits of harmonizing the incentives of users with the requirements of sustainable resource use will grow as well.

This chapter has focused on economic instruments – particularly, property rights and pricing – as means of improving resource use and development. It has not given much attention to the distributional (and hence political) implications of adopting such policies. These implications are nevertheless important, if not decisive, influences on resource management policies, and warrant careful investigation. The chapter suggests at least two other important subjects for investigation. The first area is the types of property rights available for application to various resources and their implications for management. Special attention needs to be given to the legal form of property suitable for resources now managed under poorly defined rights, such as fish, wildlife, environmental, and recreational values. The other area is the institutional, organizational, and bureaucratic changes needed to accommodate a shift in policy direction from traditional command and control regulation of forest users to an increased reliance on market forces and the incentives of users to achieve resource management objectives. This issue goes beyond the familiar problems of bureaucratic inertia and resistance to change, to more complicated issues of appropriate organization, professional expertise, and divisions of responsibility between government and private resource users, some of which are discussed in other contributions to this book.

Acknowledgment
This chapter draws heavily on a report prepared for Environment Canada (Pearse and Jessee 1994).

References
BC Ministry of Forests. 1989. *Range Program Review: Final Report of the Review Task Force.* Victoria: Ministry of Forests
–. 1993a. *Annual Report of the Ministry of Forests.* Victoria: Queen's Printer
–. 1993b. *Forest Practices Code: Rules Document.* Victoria: Queen's Printer
Binkley, C.S. 1995. "A cross road in the forest: The path to a sustainable forest sector in BC." *BC Studies* 113: 39-61
Bromley, D.W. 1989. *Economic Interests and Institutions: The Conceptual Foundations of Public Policy.* Oxford: Basil Blackwell
Brundtland, G.H. 1987. *Our Common Future.* Oxford: World Commission on Environment and Development, Oxford University Press
Campbell, R.S., P.H. Pearse, and A. Scott. 1972. "Water allocation in British Columbia: An economic assessment of public policy." *University of British Columbia Law Review* 7: 247-92
Canada's Green Plan. 1990. *Report on the Green Plan Consultations.* Ottawa: Government of Canada
Cassils, J.A. 1991. *Exploring Incentives: An Introduction to Incentives and Economic Instruments for Sustainable Development.* Ottawa: National Round Table on the Environment and the Economy
Coase, R.H. 1960. "The problem of social cost." *Journal of Law and Economics* 3: 1-44
Commission on Resources and Environment (CORE). 1994. *Finding Common Ground: A*

Shared Vision for Land Use in British Columbia – Appendices. Vol. II. Victoria: Queen's Printer

Demsetz, H. 1967. "Toward a theory of property rights." *American Economic Review* 57: 347-59

Government of Canada. 1992. *Economic Instruments for Environmental Protection.* Ottawa: Ministry of Supply and Services

Haley, D. 1996. "Paying the piper: The cost of the British Columbia *Forest Practices Code.*" Paper prepared for the conference Working with the Forest Practices Code, Vancouver

–, and M.K. Luckert. 1990. *Forest Tenures in Canada: A Framework for Policy Analysis.* Information Report E-X-43. Ottawa: Forestry Canada

Luckert, M.K., and D. Haley. 1990. "The implications of various silvicultural funding arrangements for privately managed public forest land in Canada." *New Forests* 4: 1-12

Manitoba Environment. n.d. *Harnessing Market Forces to Support the Environment.* Winnipeg: Department of Environment

Minister of the Environment. 1983. *Guidelines for Wildlife Policy in Canada.* Catalogue no. CW66-59/1983E. Ottawa: Supply and Services

Novak, M., J.A. Baker, M.E. Obbard, and B. Malloch, eds. 1987. *Wild Furbearer Management and Conservation in North America.* Toronto: Ontario Ministry of Natural Resources and Ontario Trappers Association

Olewiler, N. 1993. "Pricing the Environment: Efficient Pollution Policies: Summary." In *Competitiveness and Pricing in the Public Sector,* ed. R.W. Crawley, 89-95. Kingston: Queen's University

Opschoor H., and H. Vos. 1989. *Economic Instruments for Environmental Protection.* Paris: Organization for Economic Cooperation and Development

Organization for Economic Cooperation and Development (OECD). 1992. "Developing property rights as instruments of natural resources policy: The case of the fisheries." In *Climate Change: Designing a Tradeable Permit System,* 109-22. Paris: OECD

Pearse, P.H. 1985. "Obstacles to silviculture in Canada." *Forestry Chronicle* April: 91-96

–. 1988a. "Property rights and the development of natural resources policy in Canada." *Canadian Public Policy* 14(3): 307-20

–. 1988b. *Rising to the Challenge.* Ottawa: Canadian Wildlife Federation

–. 1990. *Introduction to Forest Economics.* Vancouver: UBC Press

–. 1992a. "Forest Tenure, Management Incentives, and the Search for Sustainable Development Policies." Paper prepared for the Conference on Forestry and the Environment, Jasper AB

–. 1992b. *Managing Salmon in the Fraser.* Report to the Minister of Fisheries and Oceans on the Fraser River Salmon Investigation. Vancouver: Department of Fisheries and Oceans

–. 1993a. "Evolution of the forest tenure system in British Columbia." Unpublished discussion paper. UBC, Faculty of Forestry

–. 1993b. "It's time to break the log jam." *Globe and Mail* June 17: A19

–, A.V. Backman, and E.L. Young. 1974. *Timber Appraisal: Policies and Procedures for Evaluating. Crown Timber in B.C. Second Report of the Task Force on Crown Timber Disposal.* Victoria: Forest Service

–, F. Bertrand, and J.W. MacLaren. 1985. *Currents of Change. Final Report, Inquiry on Federal Water Policy.* Ottawa: Environment Canada

–, and J. Jessee. 1994. *Economic Instruments for Managing the Resources of Forest Landscapes.* Vancouver: Environment Canada

–, and D.M. Tate. 1991. "Economic Instruments for Sustainable Development in Water Management." In *Perspectives on Sustainable Development in Water Management: Towards Agreement in the Fraser River Basin,* ed. A.H.J. Dorcey, 431-52. Vancouver: Westwater Research Centre, UBC

–, and J.R. Wilson. Forthcoming. "Local Co-Management of Fish and Wildlife: The Quebec Experiences." *Wildlife Society Bulletin*

Pezzy, J. 1992. "Sustainability: An interdisciplinary guide." *Environmental Values* 1: 321-62

Portney, P. 1988. "Reforming environmental regulation: Three modest proposals." *Columbia Journal of Environmental Law* 13(2): 201-16

Posner, R.A. 1977. *Economic Analysis of Law.* 2nd edition. Boston: Little, Brown and Company

Private Forest Landowners' Association (PFLA). 1995. *A Framework for the Regulation of Forest Practices on Private Forest Land.* Submission to the Ministry of Forests, Vancouver

Range Act. 1979. Revised Statutes of British Columbia, c. 355

Rich, E.E. 1967. *The Fur Trade and the Northwest to 1857.* Toronto: McClelland and Stewart

Scott, A., and G. Coustalin. 1996. "The evolution of water rights." *Natural Resources Journal* 35(4): 821-979

Stager, D. 1988. *Economic Analysis and Canadian Policy.* 6th ed. Toronto: Butterworths

Tate, D.M., S. Renzetti, and H.A. Shaw. 1992. *Economic Instruments for Water Management: The Case for Industrial Water Pricing.* Social Science Series no. 26. Ottawa: Environment Canada

Tietenberg, T. 1995. *Environmental and Natural Resource Economics.* 4th ed. New York: HarperCollins

Wiener, J.B. 1995. "Promoting market-based performance incentives in regulatory reform." Statement before the Committee on Governmental Affairs, United States Senate

Wildlife Branch. 1996. *British Columbia 1996-1997 Hunting and Trapping Regulations Synopsis.* Victoria: Ministry of Environment, Lands and Parks

Wilman, E.A. 1988. "Pricing policies for outdoor recreation." *Land Economics* 64(3): 234-41

Zhang, D., and P.H. Pearse. 1996. "Differences in silvicultural investment under various types of forest tenure in British Columbia." *Forest Sciences* 42(4): 442-48

3

Governing Instruments for Forest Policy in British Columbia: A Positive and Normative Analysis

W.T. Stanbury and Ilan B. Vertinsky

Introduction

Themes/Foci/Concerns

The current government of British Columbia has sought to transform the forest industry in a number of ways in seeking the overarching goal of environmental, economic, and social sustainability.[1] To do so, it has introduced a host of major policy initiatives[2] whose principal effects appear to be the following:

- reductions in the allowable annual cut (AAC);
- elaborate and stringent regulations on forestry practices;
- reductions in the size of the land base that may be used for industrial forestry;
- increases in the price of wood taken from Crown land, both direct (stumpage) and indirect (reforestation requirements); and
- changes in land use policy-making, i.e., extensive use of multi-stake-holder advisory processes.

The government is trying to achieve a host of major changes in only a few years. It maintains that the many initiatives are part of a comprehensive and coherent plan; however, several initiatives appear to be more immediate responses to external pressures or to the public's reactions to earlier initiatives. External pressures include boycotts in important foreign markets (Stanbury, Vertinsky, and Wilson 1994; Stanbury and Vertinsky 1996, 1997).

The full effects of the policy initiatives, both economic and ecological, have not yet been felt (Price Waterhouse 1995; Peterson 1996; Haley 1996). The biggest unknown (in scale if not direction) is the implications for forest policy (and the industry) of the settlement of Native land claims.

The number of policy goals or objectives has expanded enormously,[3] while the number of types of instruments (means to achieve them) has increased very little. The dominant means to achieve many of the new goals is "command and control" regulation (e.g., the Forest Practices Code), and the zoning and land use regulations required to implement the various regional land use plans (e.g., the Vancouver Island Land Use Plan; see CORE 1994).[4]

In the rush to achieve new environmental goals, it appears that insufficient effort has been made to ensure that these goals are being pursued in a cost-effective manner. The province has provided almost no information on the use of cost-benefit analysis, for example, to assess proposed major changes. By comparison, see van Kooten (1996).

Both the important forest policy issues and the recent policy responses launched by the BC government raise difficult problems for policy analysts. In this chapter, these problems are examined by analyzing the nature of governing instruments. This introductory part looks at several problems associated with instrument choice. The following part describes and analyzes several important instruments used by the provincial government to achieve its policy objectives relating to forests and the forest industry. The next part considers the politics of instrument choice. The discussion of the politics of instrument choice continues with a brief examination of the new Forest Practices Code, a practical example of "regulation" as a governing instrument.[5] The attractiveness of regulation is examined in relation to politicians, bureaucrats, and interest groups. The concluding part examines difficulties in the use of incentive-focused ("economic") instruments to achieve environmental objectives in forest policy.

Problems Associated With Forest Policy-Making
Making public policy for forests and the forest products industry in British Columbia (and elsewhere) is fraught with difficulty for a wide variety of reasons.

Role of the Forest. Forests have a vital role in both the local/regional economy and in the ecology of the planet.[6]

Significance for Aboriginals. Forests have particular cultural and economic significance for aboriginals. The extensive land claims made by aboriginal groups include most of the commercial forest land in BC.

Benefits Separate from Costs. Beneficiaries of efforts to protect forests are often separated (geographically and temporally) from persons bearing the greater part of the costs in terms of reduced incomes and higher unemployment in forest-dependent towns.[7]

Provincial Ownership of Forest Land. Over 90 percent of the forest land in British Columbia is owned by the province. This fact gives it legal powers associated with ownership, but it also increases the number of interdependent roles the province must play in forest policy, e.g., regulator, taxer, spender, and so on. For some observers, this ownership represents a major constraint on the attainment of allocative efficiency for the forests (Borcherding 1983; Baden and Stroup 1981; Dowdle 1981).

Non-Market Benefits. There are increasing demands that forests be used to provide a variety of benefits (besides commercial timber and fibre production) to a number of constituencies.[8] Many of these benefits are non-market in nature (which makes signalling of preferences and valuation much more difficult), and involve assumptions about the preferences of future generations.

Time Horizons. Forest policy requires dealing with a long time horizon, largely because of the age of the substantial areas of old-growth forests in BC (typically, 200 to 500 years) and the long rotation period of plantation forests (about 80 years). Yet the electoral cycle of the guardians of the province's forests is only about four years, and the time horizon of some corporate actors may be even shorter. Other corporations are prepared to take a much longer view when they invest in new mills with an expected economic life of thirty or forty years. The time horizon of forest firms depends on the security of their property rights as well as normal business risks.

Limited Knowledge. Although scientific knowledge has grown a great deal, our stock of hard scientific knowledge is modest relative to the problems with which we are confronted. As a result, the more risk-averse advocate widespread application of the "precautionary principle."

Differing Goals. Citizens and groups active in the policy debate have been sending widely differing signals to government about what actions it should take (see Table 3.1). A major problem for politicians (and the policy analysts that advise them) is that the stated goals are incommensurable. Conflicts over the use (or non-use) of forests are often intense and prolonged (Stanbury 1994b; Stanbury, Vertinsky, and Wilson 1994; Stanbury and Vertinsky 1996, 1997; Chase 1995; Dietrich 1992). These disagreements make systematic and coherent action by government difficult, if not impossible. The analysts' concept of optimization is called into question because the construction of anything similar to a social welfare function is impossible. Moreover, under political rationality, means and ends are highly interdependent, as discussed below.

Analysis of the Instrument's Ability to Achieve Forest Policy Objectives

Formulating Objectives and Selecting Governing Instruments

In formulating public policy, a distinction must be made between positive

Table 3.1

Comparative perspectives on forestry issues in British Columbia

Forest companies	Environmental groups
Commercial forests as a source of wood for lumber; fibre for pulp and paper	All forests seen as "lungs of the earth" and large store of biodiversity
Forests are a major source of jobs, profits, exports, government revenues	Preservation of a shrinking stock of a unique asset; stop "liquidating" old-growth forests
Abundant supply of processed products to increase the standard of living	Need to *reduce* consumption; maintain biodiversity
80-year cycle of cutting, replanting, growth, and maturity	Preservation of almost all remaining ancient forests for all time in their *natural* state; plantation forests are biological deserts
Benefits for the present generation and future generations through long run-sustained yield	Concern about *future* generations; indefinite time horizon; preserve old-growth/ancient forests as they are today
Maintenance of property rights in land and harvesting rights	May need to override property rights in the collective interest to protect ecosystems
Scientific regeneration, perhaps using intensive silviculture; can improve on some natural processes	Natural regeneration; natural processes are inherently benign and the ideal
Economic methods of harvesting, notably clearcuts	Limit clearcuts to 1 hectare; use selection logging to protect the environment
Fairly limited public participation in decision-making; rely on elected representatives and public servants	Broad public participation
Rely on scientific/professional expertise on forestry issues	Good ecology is largely intuitive; use science selectively when it helps protect ecosystems
Profits are the "bottom line"; emphasis on efficiency	Protection of the environment is the "bottom line"; emphasis on whole ecosystems

analysis (what is) and normative statements (what ought to be). Politicians, on behalf of citizens, provide policy goals or objectives. The production of such goals is normative, based on value judgments. Economists (in general, policy analysts) seek to ascertain the best policy to achieve the stated objectives using positive analysis.[9]

We argue that values are embedded in almost all aspects of policy making – even if they are obscure. We also argue that it is both an analytical and practical impossibility to separate policy objectives (ends) from governing instruments (means) in the analysis of most public policies. Further, we suggest that in selecting policy instruments (as well as objectives), cabinet ministers seek to maximize the probability of being reelected. In doing so, they behave according to the canons of *political rationality* rather than the implicit norms of economists who focus largely on efficiency.

Making "good" public policy is a difficult task: the world is a dynamic and uncertain place; information is not only costly to acquire (like many other goods) but its distribution is highly asymmetric; the methods of signalling preferences in the political arena are quite crude; some voices are very loud in policy making, others are absent entirely; and political rationality requires consideration of variables that are less familiar to economists and other policy analysts.

Types of Governing Instruments

The major decisions of a government in terms of policy-making can be usefully summarized by the following hierarchy:

- deciding to intervene regardless of which governing instrument is used (a decision that is bound up in the choice of goals or objectives);
- choosing the category of governing instrument (e.g., taxation); and
- selecting which form of the governing instrument to use (e.g., an excise tax of x cents per unit).

This section concentrates on the last two decisions, although they are not independent of the first one.

In general terms, governing or policy instruments are the broad means a government may use to achieve its policy objectives. The broad categories of governing instruments include:

- taxes (many types), tax expenditures (e.g., giving by not taking through preferential tax rates, tax credits, or deductions);
- user charges (sale of government services);
- expenditures on government programs (exhaustive; transfers; capital; current);
- regulation (government-made rules, backed by penalties, designed to modify the economic behaviour of actors in the private sector);[10]

- loans and loan guarantees;
- Crown corporations (or public enterprise, in which a government owns all of the voting shares);
- mixed enterprises (in which a government does not own all of the voting shares, and in which it may or may not have legal control);
- chosen instruments (when government favours a particular *private* firm or organization to achieve public policy purposes);
- public enquiries (discussed below);
- modification of private law rights and procedures;
- persuasion; and
- purely symbolic actions ("policies for show").

Within each category, an instrument can take many forms. For example, regulation[11] consists of government-imposed rules backed by penalties usually aimed at modifying the behaviour of actors in the private sector (Priest, Stanbury, and Thompson 1980). Such rules can be crafted in an almost unlimited number of ways, hence the attractiveness of regulation as a governing instrument. Furthermore, regulations or subordinate legislation (made into law by the cabinet rather than the legislature) are necessary to implement all governing instruments. Regulations have been called "the sinews of modern government" (Stanbury 1992).

Classification and Analysis of Governing Instruments
Governing instruments can be grouped into three categories: coercion-focused instruments, incentive-focused instruments, and preference-focused instruments.

Coercion-Focused Instruments

Description. The pure case of coercion-focused instruments is forced modification of the behaviour of individuals and/or organizations to that desired by government. When physical coercion is ruled out, the prototypical example of a coercion-focused instrument is the use of regulation.

Many regulatory regimes can be characterized as command and control (or command and penalty) systems. The key attributes of this form of government regulation include the following. First, the behaviour or activities of regulatees is subject to specific prohibitions or mandated requirements (legal "dos" and "don'ts"). Second, the performance of the regulatees is measured against certain standards defined in the statute or regulations. Third, penalties exist for violation of the "commands" of the regime – usually fines and prohibition orders,[12] but these may also include suspension of an operating licence or imprisonment. And, fourth, there is a tendency towards exhortation and negotiation in Canadian command and control regimes, rather than mechanistic, strict legal enforcement of the standards.

Also, government officials (or regulators) usually have considerable discretion for the exercise of their authority and use it to ensure that regulation is realistic.

Until recently, most command and control regulatory regimes in Canada were embedded in the criminal law. However, there has been a general shift away from the use of the criminal law in favour of civil law in addressing a wide variety of administrative policies, including environmental protection. BC's Forest Practices Code is a good example of this shift. It relies primarily on civil law but includes the possibility of criminal prosecutions. A key element of the new civil law approach is the use of "civil monetary penalties." These penalties – which can be as large as any criminal fine – may be imposed in proceedings in which the standard of proof is on the balance of probabilities rather than proof beyond a reasonable doubt.

Coercion-focused instruments, such as regulation, may permit more precise targeting of the behaviour to be modified to achieve the desired objective(s). In principle, if a specific type of behaviour is deemed desirable, the policymaker may mandate it and prescribe a schedule of (severe) penalties aimed at ensuring universal compliance. However, public and private costs of enforcement may be high.[13] Further, the precise prescription of behaviour may preclude desirable adaptation to local conditions and may result in low efficiency and counter-productive consequences. If the costs of compliance are high relative to the expected costs of violation, the level of compliance may not meet societal objectives. Vigorous enforcement and high penalty levels may, however, result in higher risk *premia* associated with the activities being regulated and result in lower levels of activity than are socially desired. In addition, the transaction costs associated with meeting the requirements of the regulations may also reduce activity levels below the socially desirable values. This outcome is most likely to occur when the regulations are ambiguous.

The effectiveness of coercion-focused instruments for protecting the environment is likely to be reduced by the diversity of physical environments to be managed, a complex set of objectives, high uncertainty about the nature of "optimal behaviours" and the means to verify that such behaviours take place, and high enforcement costs. The two main advantages of coercion lie in the apparent directness of control it offers to ensure compliance, and the relatively short time required to modify behaviour to meet the desired objectives.

The widespread use of coercion, however, is incompatible with basic democratic values.[14] A regulatory regime is likely to be effective if it is widely accepted as socially beneficial, thereby internalizing its prescriptions; that is, compliance becomes part of the standard operating procedures of firms, their employees, and all others affected by the regulation, largely because they believe the regulations are appropriate. Thus, the introduction of new regulations may be made more effective if accompanied by an educational program that convinces most regulatees of the need for regulation and the appropriateness of the specific rules.[15] Further, the legitimacy of new regulations depends on the extent to which stakeholders are able to participate in their development.[16]

Limitations. A command and control regulatory regime has limitations as a governing instrument:

- A regulatory regime that imposes tight constraints may ignore the fact that the marginal benefits of compliance may decline or be linear while the marginal costs of compliance may grow at an increasing rate. For example, a study of optimal schedules of harvesting in the Tangier watershed showed that if the green-up constraint was relaxed only slightly, there was a large increase in the sustainable level of harvest (Brumelle 1995). There may be circumstances, however, in which even minor violations of ecological constraints are intolerable.
- Constraints imposed by government regulation may not, in fact, contribute to the policy objectives of the regulation. For example, the constraints on harvesting imposed in the Tangier watershed to protect caribou habitat did not correspond to the critical needs of the animals (Brumelle 1995). To design an effective policy of protection, it is necessary to gain detailed information about habitat characteristics and the dynamics of animal behaviour. This type of information is rarely available and is very expensive to collect.
- There is a strong temptation to respond to problems or public concerns with regulatory actions ("there ought to be a law"). Government needs to be sure that its responses are, in fact, effective remedies to the problem at hand, and that such remedies do not create new problems. In some cases, the "cure" is worse than the "disease."
- Environmental regulations based on specific standards[17] tend to create a moving target for several reasons. Competition among environmental groups, each attempting to demonstrate a high level of "purity" or dedication (Wildavsky 1993), tends to create pressure to tighten constraints and raise standards (Stanbury 1993c). Also, better information-processing and measurement procedures tend to push up standards. For example, better technologies for the measurement of pollutants tend to induce demands for higher standards of purity, even if there is no evidence that such increases are environmentally beneficial. The prospect of moving standards and tighter enforcement generates uncertainties for regulatees, and thus increases costs of achieving environmental goals.
- Regulatory regimes are typically slow to adapt to rapidly changing economic and technological environments. Thus, their efficacy tends to be reduced over time. This finding has certainly been the case for economic regulation focusing on controlling entry, prices, and the rate of return.

Almost all studies show that such regulation imposes deadweight costs on society. In practice, such regulation has been used to redistribute income.[18]

- Regulations tend to promote attention (and public debate) that focuses on the means by which environmental goals are to be achieved rather than the goals themselves.[19] For example, pollution emission regulations based on the "best available technology that is economically feasible" reduce incentives for regulated firms to seek better means to achieve the policy's objectives (Kelly 1992: 4).

Incentive-Focused Instruments

Description. Incentive-focused instruments modify the behaviour of individuals and organizations (usually in the private sector) by altering the price signals (economic incentives) they receive. Examples include taxes (e.g., to reduce consumption of certain products or to reduce waste), subsidies (e.g., to increase employment), and fees or charges for previously free goods. Recently, there has been considerable analysis of these and other economic instruments to achieve environmental protection goals (Carlin 1992; Cassils 1991; Cox and Isaac 1987; Dewees 1980, 1992; Government of Canada 1992; Hahn and Noll 1983; Hartman and Wheeler 1995; Huffman 1994; Joeres and David 1983; Kelly 1992; Montgomery 1972; Noll 1983; OECD 1989; Palmisano 1982). Furthermore, Kelly's (1992: 18) list of economic incentives includes tradable permits/offsets, full-cost pricing for Crown-owned resources, full-cost accounting (i.e., including negative externalities), earmarked taxes/fees, and improved market information.

In the case of environmental protection, the idea is to properly price the presently non-marketed adverse effects of economic activity. Thus, effluents can be taxed at rates reflecting their true costs to society. Crown timber can be priced to reflect the loss of non-timber benefits when forests are cut down.

Government officials appreciate that, because preferences (and budgets) vary, the change in behaviour will not be the same for all persons or organizations faced with the same structure of economic incentives. Thus, incentive-focused instruments can be used if it is not deemed necessary for every person to modify his or her behaviour for the aggregate goal to be achieved. Further, it is not necessary for people to change their preferences. Indeed, incentive-focused instruments rely on the fact that preferences vary and that those people with the least intense desire for the item being taxed will be the first to modify their behaviour and buy less of the good. Here, "least intense desire" means both willingness and ability to pay.

The use of incentive-focused instruments is regarded as a logical extension of the market system and reflects most of the advantages of such a system. For this reason, many economists criticize command and control regulation and propose that incentive-based measures be substituted. Using the price system takes advantage of its enormous decentralized information-processing capability and its flexibility. The result is more efficient allocation. There are two major disadvantages of incentive-focused instruments when compared to command and control regulation: the probability that a particular set of incentives will result in a particular set of desired behaviours is lower, and incentives require a longer time frame for changing behaviour.

In some cases, the use of economic incentives involves measuring the specific consequences of the behaviours that are being promoted (or curtailed). In other cases, it is sufficient to design the system so that some payoffs (not necessarily quantified) accrue to those whose behaviour is to be altered. Thus, for example, it is possible for government to design a pricing system for environmental goods produced from forest land so that private sector managers may treat such goods on an equal footing with other sources of revenues from the forest (e.g., the sale of timber).[20] Decisions about the allocation of forest land among uses and about forest land management can be made on the basis of the expected response to different price/cost signals without need to appeal to altruism or social responsibility. The pursuit of self-interest, when prices reflect all costs and benefits to society, is compatible with the socially desired behaviour.

The use of incentives, however, is not limited to situations in which the government alters some prices in an established price system. For example, the award of individuated property rights,[21] where they do not exist, provides a broad range of incentives for owners to behave in socially beneficial ways. Awarding longer terms and more secure forms of tenure alters the time horizon over which the holder calculates the costs and benefits, providing incentives for the holder to avoid short-term strategies that are not in the public interest. Similarly, permission for forest companies to use what is now considered uneconomic wood without including it in their AAC calculations creates incentives to develop new technologies to use such wood. In these examples, the incentives flow from the conferral of rights rather than from changes in specific prices (through taxes or subsidies). They are intended to reduce myopic behaviour while triggering innovation and entrepreneurship.

We can distinguish between narrowly targeted incentives (i.e., per unit taxes or subsidies imposed to reduce or expand the supply of a particular environmental good or service), and broad incentives, usually embodied in the award of well-defined, individuated property rights, enforceable by independent adjudicators.

Narrow incentive-based instruments consist of introducing market mechanisms to create markets where markets do not exist, or to alter prices to correct market failures, most notably the existence of negative externalities. Thus, pollution emissions can be taxed to reflect the social costs they create, encouraging the emitter to reduce the volume of pollution. The creation of a competitive market process allows for decentralization of decision-making and reduces the need to use coercion. Provided no serious market failures are present, resources will be allocated efficiently. Markets permit local decision makers to choose the best means to achieve an end. They respond to prices which can be altered by government officials to reflect experience gained in the system, changes in public tastes, or new information about the environment. Using incentives instead of specific regulatory prescriptions allows decision makers to compare costs and benefits of alternative actions and make trade-offs, and thereby improve efficiency.[22] Forest companies are encouraged to use their knowledge of the particular forests they manage. Environmental goods and services are produced at the locations where the opportunity costs of their production is the lowest. Decentralization of decision-making also means lower information-processing and communications costs. While monitoring is required to ensure that incentive schemes are not abused, such costs should generally be lower than the enforcement costs of command and control regulation, because only selected outputs need be monitored, not the fidelity of the total process through which objectives are achieved.

Problems and Limitations. Incentive-focused instruments also have certain problems or limitations.[23] For example, in the case of environmental goals, the implementation of incentives requires estimating the public's demand for environmental goods and services such as non-timber benefits of forests. The methods of estimation can be divided into two categories:[24]

- The expenditure function or indirect approach relies on a relationship between private goods that are traded in the marketplace and public goods to draw inferences about the demand for the public good. The travel cost method (Clawson 1959; Freeman 1979) and hedonic pricing (Freeman 1979; Wilman 1988) are indirect methods for deriving values of non-timber benefits from information obtained in markets.
- The income compensation or direct approach uses questionnaires or surveys to directly elicit an individual's willingness-to-pay for more of a public good, or the individual's willingness-to-accept compensation to have less of the public good (e.g., to preserve old-growth forests) (Freeman 1979; Johannson 1987). Because this approach requires that individuals respond to hypothetical questions in a survey setting, it is also referred to as the contingent valuation method (Mitchell and Carson 1989) if actual values are requested, or conjoint analysis if individuals are asked to choose between multi-attribute alternatives (Keeney and Raiffa 1976). Applying multiple methods of evaluation of non-timber values may provide a robust range of values.

If reliable demand functions can be estimated, it is possible to simulate a market. Methods for estimation of demand functions are subject to disputes among professionals.

Other problems associated with incentive-focused instruments include the following:

- Incentive schemes may be subverted. For example, taxes levied for environmental protection may be unduly attractive as a means to increase revenues during periods of fiscal constraint. Similarly, incentive systems may be more easily manipulated by rent-seeking activities of stakeholders. The visibility of such opportunities is likely to encourage lobbying and pressure to use the incentives for purposes other than the one originally intended.
- Effective incentive systems ought to be fairly simple. However, experience with the stumpage system in British Columbia shows a tendency over time to increase the complexity of the system to meet the complaints or demands of various stakeholders.
- Often, it is necessary to establish within the government separate independent organizations (e.g., Crown corporations) capable of responding to price signals (e.g., the separation of management functions in New Zealand between commercial and non-commercial forests).
- If incentive-based instruments are relied on too heavily, they may create an economic incentive to engage in avoidance or even evasion.[25]
- Achieving a specific outcome precisely may be difficult using incentives.
- The use of incentives presupposes calculative rationality in a world characterized by bounded rationality and considerable non-rational behaviour. Incentives may not produce the predicted response.

Preference-Focused Instruments
Preference-focused instruments are used to modify the behaviour of individuals and organizations by means of altering the preference orderings of individuals.[26]

In a complex policy choice, involving technical and scientific uncertainties, it is possible that decision makers may have similar fundamental values, but have different preferences for outcomes, or means to achieve them. The differences in interpretation of outcomes, in terms of more fundamental values, may result from differences in the paradigms or models each decision maker uses to observe and interpret reality rather than from a conflict in values. Similarly, inconsistencies may be present when deciding on preferences among policies. For these reasons, it seems legitimate for stakeholders (including governments) to attempt to change the paradigm used by decisionmakers in making choices. This type of influence will be parallel to educational benefits often attributed to advertising.

Indeed, when provided with new data and scientific information, a decision maker may select a new, more appropriate paradigm.[27]

Few would argue against government strategies to influence citizens using verified, scientifically based information (where social benefits outweigh disseminating costs). However, the danger exists that an attempt to change people's preferences may involve disinformation or the use of affective messages. Indeed, the use of (overt) propaganda by governments in democratic societies is not acceptable (Trebilcock et al. 1981: II-2).[28]

Public participation processes that encourage open debate and access to a variety of information sources and opinions are an important vehicle for informing all stakeholders not only about the forest system to be managed, but also about the values of other stakeholders (Fenge 1995; Scott 1995). This public participation was certainly part of the Commission on Resources and Environment process in British Columbia.[29] Consensus that may emerge from such processes may provide a basis for voluntary compliance with public policy. Such consensus may, however, be difficult to reach when worldviews are irreconcilable.[30]

Combining Incentive-Focused and Preference-Focused Instruments
An important class of government instruments, which incorporates elements of both classes of instruments (i.e., changing people's worldview and preferences and the provision of incentives), is one associated with the awards of bundles of property rights. The award of individuated property rights may shift the perspective of recipients and expose them to new information and new experiences.[31] These rights may not only change the specific calculations in choosing actions, but also the paradigms used for calculation, e.g., awards of long-term timber harvesting rights may shift the view of firms with respect to their forest stewardship obligations. Indeed, if the grant of long-term tenure rights is considered valuable, but it is seen to be conditional on continued public acceptance of the holder's behaviour, those given such rights are likely to anticipate public demands and attempt to satisfy them to prevent loss of these property rights.[32] People may not only respond to new sets of constraints and prices associated with ownership, but they may also change their preferences. This outcome may be especially true when the award of individuated property rights also involves acceptance of obligations such as consulting with affected interests, being subject to independent audits, and so on. Exposure to information and the views of others deemed to be legitimate stakeholders is likely to affect the preference ordering of *all* participants (individuals and organizations). The direction of such change, however, depends on the nature of the interactions, the predisposition of stakeholders, the economic and social environments in which the interactions take place, and the availability of mediating institutions.

The design of preference-focused instruments is complex, their implementation expensive, and their consequences uncertain. Despite these characteristics and concerns, the size of the stakes and the uncertainties involved in using only the other two categories of instruments to achieve a sustainable development path suggest that preference-focused instruments are likely to be used by governments in situations involving what have been called "wicked problems" (i.e., problems involving uncertainties about both values and the consequences of alternative solutions).

Governing Instruments for Forest Policy in British Columbia
The BC government uses all the broad categories of governing instruments in forestry policy-making. However, regulation tends to be the government's instrument of choice.

Coercion-Focused Instruments
There appear to be two main reasons for the province's preference for coercion-focused instruments. The first relates to the province's ownership of over 90 percent of the forest land in BC. It has (so far) chosen not to directly engage in harvesting and processing of timber/fibre or reforestation either through a line department or by means of a Crown corporation. Rather, the province has leased harvesting rights to privately owned firms by means of complex administered contracts.[33] Management of Crown-owned forest lands is accomplished through a complex regulatory regime which has a number of parallels to traditional public utility regulation (Stanbury and Vertinsky 1985). In addition to this type of direct or economic regulation, the province attempts to achieve its environmental protection goals by means of regulation rather than by the use of economic incentives. A further reason for the extensive use of the regulation instrument relates to the politics of instrument choice.

Regulating the Allowable Annual Cut. Perhaps the most important regulatory parameter affecting the forest industry is the AAC. It must be set at least once every five years by the chief forester under s. 7 of the *Forest Act*. It indicates that range of factors must be considered. The AAC is the focus of a number of the province's recent policy initiatives established in the name of sustainability.

The Chief Forester has discussed the relationship between the AAC and the goal of sustained yield[34] (Chief Forester, 1991, Appendix III):

The Chief Forester selects the AAC for each of thirty-six Timber Supply Areas.[35] The choice results from the consideration of social, economic, philosophical, and technical issues – not from a numerical calculation based on technical or biological factors. The appropriate AAC reflects the natural capabilities of the land base and "prevailing social values." The upper limit of the productive capacity of the forest in terms of the volume of timber which may be cut is often described as the long run-sustained yield (LRSY): "In BC, this is a theoretical upper limit of what could be harvested on a regular basis if a number of conditions (such as appropriate age distribution, complete accessibility, no uncertainty in stand growth, etc.) are met."

In general terms, the LRSY depends upon the maximum mean annual increment (average increase in the volume of wood) of each stand which is associated with a particular period of rotation (period between harvests). Changes in the rotation period may increase or decrease the maximum achievable growth rate and thus affect the LRSY. The Chief Forester (1991, Appendix III) notes:

> Allowable annual cut levels are currently set above the LRSY in most TSAs. It is possible to harvest at these levels because of the abundance of old, high-volume stocks of timber. However, this will eventually lead to a "falldown," which can take a variety of forms, differing in timing and severity. It is the Forest Service's intention that the falldown be regulated so that AACs drop gradually until they reach the long run sustained yield.

The LRSY is affected by a variety of factors, including withdrawals from the land base (e.g., to create parks), and changes in management practices (e.g., more or less intensive silviculture).[36]

Forest Practices Code. The Code is a superb example of command and control regulation. Its focus is on the stewardship of the province's forest land and its multiple uses. With the Code, the province moved beyond long run-sustained yield, in terms of the volume of wood harvested, to the concept of "sustainable use of forests."

The Code is a multi-component regulatory regime, integrating many existing regulatory provisions,[37] and greatly extending the scope and detail of the regulation of timber harvesting in BC.[38]

Incentive-Focused Instruments
In BC, incentive-focused instruments are rarely used to achieve environmental objectives. However, such instruments are used to promote other objectives. For example, a firm's investment in silviculture can be used by the Ministry of Forests in the calculation of the firm's AAC. As well, to increase use of what was deemed uneconomic wood, the government has allowed the use of small diameter trees, considered uneconomic to harvest, without reduction in the AAC. These instruments also help the evolution of TFL tenures by creating incentives for forest product firms to increase investment in their holdings and encouraging more "socially responsible

forest stewardship" practices (using the implied threats of non-renewable TFLs) (Zhang 1994).

Taxation/Rent Collection. There has been a dramatic increase in stumpage and royalty revenues collected from the forest industry over the past decade (Schwindt and Heaps 1996). As owner of over 90 percent of commercial forest land in BC, the province has erected an elaborate taxation regime to collect the rents accruing to it.[39] The province made substantial changes to the stumpage system in 1987 (changing the method of calculating stumpage and shifting responsibility for reforestation to the companies), and again in 1994 (increasing revenues by about $400 million per year to finance the Forest Renewal BC initiative). Rates were reduced somewhat in 1998. Stumpage is an example of a broad incentive-type instrument. The aim is waste reduction: high stumpage increases the utilization of all fibre extracted by logging and reduces waste. The policy may also encourage efforts to create greater value added within BC. Such differentiated products will be less price sensitive.

Preference-Focused Instruments
During its first term under Mike Harcourt (October 1991-January 1996),[40] the New Democratic Party made use of multi-stakeholder advisory processes (Fenge 1995; Scott 1995). The most conspicuous example was that of the Commission on Resources and Environment. Its statutory mandate specified that it was "to ensure that resource and environmental management involves greater public participation, and that the ... plans achieve economic, social and environmental balance" (CORE 1994: 5). It is difficult to determine whether an advisory process or a shared decision-making process was intended.[41] The various advisory processes appear aimed at resolving forest policy conflicts and closing the gap in the preferences of the various stakeholders.

Trebilcock et al. (1982: Ch.4) notes that advisory processes fall within the category of governing instruments called public inquiries. This category includes royal commissions, task forces, legislative committees, regulatory and advisory agencies, and departmental and interdepartmental studies. Advisory processes can be viewed either as an effort to search for a solution to a particular policy problem, or as a policy instrument.[42] Trebilcock et al. (1982: 37) favour the latter because "commonly the subject under study is one to which there is no solution in a technocratic sense and to which no final answers will be given or found." Thus, an advisory process is "in large part ... a form of response to the policy problem and not merely ... a means to search for a response to the policy problem" (Trebilcock et al. 1982: 37). As a result, the public inquiries instrument (including advisory processes) "emphasizes the inadequacy of the means-ends dichotomy when speaking of the major policy issues on the public agenda." As a policy instrument, however, it can serve as a mechanism to inform the public or attempt to alter the points of view of some parts of the public.

To be legitimate, inquiries must allow some type of participation by all or most of the key stakeholders, although some interests may not be organized and thus may not be represented. But who is a stakeholder? CORE (1994: 260) defined a stakeholder as "someone who will be affected by a decision or can block or undermine any agreement if they are not given the opportunity to participate in the decision making process." If the stakeholders can reach a consensus,[43] the implication is that government will implement it. If a government selects stakeholders in part on the basis of who might be able to "block or undermine" the decision in question, it will invite strategic behaviour by interest groups.

One of the problems with multi-stakeholder processes is that they tend to carry the implication that all interests are equally important. Fourteen interest sectors were represented in the CORE process for Vancouver Island: conservation, agriculture, fisheries, outdoor recreation, forest employment, forest industry independents, forest managers/manufacturers, general employment, local government, mining, provincial government, social and economic sustainability, tourism, and youth.[44] Further, the nature of the interests represented are not commensurable. For example, compare the interests of a forest company executive to those of the animal welfare activist to those of the person speaking for a particular spiritual value requiring that the forests remain untouched by humans.

The effects of advisory processes, such as those employed by CORE, are difficult to discern. The government's stated objective was to avoid more valley-by-valley fights between environmental groups and forest companies. The lengthy multi-stakeholder consultation process before the announcement of the cabinet's decision on Clayoquot Sound in April 1993 did not prevent record civil disobedience in summer and fall 1993 (Stanbury 1994b). Nor did the CORE process and government decision on the Vancouver Island Land Use Plan prevent a massive one-day protest in Victoria a few weeks later. These observations, however, do not provide a proper assessment of BC's multi-stakeholder processes.

Politics of Instrument Choice[45]

The conventional model of policy formulation attempts to separate the choice of objectives (ends) from the choice of instruments (means).[46] This model is naive – indeed, misleading – for several reasons:[47] in the real world, means and ends are interdependent; ministers have different goals than do policy analysts, and they have the authority to choose both instruments and goals; and political rationality dictates that ministers pay equal attention to the choice of instruments and the specification of policy goals. Trebilcock et al. (1982) focus on public choice in instrument choice.

Assumptions

These assumptions follow Trebilcock et al. (1982).

- Politicians operate in a world of costly and asymmetric information. Thus, the political environment is characterized by considerable uncertainty.
- There is competition among political parties for votes so as to gain/retain office.
- The dominant operational goal of ministers is to maximize the probability of re-election.
- Competing parties make promises of policies in exchange for political support.
- In shaping proposed policies, politicians rationally focus on the appeal of these policies to marginal voters in marginal ridings.
- An ideology may enable voters to form reasonable predictions about how a party, if elected, will respond to a wide range of problems and issues. However, an ideology may inhibit the ability of a party to adapt to the dynamics of an uncertain world.
- Ministers are assisted by a professional public service. Senior officials have some measure of autonomy and thus exercise discretion in varying degrees.

Political Rationality

Why does political rationality dictate that ministers[48] focus their limited time and energy on the choice of instruments as well as on the specification of the objectives of public policies? In general, the answer lies in the overriding operational objective of ministers: maximizing the probability of re-election.[49] This objective involves, among other things, focusing perceived benefits on marginal voters in marginal ridings (while providing sufficient benefits to traditional supporters so as to prevent them from shifting to a rival party). It also means trying to ensure that the costs of public policies are less apparent and diffused among infra-marginal voters.

Ministers are engaged in the rational pursuit of sufficient votes to stay in power, thus they know that means and ends cannot be separated in the formulation of public policies for several reasons. First, stated policy objectives (e.g., putting 12 percent of BC's land area into protected areas)[50] are the means to more final objectives (e.g., sustainability), but they are also a possible means for re-election. Not only does re-election generate much utility for ministers, but it too is a necessary condition to obtain other higher-order policy goals. In general, "the process of translating objectives or ends into more ultimate objectives involves an almost infinite regress" (Trebilcock et al. 1982: 25). The organization of a means-ends chain can produce a hierarchy of objectives. "Through the hierarchical structure of ends, behaviour attains integration and consistency, for each member of a set of behavioural alternatives is then weighted in terms of a comprehensive scale of values – 'the ultimate ends'" (Simon 1976: 63).

Second, the objectives (ends) and instruments (means) embodied in public policies are almost always *interdependent*. The choice of instruments is not a neutral exercise because virtually all instruments affect policy objectives other than the primary policy objective. Instruments are value-laden. For ministers, the choice of instruments is far more than a matter of selecting one that achieves the objective at the lowest cost.

Lindblom (1959: 82) emphasizes that choosing goals and choosing instruments are not separable processes, because, among other things, "there is no practicable way to state marginal objectives or values except in terms of particular policies." Thus, in practice, policymakers have to choose directly among alternative policies that offer different combinations of values.[51] Thus, Lindblom (1959: 83) asserts that "one simultaneously chooses a policy to attain certain objectives and chooses the objectives themselves." At the margin, decision-making concerning values can only be done in the process of choosing among alternative policy instruments.

Three Factors That Influence the Ministers' Choice of Instruments
Trebilcock et al. (1982: 27) argue that ministers will be guided by the calculus of maximizing the probability of re-election in determining simultaneously both objectives and instruments. They then argue that political rationality will result in ministers taking three main factors into account in selecting among potentially substitutable governing instruments:

- legal constraints, including constraints imposed by the Constitution and international agreements;
- the impact (actual or potential) of different instruments on marginal and infra-marginal voters, particularly the distribution of costs (preferably focused on infra-marginal voters) and benefits (preferably focused on marginal voters); and
- the effect of different instruments on a world characterized by costly and/or asymmetric information. (For example, some instruments make it easier to convey information to marginal voters and disguise it from infra-marginal voters.)

At the same time, in selecting among alternative instruments, ministers are aware that the selection is made in the context of a dynamic world. Thus, "choices of objectives and instruments in functionally unrelated areas are politically independent," i.e., they may be used to counteract voter reactions to previous policy choices (Trebilcock et al. 1982: 32).

Furthermore, Trebilcock (1994: 7) argues that three sets of considerations should bear on the question of instrument choice: technical or efficiency considerations; distributional considerations (who should pay for what); and political considerations. The last set of considerations "can be viewed as a constraint on the application of efficiency and distributional considerations in choosing [governing] instruments."

Instrument Choice: The Case of BC's Forest Practices Code[52]

Our Approach

This section extends and applies the public choice perspective on the choice of governing instruments described earlier. Our analysis differs from that of Trebilcock et al. (1982) in several ways:

- We recognize that the cabinet of a government with a substantial majority is wise to take into account interests of key stakeholders in selecting governing instruments and designing their details.
- We examine a major piece of command and control regulation (the Forest Practices Code) from four different perspectives: the cabinet, public servants, environmental interest groups, and the firms subject to it.
- We appreciate that, in practice, the linkage between instrument choice and marginal voter politics (emphasized by Trebilcock et al. 1982) is quite loose. Therefore, we focus on the implications of selecting command and control regulation for more immediate objectives (i.e., the ultimate goal of re-election for the party in power).

The Cabinet's Perspective

In the Westminster (or Parliamentary) model of democratic government, the centre of power is the prime minister or premier and the cabinet (chosen by him or her) (Hogg 1985). Cabinet has the legal authority and practical power to be the sole decision maker in selecting the governing instrument(s) used to implement policy. However, it often finds it prudent to consider the concerns of other groups.

The Forest Practices Code, as a governing instrument, aimed to achieve the following cabinet objectives:

- The BC cabinet wanted to be seen to be "getting tough" with tenure holders by forcing them to make their timber-harvesting practices more sensitive to environmental considerations.
- The Code contains over 300 offences with heavy penalties for breach (up to $1 million). This power enables cabinet to use the Code as example of its commitment to protecting the environment, and restraining the activities of forest companies (both of which are important to the party's "green" supporters).
- The Code is a major part of the comprehensive response by the province to pressures exerted by domestic environmental interest groups which have linked up with foreign counterparts to exert direct economic pressure on BC producers (Stanbury, Vertinsky, and Wilson 1994; Stanbury and Vertinsky 1996, 1997). In summary, the Code provides an antidote to future threats to boycott BC's forest products.
- The Code is a tool for central control of a major industry by the provincial

government. It relies on numerous detailed government directives (dos and don'ts) rather than economic incentives and competitive market forces.[53] It makes extensive use of public servants who work within a policy framework established by the cabinet.

- The Code meets the cabinet's desire to increase its discretionary control over the forest industry. In particular, the Code is designed to provide discretion for the Minister of Forests. The operative criteria need not be official criteria, they need not be disclosed, and they may change over time to reflect the political concerns of the moment.

   Most of the operative details of the Code lie in its Regulations, Standards, and guidebooks, although the parent statute (Forest Practices Code of British Columbia Act) is long and detailed. Changes to the Regulations (and closely related Standards) are made entirely by a committee of the cabinet. By comparison, the federal government requires advance notice of a new regulation in the annual regulatory plan, consultation with affected interests, the preparation and publication of a Regulatory Impact Analysis Statement, and pre-publication of the draft regulations followed by at least a 30-day comment period (Stanbury 1992).

- The cabinet did not make public any estimate of the social costs and benefits of the Code. Van Kooten (1996) concludes that even under the most optimistic scenario, the net social cost of the Code will be $297 million per year. In present value terms, this amounts to $7.4 billion. In other words, the Code will result in a net reduction in the wealth of BC residents of over $7 billion.

- The government believed prior environmental protection regulations had not been effectively enforced: too few violations were pursued and economic penalties were too low. Thus, a government newspaper advertisement emphasized that the Code would result in "tough enforcement and heavy penalties," including fines of up to $1 million per day. Emphasis on potential penalties allows the government to use the political advantages of command and control regulation.

- The plethora of rules, plus plenty of new "eager beaver" inspectors in the field, put the forest companies on the defensive from the word go. As a recent newspaper report noted, "Loggers are more concerned that they may inadvertently make a mistake; do something that is forbidden by the metre-high pile of documents. It has bred insecurity. Companies are worried about fines and loggers are worried about their jobs" (*Vancouver Sun*, 15 June 1996, B7).

This is exactly the frame of mind a government with a strong penchant for central planning wants to inculcate in its regulatees. It maximizes the discretionary power of MOF officials and their political masters.

Public Servants' Perspective on the Forest Practices Code
The following outlines the public servants' perspective on the Code.

- The Code will generate additional employment for public servants. As Haley (1996: 5) notes, the Code is "very labour intensive."[54]
- Public servants have a central role in administering and enforcing the Code. Under the Code, even lower- to middle-level officials can act as investigators and judges through their very extensive power to invoke administrative penalties. The staff of three ministries can invoke administrative penalties: monetary penalties (fines), stop-work orders, remediation orders, and seizure and sale of timber. These penalties are subject to fairly elaborate review and appeal processes.
- Despite the very large number of specific offences, the great scope of the Code provides public servants with considerable discretion in its application. However, much of the discretion is attributable to vagueness, for example, the definitions in the Code of "sensitive areas" and "visual quality objectives."
- The guidebooks assist the Ministry of Forests district managers in administering the Code, by providing "'how-to' guides [which] detail the Code's recommended procedures, processes and results. Within the restrictions and limits set by the Code, guidebooks give information on how to make site-specific interpretations and modifications" (BC Ministry of Forests 1995: inside cover page). Thus, the guidebooks make overt what is often contained within confidential internal documents or only in the unwritten rules of thumb adopted by lower-level public officials. This detail reduces the chance of arbitrary application of the Code and discrimination among regulatees. However, the guidebooks provide a more flexible means of shaping the behaviour of regulatees than do the statute or regulations. Whether a guidebook provides an authoritative interpretation of the formal legal provisions in the Act and Regulations will depend on the courts. However, such challenges are costly for regulatees.
- Extensive command and control regulation such as the Forest Practices Code is the antithesis of using economic incentives and market forces to achieve public policy objectives. Most public servants appear to believe that a comprehensive regulatory regime in which cabinet ministers follow their advice closely is the best of all possible worlds. Why? Because public officials can be responsive to the need for change, and they can change uncertainty into orderliness. Also, regulation is the most "natural" function of government (along with spending and taxing) from the perspective of public servants. Using it, they can give the illusion that the will of their political masters can be directly translated into the desired types of behaviour by firms in the private sector. Even better, most of the costs of regulation are hidden.

- The Code meets the strong but inchoate need of senior public servants for rational comprehensive planning to achieve the new goal of sustainable forestry. Moreover, because the key operative requirements are incorporated into the regulations, officials can continuously adjust the regime in light of experience (provided cabinet approves new regulations or amends existing regulations). However, the edifice of regulation may result in gridlock. Keith Purchase, president and CEO of Timber West Forest Ltd., has said that the highly bureaucratic process under the Code "slowed harvest approvals to a snail's pace" in 1995. An estimated 14 million cubic metres of timber went unharvested "as a result of increased bureaucracy" (Price Waterhouse 1996b: 2).

Environmental Interest Groups' Perspective

It is hazardous to offer generalizations about the views of even the handful of major environmental interest groups that have focused on forestry issues in BC.[55] While sharing some objectives, their perspectives and tactics vary considerably. Further, as a matter of strategy, their leaders rarely offer detailed criticisms of major government initiatives. Instead, they tend to focus on a few key issues that provide ease of communication through the news media.

A massive command and control regulatory initiative, like the Code, is likely to be attractive to environmental groups for several reasons.

- As noted, the Code contains hundreds of specific dos and don'ts, giving content to the vague term "sustainable forestry."[56] Environmentalists tend to be suspicious of the use of economic incentives and market forces to achieve environmental goals. Regulation is their instrument of choice – largely because most environmentalists appear to believe that market failures stem from the moral failure of profit-seeking individuals/firms. Hence, they do not agree that negative externalities, for example, can be attributed largely to the mis-pricing of key inputs and outputs (which causes prices to fail to reflect true social costs), and that the errors in prices can be corrected by assigning a complete set of individuated property rights in what are called "common property resources" (including the air and large bodies of water). Command and control regulation is likely to be approved by environmental groups because it is based on standards set by law enacted by the legislature exercising a visible hand in light of popular political pressure.
- The BC environmental groups (notably, Greenpeace and Sierra Legal Defence Fund) have long argued that the Ministry of Forests has failed to enforce the previous regulations designed to protect the environment, and that the fines or other penalties were grossly inadequate. The Code provides specific means of enforcement. While the Ministry of Forests is the key government agency responsible for enforcing the Code, the visibility and hence political symbolism of the new regime is enhanced by

two related agencies: the Forest Practices Board (FPB) and the Forest Appeals Commission. The Practices Board oversees the activities of the two protagonists (industry and government) with respect to the Code. The Appeals Commission is an independent quasi-judicial tribunal responsible for hearing appeals following the Administrative Review process or FPB decisions.[57] Furthermore, the Code provides environmental groups with an opportunity to engage in private enforcement. In light of the previous failure of federal and provincial officials to pursue many cases brought to them by environmental groups, this provision should be well received.

- Command and control regulation like the Code appears to make it clear that regulatees have no right to pollute or otherwise despoil the environment. In contrast, a tax or charge on harmful forest practices seems to suggest that the firms have a "right to damage the environment" so long as they can pay the tax or charge.
- Environmental groups should welcome the comprehensiveness of the Code as a regulatory regime, because they tend to favour political and bureaucratic control over the activities of private firms.
- Environmental groups favouring an end to harvesting prefer aggressive enforcement of the Code as a vehicle to make at least some logging uneconomic and to reduce the actual level of harvest by "regulatory gridlock." Note that the Clayoquot Rainforest Coalition, led by RAN (1995), has as its stated central objective the end to all commercial logging of old growth in BC (ancient temperate rain forests) (Stanbury and Vertinsky 1996, 1997).[58] Because old growth accounts for 44 percent of all BC's commercial forest lands, and a much higher fraction of the coastal region, achievement of this goal would result in the collapse of the forest industry.

Forest Firms' Perspective

The following outlines the perspective of the forest firms concerning forest policy in BC.

- To meet the requirements of the Code, forest companies emphasize that they have incurred high compliance costs.[59]
- The costs of meeting environmental objectives are greater than they need to be because of the use of a less efficient instrument, namely command and control regulation.[60] However, companies may underestimate the difficulty of using incentive-focused instruments in this context.
- Costs of complying with the Code are largely hidden, but in the first instance, they are paid by the forest companies as they are price-takers in almost all their markets. Note that if stumpage was actually an efficient rent collection device, part of the cost would be borne by provincial taxpayers in terms of lower rents from the forests they own.

- To the extent that command and control regulation has a high symbolic value, it may help to reduce the threat of boycotts advocated by both domestic and foreign environmental groups. In practice, it appears that victories for environmental groups may lead to an escalation of demands, or demands for new types of environmental protection.[61]
- The Code delegates considerable discretion to middle- and lower-level Ministry of Forests officials. This delegation increases uncertainty for the firms, thereby increasing their costs. On the other hand, the Code does provide a wide range of informal guidelines and practices, and so, to the extent that these fetter the discretion of Ministry officials in processing harvesting applications, the Code has reduced uncertainty.
- The Code creates many requirements, the violation of which could result in high penalties. Further, the Minister made it clear that the firms' performance will be more closely monitored and that every violation will be pursued.
- The province does not acknowledge that the Code will result in extensive takings of forest companies' tenure rights. At the very least, the Code will reduce the AAC by 6 percent over the first decade (Peterson 1996). No compensation will be paid for this "expropriation" of the firms' property.

Conclusion

Our analysis indicates that the BC government has greatly expanded the use of command and control regulation in an effort to protect the environment in the context of the harvesting of timber. This expansion runs counter to a large body of literature advocating economic incentives instead of regulation primarily on the grounds of efficiency, but also in terms of effectiveness.[62] Even the federal Department of Environment supported publication of an extensive study advocating incentive-focused instruments (Government of Canada 1992).

"If They're So Great, Why Aren't They Being Used?"[63]

The first thing to note is that almost all the extensive literature on the application of incentive-focused instruments to environmental protection relates to the problem of pollution control, i.e., the emission of noxious substances into the air, water, or land. There is virtually no discussion of how to apply economic incentives to environmental protection in the forest industry, although there are a few examples of economic instruments being employed for environmental protection (tradable emission permits, "bubbles," carbon-related taxes, higher user charges for refuse collection disposal). Overall, however, regulation is clearly the instrument of choice, even for pollution control.

Our reading of the history of government intervention in Canada is that governments generally focus on redistributing income to politically salient groups/regions/sectors and that substantial *reductions* in economic effi-ciency result from these actions. The evidence suggests that political rationality does not require politicians to pay much attention to efficiency. Further, when an important policy issue is also an emotional one – as activists have made environmental protection in BC – there is even less inclination to count the costs of even extensive government action.

Practicable incentive-based instruments for environmental protection in the forest industry have not been developed because of the nature of the problem being confronted. Here are several characteristics of the problem:

- The apparent overarching goal of sustainable forestry is not clearly defined in the Code or other government documents. Some goals of BC's forest policy are aesthetic, e.g., the visual quality requirements. Such goals are difficult to define and measure. In contrast, goals to reduce the volume of certain air or water pollutants differ because of their (fairly) well-defined adverse consequences. This fact makes it much easier to create incentive-based instruments to deal with pollution.
- The "production function" for sustainable forestry is highly uncertain. Knowledge of the shape of environmental damage or benefit functions related to activities regulated by the Code is limited. The present perceived state of knowledge seems to be limited to stating a number of dos and don'ts, e.g., riparian zones, constraints on the size and spacing of clearcuts, road construction requirements, and "green-up" requirements, and to ensuring certain minimum habitat requirements. However, the state of knowledge is not sufficiently advanced to deal with the interdependencies among the specifically regulated activities.
- The regulations in the Code are aimed at protecting streams, preventing erosion, limiting adverse effects on uncut regions of forest during cutting, constraining the methods of logging, and ensuring that the habitat of certain creatures is protected. How these combine to achieve "sustainable forestry" is more an act of faith than science. But it must be emphasized that relevant knowledge has been generated in recent years. However, there is no evidence that the Code will be able to ensure environmental protection.
- One of the most frequently articulated concerns in forest policy is "biodiversity." Yet, as Kimmins (1996: 4) explains, "because there are so many measures of biodiversity,[64] such a broad, general statement of intent [to manage for biodiversity] provides no basis for deciding how a particular forest should be managed."
- The "achieve sustainable forestry" problem is quite different than the "reduce pollutant X below level Y" problem. The latter policy problem is to reduce or eliminate a specific negative externality. What is to be done is very clear – even if exactly how to do it most efficiently is much more complex. The "achieve sustainable forestry" policy problem is framed in the positive – i.e., we wish to induce behaviour which will contribute to

the goal. But many behaviours appear to be consistent with the goal. Behaviours adverse to the goal cannot be simply defined as the opposite of those consistent with it (as can be done in the case of pollution). That is why the Code requires certain behaviours and prohibits others.

- Devising incentive-based instruments to promote sustainable forestry is more difficult than devising such instruments to prevent effluents despoiling the air, water, or land. It is possible to imagine using environmental damage charges for a number of harvesting activities, but it is much more difficult to see how the level of the charges could be determined in a scientific fashion. Worse, it is not at all clear that the behaviour induced by the charges will, in fact, result in "sustainable forestry" or even achieve one of the many sub-goals under that broad concept. By comparison, a charge on certain effluents will probably cause the polluting companies to reduce the volume of emissions subject to the charge.

- The historical context in which the Code was created is also important in understanding why incentive-based instruments have not been developed. First, it is evident that distrust of the companies' willingness or ability to protect the environment had achieved new heights in BC. Even strong price signals were not seen as sufficient to absolutely prevent certain disastrous practices or to mandate certain beneficial activities. Further, with absolute requirements, it was believed that enforcement would be easier – even if more Ministry of Forests employees were needed to monitor performance. Second, the industry was under serious threat in some of its foreign markets. The new government actions had to be seen to respond directly to the most intense concerns of the most notable environmental groups. For these groups, incentive-type instruments would simply not do.

In many ways, the Code is attempting to find ways of allowing a structural change in a forest landscape (namely, harvesting) while trying to retain a complex set of features of that landscape. At the same time, the method of harvesting must be economically viable. So far that has meant large scale operations.

Economists and the Design of Economic Instruments

The major obstacle to greater use of economic instruments is the politicians' belief that incentive-focused instruments will be less effective in achieving key objectives (ultimately, their re-election). Furthermore, other (technical) problems merit more attention from economists.

The main tasks for economists include: identifying the bundle of goods and services to be produced by the forest (including non-use benefits); estimating demand functions for non-market goods included in the above bundles; and designing a pricing system that will provide socially optimal trade-offs. In addition, economists may identify institutional arrangements that promote efficient behaviour in the sense of maximizing social welfare. These arrangements largely involve the design and allocation of certain property rights.

Economists must increase efforts to show environmentalists that it is not in the latter's interest to advocate inefficient instruments. There is no virtue in achieving necessary improvements in green goals at higher than the lowest attainable costs. On the desire to minimize waste, surely ecologists and economists are at one.

Economists and Preference-Focused Instruments

Economists have not been successful in analyzing preference-focused governing instruments. Economists deal with prices, technologies, information, resource endowments, and allocation of resources, but not with preferences. Preferences are taken as givens.[65] The economist also ignores the interactive relationship that exists between consumption and preferences. While consumption is determined in part by preferences, preferences are moulded in part by experiences with consumption (i.e., what we want reflects in part what we consume).

Ecologists argue that sustainable development can be achieved within a market economy only if preferences are modified (not just the behaviour in response to higher prices reflecting all social costs). In particular, they argue for delinking individual and social welfare from traditional material consumption. The greening of corporate behaviour is becoming more important in explaining changes in forestry practices. It is largely the result of the greening of consumer preferences (often influenced by environmental groups). Our research indicates that some consumers are prepared to make purchasing decisions about wood or paper products on the basis of their perception about the forestry practices of the producer (Stanbury, Vertinsky, and Wilson 1994). With growing consumer preferences for green products, the role of government may be more one of monitoring and certifying practices, and informing and educating the public, and less one of protecting the environment through command and control regulation.

Finally

While change may be a constant, today the makers of forest policy are condemned, in the words of the Chinese proverb, "to live in interesting times." Uncertainty is pervasive, and uncertainty creates fear. Fear creates resistance to change. Change often has high costs, both psychological and economic. But resistance to change is more than the fear of the new, or a hankering for the security of the familiar. It is also shaped by the idea that large costs will be incurred, and the majority of the people who will bear them will not be better off. That is why good analysis which helps us "look before we leap" is so important.

Acknowledgments
This study was funded in part by a grant from the Networks of Centres of Excellence (Sustainable Forest Management). We are indebted to Mabel Yee for superb word-processing services. The editor of this edition required a number of changes to shorten the chapter, which reduced the detail on theoretical issues and the support for a number of our arguments.

Notes
1 Generally, see Bishop (1993); Common and Perrings (1992); Daley (1990); Jansson et al. (1994); Jereon et al. (1994); and Robinson et al. (1990). The characteristics of the "sustainable use of forests" are described in the preamble to the *Forest Practices Code of British Columbia Act.*
2 Price Waterhouse (1995) contains a useful sketch of a number of the major policy initiatives: the Protected Areas Strategy; CORE process leading to a very detailed land use plan for several areas (Vancouver Island, Cariboo-Chilcotin, East and West Kootenay areas); *Forest Practices Code of British Columbia Act* (and Regulations); Revision of the AAC for all Timber Supply Areas; Forest Renewal Plan; and the series of decisions relating to Clayoquot Sound. Three major items are likely to dominate the NDP's forest policy agenda in its second term in government following the general election in May 1996: changes in the tenure system; efforts to get tenure holders to provide more jobs per 1,000 cubic metres of wood harvested; and the settlement of Native land claims which will require the transfer of forest land to aboriginal peoples.
3 See the objectives set out in the *Ministry of Forests Act*, s. 5.4, *Forest Act*, s. 2.4, and *Forest Practices Code of British Columbia Act.* CORE's "Land Use Charter" for BC (1992) included some forty-five policy goals.
4 The multi-stakeholder advisory process undertaken by CORE could be described as a governing or policy instrument.
5 For a more in-depth analysis of the Forest Practices Code, see Cook, infra, Ch. 9.
6 Old-growth forests are storehouses of biological diversity (see Maser 1994).
7 For example, when German environmental groups insisted in 1993 that BC not log any more of its old-growth forests (see Stanbury 1994b), some BC residents argued that the Germans should purchase the land to be logged and turn it into a park or wilderness area. This argument also arose in the threats of boycotts of MacMillan Bloedel in California (see Stanbury and Vertinsky 1996, 1997).
8 This requirement is now reflected in the preamble to the *Forest Practices Code of British Columbia Act.* See Cook, supra, note 5.
9 Economists (or anyone else) cannot say anything in a *scientific* fashion about what goals ought to be pursued through government action or what weights should be assigned to each. It is argued that the scientific expertise of the economist (policy analysts in general) can usefully be brought to bear on the second task – that of ascertaining the "best" policy instruments to achieve the stated objectives.
10 Tenure arrangements under which companies acquire timber from the Crown as a form of regulation are included here. Generally, see Haley and Luckert, infra, Ch. 6.
11 The distinction between regulation (the governing instrument) and regulations (subordinate legislation) is discussed in Stanbury (1993d).
12 One could describe such penalties as "incentives" to comply. The objective is to get everyone subject to the regulation to comply. True incentive-based instruments use economic incentives that reflect closely the social costs resulting from certain types of behaviour requiring modification.
13 The costs are likely to be high if there is little acceptance of the goals of the policy and/or the penalties are deemed excessive relative to the social harm to be prevented. In general, efficient compliance strategies require that both the size of the penalty and the likelihood of being of caught and convicted be taken into account.
14 This incompatibility can be seen in the reaction to the federal government's gun registration law. See "Fighting Back," *Maclean's* (5 June 1995): 14-23.
15 In some cases, however, more information may actually increase opposition.
16 The BC Ministry of Forests publicity material concerning the Forest Practices Code has strongly emphasized the extensiveness of the consultation process that resulted in the Code. The material in May 1995 described "the public's continuing role" in the operation

of the Code in terms of strategic planning, operational planning, complaints, and technical working committees on new guidelines.

17 Standards are specified as a separate part of the Forest Practices Code. See Cook, supra, Ch. 9.

18 An example is the agricultural products supply management scheme (see Stanbury and Lermer 1983).

19 For example, there was almost no debate in BC over the extent to which the Protected Areas Strategy (to put 12 percent of BC's land base into parks) will contribute to various environmental protection goals, e.g., biodiversity, maintenance of habitat, and so on.

20 The point is for government to create prices for environmental goods for which there is no market at present.

21 This approach is in contradistinction to the problem of common property resources – even when ownership is overtly vested in government.

22 Kelly (1992: 13) puts it this way: "Economic instruments reinforce thrift, quality, efficiency, knowledge, investment and innovation."

23 The central problem is that the instruments are difficult to apply to environmental problems other than the control of air or water pollution. Even in that context, there are problems in translating the theory into practice. See, for example, Carlin (1992); Hahn and Noll (1983); Joeres and David (1983); Noll (1983); Hartman and Wheeler (1995).

24 These points were taken from Vertinsky et al. 1994: 42-43.

25 Federal and provincial tobacco taxes in Ontario and Quebec (and some other factors) resulted in almost 40 percent of the cigarettes consumed in those provinces being supplied by smugglers before the taxes were reduced in 1994. Similarly, high federal and provincial taxes on distilled spirits have resulted in one-quarter of the volume consumed in Canada coming from smuggling, according to a spokesman for the Association of Canadian Distillers.

26 Note that economists usually begin by assuming that each person has a complete, well-ordered, and stable preference ordering (which is exogenously determined in almost all of their models). This conception of the world is convenient since it allows economists to focus their inquiries about the influence of prices upon behaviour and thus develop tractable models. Few deny, however, the power of advertising to change behaviour, and not only by virtue of providing information to decision makers but also by changing their valuation of outcomes, i.e., their preferences.

27 An example of such a shift of paradigm can be found in a recent change in forestry practices in Sweden (see National Board of Forestry 1992).

28 This point does not mean that public money is not used for government propaganda. Rather, it must be disguised and re-labelled as the "provision of information" about policy issues, options being considered by government, or government programs already in place. Stanbury and Fulton (1984) note that suasion by political leaders can involve leadership in the best possible sense of that term.

29 The Commission on Resources and Environment (CORE) was created in January 1992 and ended in March 1996.

30 The CORE process did *not* produce a consensus regarding land use planning, but it probably served to sharpen and narrow the issues for the BC cabinet that had to make the final decisions.

31 For example, it is common to award stock options to the top managers of widely held firms. The objective is to get managers to think and act like owners.

32 It is possible that some tenure holders will seek to persuade the government and other stakeholders to make the conditions less onerous and/or to create the impression that the public demands are being met when, in fact, they are not. Rent-seeking or other forms of self-interested behaviour are ubiquitous and NDP governments are no exception.

33 "The *Forest Act* specifies a number of forms of agreement by which the Forest Service may authorize timber harvesting. The number of timber harvesting agreements in the province is approximately: 190 forest licences; 35 tree farm licences; 270 timber licences outside of tree farm licences, and 400 timber licences within tree farm licences; 140 timber sale licences (excludes timber sold competitively under the Small Business Forest Enterprise Program); 10 pulpwood agreements; 500 woodlot licences; 1,500 road permits; 900 licences to cut; and, 125 Christmas tree permits" (BC Ministry of Forests 1994: 15).

34 While sustained yield management is said to be the most fundamental guiding principle of BC forest policy for over half a century, that principle is not specifically spelled out in either the Forest Act or Ministry of Forests Act. However, s. 6 of the *Forest Act* states that the Minister may "designate as a public sustained yield unit Crown land that is not a tree farm licence area." But the term public sustained yield unit is not defined in s. 1, the interpretation section. Further, s. 28 of the *Forest Act* spells out the terms and conditions for a tree farm licence (the most important form of tenure). The Chief Forester, in approving the holder's working plans, must determine that the AAC "may be sustained from the tree farm licence area," having regard to five specific and one general factor. The word "sustained" is not defined, but the widely held assumption is that it means sustained indefinitely or in perpetuity.

35 This selection is a statutory obligation under the Forest Act.

36 More generally, see Dellert, infra, Ch. 11.

37 Previously, forest practices in the province had been governed by 6 federal and 20 provincial acts, approximately 700 regulations, and 3,000 separate guidelines. This number has made consistent administration difficult.

38 A very useful and more detailed discussion can be found in Cook, supra, Ch 9. See also Cassidy 1994; BC Ministry of Forests 1993, 1994, 1995, 1996. Haley (1996) estimates that the total costs of the Code amount to $2.1 billion annually. This is more than the province has ever collected in any year in stumpage and royalties from the industry (see Schwindt and Heaps 1996). However, van Kooten (1996) estimates the *net* social cost of the Code is closer to $300 million per year.

39 See Scarfe, infra, Ch. 8.

40 Mr. Harcourt was replaced by Glen Clark as premier in January 1996. Mr. Clark led the NDP to re-election in May 1996 with a reduced majority.

41 CORE (1994: 259) defines "shared decision-making" as "a framework approach to participation in public decision-making in which, on a certain set of issues, for a defined period of time, those with authority to make a decision and those affected by that decision are empowered jointly to seek an outcome that accommodates rather than compromises the interests of all concerned."

42 The provincial government may have a number of objectives in using multi-stakeholder processes: to engage in consultation; achieve consensus (among organized/vocal interests); to legitimize its decisions; to respond to vocal demands for greater direct participation in the policy process; to demonstrate to a wider public the complexity and intractability of decisions involving deep-seated value conflicts; to create delay, so as to allow conflicts to cool; and to try to channel overt conflict into a more manageable form.

43 The representatives of fourteen interest sectors on Vancouver Island were "unable to reach consensus on recommendations on the location of specific land use designation boundaries" (CORE 1994: 99). Thus, CORE Commissioner Stephen Owen stated in the report presented to the cabinet in February 1994.

44 Representatives of First Nations were not at the Vancouver Island regional negotiation table. In general, the province met with them on a government-to-government basis under a protocol of 20 August 1993.

45 For more discussion on the theories of instrument choice, see Trebilcock et al. (1982), Becker (1958), Doern and Wilson (1974), Baxter-Moore (1987).

46 A number of authors have offered theories of instrument choice by government: Becker (1958) offers both a positive theory of technical efficiency in instrument choice and a normative argument for technical efficiency; Doern and Wilson (1974) array all governing instruments on a scale measuring the degree of coercion; Baxter-Moore (1987) categorizes governing instruments from non- to highly intrusive. None of these theories reflects the public choice perspective.

47 Note that the theory propounded by Trebilcock et al. (1982) is an exception. This theory focuses on a public choice approach to instrument choice. Generally, see Mueller (1989).

48 While we focus on ministers (the locus of power in the Westminster model of representative government), politicians on the opposition benches and candidates in elections will also focus their promises on both the means and the ends of proposed policies.

49 This point ignores the possibility that certain actions by a minister may help that minister personally while harming the party's chance to retain power.

50 The Protected Areas Strategy (PAS) is an interagency initiative to develop a provincial system of areas protecting conservation, recreation, and cultural heritage values. "The two goals of PAS are to protect viable representative areas, and special features. The target is to designate 12 percent of BC as protected areas by the year 2000" (BC Ministry of Forests 1994: 2).

51 Indeed, it may be useful to define a "policy" as a bundle consisting of a particular combination of values embodied in certain policy objectives and instruments.

52 For an in-depth analysis of the Forest Practices Code, see Cook, infra, Ch. 9.

53 The paper associated with the Code, when stacked up, exceeds six feet in height. See the photograph in the *Vancouver Sun*, 15 June 1996, B1.

54 See Cook, infra, Ch. 9.

55 These interest groups include Western Canada Wilderness Committee, Friends of Clayoquot Sound, Sierra Club of Western Canada, Sierra Legal Defence Fund, Greenpeace Canada (Vancouver office), BC WILD (a coalition), and the Valhalla Society.

56 As with the term "sustainable development," there are conflicting views about the meaning of "sustainable forestry." The publications of the BC Ministry of Forests do little to spell out what the term means, other than the implementation of the Code and the reduction in the AAC under the Timber Supply Review. For the Rainforest Action Network (RAN), "sustainable forestry" means the end to harvesting of any old growth in BC. More generally, see Maser 1994; Aplet et al. 1993; Noss 1994; Bormann et al. 1994; and M'Gonigle and Parfitt 1994.

57 See Cook, infra, Ch. 9.

58 RAN's most recent position regarding BC forests is described in Hamilton (1997: D3).

59 See Cook, infra, Ch. 9.

60 Forest companies are reportedly most concerned with the Code's methods of regulation (which focus on controlling processes rather than outcomes), and the huge increase in administrative costs (including delays in the permit approval process). Some executives have said that they do not want the environmental protection regulations weakened (see *Vancouver Sun*, 15 January 1997: D1, D6).

61 For example, shortly after the BC government said that it would require forest companies logging in Clayoquot Sound to meet all of the recommendations in the final report of the Scientific Panel for Sustainable Forest Practices in Clayoquot Sound (1995), representatives of some environmental groups demanded that these requirements be applied throughout the province. Thus, they ignored the scientific advice that to achieve the biodiversity objective, forest management must be region-specific, landscape-specific, and site-specific (Kimmins 1996: 4).

62 See the bibliography in Kelly 1992; OECD 1989; Government of Canada 1992; and Hartman and Wheeler 1995. The latter is a particularly useful survey of the use of economic incentives in the control of pollution.

63 This mordant question is asked by Kelly (1992: 19).

64 Kimmins (1996: 4-5) refers to genetic, species, structural, and functional diversity. He notes that each can be applied at various spatial and temporal scales.

65 J.K. Galbraith is an exception. He argues that large corporations, through advertising, can shape preferences.

References

Aplet, G., N. Johnson, J.T. Olson, and V.A. Sample. 1993. *Defining Sustainable Forestry.* Washington, DC: Island Press

BC Ministry of Forests. 1993a. *Proposed Forest Practices Rules for British Columbia.* Victoria: Ministry of Forests

–. 1993b. *The British Columbia Forest Practices Code Discussion Paper.* Victoria: Ministry of Forests

–. 1994. *Five-Year Forest and Range Resource Program, 1994-1999.* Victoria: Ministry of Forests

–. 1995. "Summary of *Forest Practices Code* Guidebooks." Victoria: Ministry of Forests

–. 1996. *Forest Practices Code: Timber Supply Analysis.* Victoria: Queen's Printer

Baden, J., and R.L. Stroup, eds. 1981. *Bureaucracy vs. Environment: The Environmental Costs of Bureaucratic Governance.* Ann Arbor: University of Michigan Press

Baxter-Moore, N. 1987. "Policy Implementation and the Role of the State: A Revised Approach to the Study of Policy Instruments." In *Contemporary Canadian Politics*, ed. R.J. Jackson, Doreen Jackson, and N. Baxter-Moore, 335-55. Scarborough: Prentice-Hall

Becker, G.S. 1958. "Competition and democracy." *Journal of Law and Economics* 1: 105-8

Bishop, R.C. 1993. "Economic efficiency, sustainability, and biodiversity." *Ambio* 22: 69-73

Borcherding, T.E. 1983. "Toward a Positive Theory of Public Sector Supply Arrangements." In *Crown Corporations: The Calculus of Instrument Choice*, ed. J.R.S. Pritchard, 99-184. Toronto: Butterworths

Bormann, T. et al. 1994. "A Framework for Sustainable-Ecosystem Management." Forest Service General Technical Report PNW-GTR-319. Portland: USDA

Brumelle, S. 1995. "A TABU Search Algorithm for Finding a Good Forest Harvest Schedule Satisfying Green-up Constraints" Vancouver: Forest Economics and Policy Analysis Research Unit, UBC

Carlin, A. 1992. "The United States Experience with Economic Incentives to Control Environmental Pollution." No. 230-R-92-001. Washington, DC: Environmental Protection Agency Office of Planning and Evaluation

Cassidy, P. 1994. "BC *Forestry Code* adopting suspect regulatory approach." *Environment Policy and Law* (December): 135-47

Cassils, J.A. 1991. "Exploring Incentives: An Introduction to Incentives and Economic Instruments for Sustainable Development." Victoria: Socio-Economic Impact Committee of the National Round Table on the Environment and the Economy

Chase, A. 1995. *In a Dark Wood: The Fight Over Forests and the Rising Tyranny of Ecology*. Boston: Houghton Mifflin

Clawson, M. 1959. *Methods of Measuring the Demand For and Value of Outdoor Recreation*. Reprint 10. Washington, DC: Resources for the Future

Commission on Resources and Environment (CORE). 1994. *Vancouver Island Land Use Plan*. Vol. 1. Victoria: CORE

Common, M., and C. Perrings. 1992. "Towards an ecological economics of sustainability." *Ecological Economics* 6: 7-34

Cook, T. 1998. "Sustainable Practices? An Analysis of BC's *Forest Practices Code*." In *The Wealth of Forests: Markets, Regulation, and Sustainable Forestry*, ed. Chris Tollefson, 204-31. Vancouver: UBC Press

Cox, J.C., and R.M. Isaac. 1987. "Mechanisms for incentive regulation: Theory and experiment." *Rand Journal of Economics* 18(3): 348-59

Daley, H.E. 1990. "Toward some operational principles of sustainable development." *Ecological Economics* 2: 1-6

Dellert, L. 1998. "Sustained Yield: Why has it Failed to Achieve Sustainability?" In *The Wealth of Forests: Markets, Regulation, and Sustainable Forestry*, ed. Chris Tollefson, 255-77. Vancouver: UBC Press

Dewees, D. 1992. "Taxation and the Environment." In *Taxation to 2000 and Beyond*, ed. R. Bird and J. Mintz. Toronto: Canadian Tax Foundation

Dewees, D.N. 1980. "Instrument choice in environmental policy." Paper presented at University of Toronto Law and Economics, Toronto

Dietrich, William. 1992. *The Final Forest*. New York: Simon and Schuster

Doern, G. Bruce, and V.S. Wilson. 1974. *Issues in Canadian Public Policy*. Toronto: McMillan

Dowdle, B. 1981. "An Institutional Dinosaur with an Ace: Or, How to Piddle Away Public Timber Wealth and Foul the Environment in the Process." In *Bureaucracy vs. Environment: The Environmental Costs of Bureaucratic Governance*, ed. J. Baden and R.L. Stroup. Ann Arbor: University of Michigan Press

Fenge, T. 1995. "Multi-Stakeholder Processes: Interest Group Politics versus the Public Good." *National Round Table Review* winter: 4

Freeman, A.M. 1979. *The Benefits of Environmental Improvement*. Baltimore: Johns Hopkins University Press

Government of British Columbia. 1993. *A Protected Areas Strategy for British Columbia*. Victoria: Queen's Printer

–. 1994. *British Columbia's Forest Renewal Plan*. Victoria: Government of BC

Government of Canada. 1992. *Economic Instruments for Environmental Protection*. Ottawa: Minister of Supply and Services Canada

Hahn, R.W., and R.G. Noll. 1983. "Barriers to implementing tradable air pollution permits: Problems of regulatory interactions." *Yale Journal on Regulation* 1(1): 63-91

Haley, D. 1996. "Paying the piper: The cost of the British Columbia *Forest Practices Code*." Paper prepared for the conference Working with the *Forest Practices Code*, Vancouver

–, and M. Luckert. 1995. "Policy Instruments for Sustainable Development in the British Columbia Forest Sector." In *Managing Natural Resources in British Columbia: Markets, Regulations, and Sustainable Development*, ed. A. Scott, J. Robinson, and D. Cohen, 54-79. Vancouver: UBC Press

–. 1998. "Tenures as Economic Instruments for Achieving Objectives of Public Forest Policy." In *The Wealth of Forests: Markets, Regulation, and Sustainable Forestry*, ed. Chris Tollefson, 123-51. Vancouver: UBC Press

Hamilton, G. 1997. "Thorn in MB's side eases his campaign." *Vancouver Sun* 24 January: D3

Hartman, R.S., and D. Wheeler. 1995. "Incentive Regulation: Market Based Pollution Control for the Real World." In *Regulatory Policies and Reform: A Comparative Perspective*, ed. C.R. Frischtak, 210-335. Washington, DC: World Bank

Hogg, P.W. 1985. "Responsible Government." Abridged from P.W. Hogg, *Constitutional Law of Canada*, 2nd ed., 189-213. Toronto: Carswell

Huffman, J.L. 1994. "Markets, regulation and environmental protection." *Montana Law Review* 55: 425-34

Jansson, A.M., et al. 1994. *Investing in Natural Capital: The Ecological Economics Approach to Sustainability*. Washington, DC: Island Press

Joeres, E., and M. David, eds. 1983. *Buying a Better Environment: Cost-Effective Regulation Through Permit Trading*. Madison: University of Wisconsin Press

Johansson, P.-O. 1987. *The Economic Theory and Measurement of Environmental Benefits*. New York: Cambridge University Press

Keeney, R.L., and H. Raiffa. 1976. *Decisions With Multiple Objectives: Preferences and Value Tradeoffs*. New York: John Wiley and Sons

Kelly, M. 1992. "Market Correction: Economic Incentives for Sustainable Development." Working Paper 4. Montreal: Institute for Research on Public Policy, National Round Table on the Environment and the Economy

Kimmins, J.P. 1996. "Biodiversity and its relationship to ecosystem health and integrity." *Policy Options* November: 3-7

Lindblom, C.E. 1959. "The Science of 'Muddling Through.'" *Public Administration Review* 19: 79-88

M'Gonigle, R.M., and B. Parfitt. 1994. *Forestopia: A Practical Guide to the New Forest Economy*. Madiera Park, BC: Harbour Publishing

Maser, C. 1994. *Sustainable Forestry: Philosophy, Science and Economics*. Delray Beach, FL: St. Lucie Press

Mitchell, R.C., and Carson, R.T. 1989. *Using Surveys to Value Public Goods: The Contingent Valuation Method*. Washington, DC: Resources for the Future

Montgomery, R. 1972. "Markets in licenses and efficient pollution control programs." *Journal of Economic Theory* 5(3): 395-418

Mueller, D. 1989. *Public Choice II*. Cambridge: Cambridge University Press

National Board of Forestry. 1992. *A Richer Forest*. Jonkoping, Sweden: National Board of Forestry

Noll, R.G. 1983. "The Feasibility of Tradable Emissions Permits in the United States." In *Public Sector Economics*, ed. J. Finsinger, 189-225. New York: St. Martin's Press

Noss, R.F. 1994. "A Sustainable Forest is a Diverse and Sustainable Forest." In *Clearcut: The Tragedy of Industrial Forestry*, ed. B. Devall, 54-79. San Francisco: Sierra Club Books

Organization for Economic Cooperation and Development (OECD). 1989. *Economic Instruments for Environmental Protection*. Paris: OECD

Palmisano, P. 1982. "Have Markets for Trading Emission Reduction Credits Failed or Succeeded?" Working Paper 2. Washington, DC: United States Environmental Protection Agency, Office of Policy and Resource Management

Peterson, L. 1996. *Forest Practices Code – Timber Supply Analysis*. Victoria: Ministry of Forests

Price Waterhouse. 1995. *Analysis of Recent British Columbia Government Forest Policy and Land Use Initiatives*. Study for the Forest Alliance of BC

–. 1996a. *The Forest Industry in British Columbia, 1995*. Vancouver: Price Waterhouse

–. 1996b. *Report of the 9th Annual British Columbia Forest Industry Conference – 1996*. Vancouver: Price Waterhouse

Priest, M., W.T. Stanbury, and Fred Thompson. 1980. "On the Definition of Government Regulation." In *Government Regulation: Growth, Scope, Process*, ed. W.T. Stanbury, Ch.1. Montreal: The Institute for Research on Public Policy

Rainforest Action Network (RAN). 1995. *Ten Years of Rainforest Action*. San Francisco: RAN

Robinson, J., G. Francis, R. Legge, and S. Lerner. 1990. "Defining a Sustainable Society: Values, Principles and Definitions." *Alternatives* 17(2):36-46

Scarfe, B.L. 1998. "Timber Pricing Policies and Sustainable Forestry." In *The Wealth of Forests: Markets, Regulation, and Sustainable Forestry*, ed. Chris Tollefson, 186-203. Vancouver: UBC Press

Schwindt, R., and T. Heaps. 1996. *Chopping Up the Money Tree: Distributing the Wealth British Columbia's Forests*. Study for the David Suzuki Foundation, Vancouver

Scientific Panel for Sustainable Forest Practices in Clayoquot Sound. 1995. *Sustainable Ecosystem Management in Clayoquot Sound*. Victoria: Queen's Printer

Scott, S. 1995. "Multistakeholder processes: A panel discussion." *National Round Table Review* winter: 5-15

Simon, H.A. 1976. *Administrative Behavior*, 3rd ed. New York: Free Press

Stanbury, W.T. 1992. *Reforming the Federal Regulatory Process in Canada, 1971-1992*. Study prepared for the Subcommittee on Regulations and Competitiveness of the House of Commons Standing Committee on Finance. In *Minutes and Proceedings of the Standing Committee on Finance* 23: A1-293

–. 1993a. *Business-Government Relations in Canada: Influencing Government*. Toronto: Nelson Canada

–. 1993b. "A skeptic's guide to the claims of so-called public interest groups." *Canadian Public Administration* 36(4): 580-605

–. 1993c. *Regulating Water Pollution by the Pulp and Paper Industry in Canada*. Vancouver: Forest Economics and Policy Analysis Research Unit, UBC

–. 1993d. "Regulation, Regulations and Regulatory Reform." In *Readings and Canadian Cases in Business, Government and Society*, ed. M. Baetz, 112-30. Toronto: Nelson Canada

–. 1994a. "Holding Governments Accountable: Insights From Efforts to Reform the Federal Regulation-Making Process." In *Policy Making and Competitiveness*, ed. B. Purchase, 67-96. Kingston: Queen's University School of Policy Studies

–. 1994b. *Protest Behaviour and Other Efforts to Influence Forest Policy in British Columbia: The Case of Clayoquot Sound*. Vancouver: Forest Economics and Policy Analysis Research Unit, UBC

–, and J. Fulton. 1984. "Suasion as a Governing Instrument." In *How Ottawa Spends 1984: The New Agenda,* ed. A.M. Maslove, 282-324. Toronto: Methuen

–, and G. Lermer. 1983. "Regulation and the redistribution of income and wealth." *Canadian Public Administration* 26: 378-401

–, and I.B. Vertinsky. 1985. "The nature of regulation of the British Columbia forest industry." *UBC Business Review* 1: 45-52

–, and I.B. Vertinsky. 1996. "Analysis of Boycotts as a Tactic Employed by Environmental Groups" Vancouver: Forest Economics and Policy Analysis Research Unit, UBC

–, and I.B. Vertinsky. 1997. "The use of the boycott tactic in conflicts over forestry issues: The case of Clayoquot Sound." *Commonwealth Forestry Review* 76(1): 18-24

–, I.B. Vertinsky, and B. Wilson. 1994. "The Challenge to Canadian Forest Products in Europe: Managing a Complex Environmental Issue." Vancouver: Forest Economics and Policy Analysis Research Unit, UBC

Trebilcock, M.J. 1994. *The Prospects for Reinventing Government*. Toronto: C.D. Howe Institute

–, and D.G. Hartle. 1982. "The choice of governing instrument." *International Review of Law and Economics* 2: 29-46

–, D.G. Hartle, J.R.S. Pritchard, and D.N. Dewees. 1981. *On The Choice of Governing Instrument: Some Applications.* Technical Report 12. Ottawa: Economic Council of Canada

–, D.G. Hartle, D.N. Dewees, and J.R.S. Pritchard. 1982. *The Choice of Governing Instrument.* Ottawa: Minister of Supply and Services

van Kooten, G.C. 1996. "Economic evaluation of British Columbia's Forest Practice's *Code*: Where are the benefits?" Vancouver: Forest Economics and Policy Analysis Research Unit, UBC

Vertinsky, I.B., S. Brown, H. Schreier, W.A. Thompson, and G.C. van Kooten. 1994. "A hierarchical-GIS-based decision model for forest management: The systems approach." *Interfaces* 24(4): 38-53

Wildavsky, A. 1993. *The Rise of Radical Egalitarianism.* New Brunswick, NJ: Transaction Books

Wilman, E.A. 1988. "Modeling Recreation Demands for Public Land Management." In *Environmental Resources and Applied Welfare Economics*, ed. V.K. Smith, 165-90. Washington, DC: Resources for the Future

Zhang, D. 1994. *Implications of Tenure for Forest Land Value and Management in British Columbia.* PhD dissertation. UBC

4
Compliance and Constraint: Economic Instruments in Context
Rod Dobell

Introduction

This chapter suggests a broad, admittedly rather abstract but non-technical, framework or context within which the more concrete discussion of other chapters can be set. It serves as a reminder that in practice the choices to be made in the design of institutions and the selection of governing instruments are less categorical, more pragmatic, and more adaptive than theoretical comparisons often imply.

The argument is simple in essence, though sometimes cumbersome in detail. In the next section, "An Ecological Frame and Context," it is suggested that there are two elements of transition to what Herman Daly has called a "full world"[1] that have not been adequately taken into account in discussion of governance and regulation generally, and of forests in particular. In this "full world," forests must be seen as part of a complex natural system (the biosphere), which in its totality forms a common heritage for humankind rather than simply a separable body of appropriable resources. Human activities impinging on this resource system must therefore be managed through institutions that reflect necessary constraints of stewardship. Moreover, in policy analysis both equity and the maintenance of social and natural capital are identified as important human concerns. Conventional goals of economic efficiency, which are focused on optimal use of physical capital and priced resources, must therefore be supplemented in the decisions of these management institutions by goals of social equity, conservation of social capital, and sustainability of natural capital. These resources are mostly not currently priced or easily valued.

These goals may be pursued in processes at various levels of aggregation or delegation, and compliance with commitments undertaken may be assessed at various levels. Ultimately, however, such goals can only be realized through local action, with the exercise of individual discretion in diverse and changing circumstances "on the ground." This chapter's third section, "Discretion and Compliance," outlines a variety of institutional structures through which social objectives and market preferences can be linked to specific characteristics of the resource system in such a way that individual compliance with policies to achieve social purposes can be promoted. In this setting, economic incentives are seen as only one part of more general flows of information affecting individual behaviour.

The fourth section, "Economic Instruments and Voluntary Compliance Mechanisms," examines the special case of economic instruments in this light, but focuses more specifically on the flexible use of regulatory instruments through appeal to "voluntary compliance mechanisms" (VCMs). The hard distinction between "command and control" and "free markets" is seen as overly simplistic in this setting: almost every institutional arrangement can be viewed as a mix of hierarchy, contract, market, and custom. Economic instruments, usually contrasted directly with "command and control," attempt to align the interests of agents with the interests of the public owner by harnessing self-interest through the secure promise of appropriate financial returns to good management. In other words, economic instruments appeal to self-interest by securing claims on future income streams.

In a complex and uncertain "full world," however, that sort of security of tenure is no longer feasible. Property rights associated with resource tenures, just as all other claims to an income stream, must be seen as subject to changing concepts of socially appropriate use, and subject to adjustment in light of the evolution of a congested world[2] characterized by inherent and possibly irreducible uncertainty. The price, financial or administrative, of access to increasingly scarce ecological space is bound to rise. So also is the competition for, and the value of, claims upon it. The fifth section, "Adaptive Institutional Design" suggests that it will be necessary to turn increasingly to more flexible notions of institutional design which recognize that ultimately the most powerful market mechanisms are "sunshine and scrutiny"[3] – information flows shaping the ongoing decisions of resource managers through the pressures brought by informed stakeholders (consumers, employees, shareholders, neighbours) and other groups. Resource managers less able to rely on certainty of property rights as such must be able to rely more on procedural assurances and effective consultative processes governing adjustments in those rights as the costs of adaptation to changing circumstances are shared among all the players in civil society.

Given this discussion, two points stand out:

- In a changing, complex world, it is not possible to aim at certainty in property rights in natural capital or resource systems themselves any more than it is in other claims on jobs, markets, or other income streams in a rapidly changing economy.
- The key economic instrument ultimately is information to assure accountability for delegated responsibilities, and to enable owners, con-

sumers, and other citizens to adapt their conduct, and their delegation of authorities, in light of their assessment of the actions of their institutional and individual agents.

These facts lead to two overriding challenges:

- The need to define acceptable processes for continuous monitoring of the exercise of discretion, and adaptive revision of the terms or constraints surrounding tenures and property rights.
- The need to establish feedback mechanisms and linkages that assure adequate flows of information for these purposes.

In summary, the basic conclusions are straightforward. In the management of forest-related activities, it is now widely recognized that forests are more than fibre. As components of a regional and global ecosystem, which includes human institutions and activities, forests are part of a complex and uncertain dynamic system with vast diversity and possibly increasing variety, in which there is growing demand for greater participation in decisions involving multiple objectives, multiple interests, and multiple agents. In the face of this complexity and recognized uncertainty, the search for optimal allocation of resources and economic efficiency may well be scaled back to the pursuit of more general qualitative merits: adaptability, resilience, stability, sustainability, equity, accountability, and contestability. In particular, in the management of renewable resource systems, some version of an overriding objective of sustainability or stewardship may well be adopted.

In exploring the question of policy or instruments in pursuit of this goal, it may be useful to think in terms of three slogans:[4]

- sustainable by intent, through covenants negotiated by epistemic communities operating at international level, expressed in codes or ethical frameworks;
- sustainable by design, through inter-institutional arrangements reflecting principles of subsidiarity in action from nation state to the local level; and
- sustainable by action, through discretionary decisions of individuals on the ground.

International agreements may provide direction; institutional design and inter-institutional arrangements may provide information and incentives. Ultimately, however, it is immediate peer pressure, tradition, and individual moral codes that determine how information and incentives lead to individual action in specific, individual, and changing circumstances. And it is the aggregate of individual actions that leads to change in the state of the system and determines, ultimately, whether the responsibility for stewardship is met. Continuing scrutiny, by non-government organizations and individuals, of these ongoing actions in the implementation of policy intentions and compliance with regulations is crucial.

In this chapter, the focus is on the issue of institutional design, and the choice of governing instruments to link international and national objectives with individual and community interests and incentives. With a focus on compliance at the workface, in the woods, both centralized regulation and economic instruments seem likely to be somewhat too unidimensional to capture all necessary aspects of the multiple objectives to be pursued in ongoing action in a multi-stakeholder setting. The interests of the public owner seem unlikely to be aligned well with the interests of private agents solely through price signals and market transactions. A more nuanced, iterative, and continuing search for communicative rationality[5] – which is to say, simply, a continuing process of broad public discussion and debate that seeks consensus on acceptable action and adaptation to a changing natural and social context – seems essential.

In this collection, this chapter sketches some theoretical considerations relevant to these general propositions about governance and serves as background to more focused discussion of market approaches and economic instruments in British Columbia's forest policies.

An Ecological Frame and Context: Social Capital and Natural Capital

The evolution of the human economy has passed from an era in which man-made capital represented the limiting factor in economic development (an "empty" world) to an era in which increasingly scarce natural capital has taken its place (a "full" world) ... But few until now have recognized that we have not only reached an economic turning point, we have passed it.[6]

In this section, a general ecological frame for the discussion is sketched. Human activities are set in the context of the natural systems of biosphere and geosphere. The natural capital corresponding to the value of the natural systems of the geosphere and biosphere, and the social capital corresponding to the cultural traditions and community institutions that enable people to work together effectively, are both emphasized in this context.

As the introductory quotation from Herman Daly suggests, the size of human populations and the scale of human activities have grown so great that humankind must now be recognized as "a mighty geological force" with the power to alter the geosphere,[7] and as a dramatic evolutionary force in competition with all other species in the biosphere.[8] Humankind is the product of a long process of co-evolution with all these other species, within the context of the natural systems of the biosphere, and hence is intimately interdependent with them. Survival of the human species seems likely to be crucially dependent on the sustainability of the biosphere overall. As a result, it becomes necessary to constrain the overall impact of the human species on the biosphere, and the extent of competition with the other species with which humanity is crucially interdependent. For this purpose, it is essential to manage the competition of individual humans with one another for claims on and shares in access to the increasingly scarce and valuable resources of the global commons.

A similar transition, it has been argued, has occurred unrecognized in the sphere of international organization. "The world is not now passing through a transition to some new form of world order, but rather ... it has already undergone such a transition and is, today, well ensconced in a new order."[9] This new order involves the rise of substantial capacity of civil society at the global level to influence the negotiation of international covenants bearing on the management of resources perceived as a common heritage, in which decisions are seen as having global consequences.[10] It also involves a similar increase in the involvement of non-government organizations and other groups or interests in inter-organizational processes of policy formation and negotiated implementation at all levels of government or community activity,[11] and a corresponding decline of confidence in the traditional institutions of representative democracy.[12]

Because trees are not migratory, and it is easy to establish title to specified tracts of forest land, it may be asked why any of this discussion is relevant. Are there any concerns with ownership and production of forest and forest products that cannot be handled by normal market mechanisms?

Although it is true that trees are not common pool resources, it is important to recognize that forests play three possibly key roles in the global commons: they form part of the habitat for an intricate and diverse biological web of overlapping and interdependent resource systems; they are part of a global carbon cycle critical to life; and they, as forest wilderness, may form part of a unique and irreplaceable endowment important to humans simply through its existence. Thus, it may be possible to imagine ownership and commercial traffic in the resource units (trees or lumber), while denying the acceptability of ownership or appropriation of elements of the forest resource ecosystem itself – a heritage which owes little for its creation to the energy or ingenuity of humankind in general or individual investors in particular.

Forests, then, have elements of public goods about them because the ecological services they perform within the carbon cycle or as habitat are not "excludable"[13] and neither are the wilderness option values they offer. Forests may be "congested" in the sense that some uses will generate "externalities" – one user's presence will impose unpriced costs on other users of the habitat.

The scale of human numbers and the extent of industrial activity now are sufficiently great, and press sufficiently against the capacity of the natural systems of the ecosphere, that significant interdependence and "externalities" are associated with almost all resource harvesting and waste disposal activities. Individual decisions to exploit particular resources or claim a role in particular resource systems will create impacts on others which will not generally be reflected in price systems or taken into account in those decisions. A social interest in these individual decisions, and a possible social responsibility to influence them in the interests of others, results from this fact of "congestion" as well as from associated problems of information asymmetry which make it difficult for consumers or citizens to appraise fully the significance of decisions of resource managers, or even to know the extent of their implementation of negotiated agreements on compliance with codes.

Thus, although we may wish to concentrate discussion on market mechanisms involving the institutions of the formal economy based on explicit prices and individual transactions reflecting voluntary exchange, we must recognize that the formal economy is only one (relatively recent) set of structures among a much vaster set of formal and informal social institutions which makes up a human community. These institutions include formal hierarchical (state) institutions – the formal apparatus of what we have come to call the government sector – as well as informal networks and kinship structures. Household and family structures are included in these informal networks as well as a "civil society" that involves organizations of various scales from small local associations to international bodies.

Within the formal economy, it has long been conventional to recognize that the allocation of human effort and current resources can be directed towards the creation of further resources (assets or property) that embody a capacity to produce goods and services in the future. This investment, or accumulation of future production potential, may take the form of building up physical capital (produced means of production), or human or intellectual capital. What has only been recognized more recently is that the capacity to produce goods and services in the future rests also on less tangible assets or resources (referred to as social capital). Accumulated traditions and rules relating to the ways in which individuals are enabled to work together effectively, and to maintain a cohesive civil society, represent stocks built up from past efforts and past decisions, which can be increased or destroyed by current actions. What also is now explicitly recognized is that the natural systems of the ecosphere are also essential to human activity, and represent a productive asset, which may be built up or drawn down as a result of human activity.[14]

Included in this sphere of human institutions and social systems are four categories now often distinguished as state or hierarchy, market, community (networks), and self-governing institutions. While considerable attention has been directed in the past towards the choice between state and market institutions, more recent discussion has focused on other elements of civil society, in particular informal networks and more formal self-governing institutions or third sector organizations.[15] This emphasis forces attention to social and institutional capital as special forms of knowledge or community assets, and to the important body of relationships outside the formal economy to which we have traditionally confined efforts at national accounting and social reporting.

Thus the context in which economic instruments must be discussed includes a social sphere of non-market institutions which may shape economic decisions, and in which stocks of social capital essential to the functioning of the system may be accumulated or drawn down. In this sphere, the distribution of income and wealth is an important determinant of social cohesion and social well-being, and hence a crucial consideration in public policy directed towards maintaining the productive capacity of the overall economy.

All of this human system functions within an ecological space in which stocks of natural capital can be identified as essential. This structure of human society, both formal economic activities and non-market social institutions, rests upon the life-support systems of the biosphere, and these in turn live within the natural systems (geological, hydrological, atmospheric, and so on) of the non-biological edifice of the planet. Humans not only draw on or harvest exhaustible and renewable resources as inputs into household and industrial activity, but they also rely on natural systems for the disposal of the waste products. The scale of these human activities has now become so great that humankind not only draws its sustenance from those systems, but acts in ways that change them significantly, possibly fundamentally. The aggregate impact of individual human decisions could put at risk the continued functioning of those basic systems. The physical environment could be changed irreversibly;[16] moreover, the accumulated draw of human activity on the same resources on which the rest of the living world depends may reach the scale of irresistible competition with other species.[17]

None of these stocks of social or natural capital is likely to be appropriately priced and represented in market exchanges. The dynamics of the processes by which social and natural capital may be accumulated, conserved, or destroyed are not well understood and are unlikely to be taken properly into account in market transactions. Questions of title to or ownership of such capital stocks are likely to be very ambiguous indeed.

There are two consequences. Because the survival of the human species, co-evolving with the rest of the ecosphere, may depend on the sustained functioning of this overall system, it becomes necessary to consider the development of institutions (effective rules of conduct) which can constrain the overall impact of humans, both as predators on other species (as in the harvesting of fish) and as competitors for common ecological space (as in the destruction of habitat through human settlement and discharges). This need for discipline and self-restraint as a direct result of the emergence of interdependence can be seen as a further stage in what Elias has called "the civilizing tendency."[18]

The same questions of scale and scarcity that give rise to a need to constrain the competition of humans with other species give rise also to the need for rules to discipline the competition of humans with others in their species in harvesting or using scarce ecological resources.

We thus begin our discussion with a presumption that there will be a need to assure social rules or incentives to govern the interactions of humans as they attempt individually to draw upon the resources of the biosphere for their own survival. These rules or incentives may be negotiated at global or international level and must ensure that the aggregate impact of the human race is contained within a scale that does not threaten the existence of the underlying natural systems on which all depends.

Because these systems are complex, interacting, and interdependent, and characterized by inherent, possibly irreducible, uncertainty, it is hard to calculate what such limits might be. In the absence of other guidelines, there is an argument for a rule of thumb expressed as an obligation of "stewardship." This obligation would entail an undertaking that human activity will be structured in such a way as to assure, so far as possible, that the capacity of the ecosphere to support such activity does not deteriorate. The allocation of resources at any time in such a way as not to reduce the options open in the future for meeting human needs is a requirement that might also be expressed as a commitment to a non-decreasing value of the total capital stock. Such allocation must take into account unpriced social and natural capital, and in particular must recognize that some forms of natural capital might be so essential to the production of future goods and services needed for life that their loss would inevitably mean a decline in the value of the earth's capital stock.

Such requirements of "sustainability" ("strong" or "weak"), and of "precautionary approaches" in decision-making, which might assure sustainability, have been extensively discussed in a long-running and sharply contested debate.[19]

The point here, however, is simply that a set of social guidelines insisting on priority for a stewardship or sustainability objective may well emerge as an overriding commitment in the design of social institutions. These guidelines would entail limits on the "rights" of individuals to claim access or authority to harvest resources as they please, or to occupy a role in ecological systems at will. In the next section, we ask how such social objectives (such as priorities attached to conservation goals, or commitments to a stewardship condition) might be reflected in and communicated through the institutions and rules of the formal economy, government sector, or civil society, in such a way as to influence individual conduct sufficiently to assure realization of overall goals. The focus there is on institutional structures, not on management processes or approaches to decision-making within them. How might institutional arrangements be structured so as to align individual interests and motivations with such social goals?

Discretion and Compliance

We start with the problem of finding institutional arrangements to serve aggregates of individual preferences assumed as given. For this purpose, in general, decentralized markets work well. Even if one cannot argue rigorously that such markets will realize an optimal allocation of resources at any given moment, market mechanisms do build in dynamic processes of competition which continually work in the direction of "better" service to the customer, better economic performance, and a presumably greater welfare within the limits of available resources.

Many important resources, however, are unpriced, and important information is often lacking or is distributed in a very asymmetric manner among agents in the market. Problems of externalities and other market failures arise,[20] and so one sees the introduction of regulation or government intervention. It can still be debated whether social goals could be achieved better through improved pricing ("free market environmentalism")[21] or through the forced imposition of centralized standards ("command and control"), either to internalize the externalities so as to improve the information going into decisions, or to substitute other constraints on the decision.

Whether this regulatory constraint imposes a required outcome, or dictates the conclusion of some intermediate decision, is a matter of choice (or negotiation) in the design of regulatory measures. In general, economic or environmental outcomes occur as a result of a long process or chain of steps, with choice of technology, choice of inputs, and operating practices all entering into the determination of consequences in terms of direct outputs, intermediate impacts, more general results or outcomes, or ultimate environmental effects. The more general (or more results-oriented) is the nature of the regulatory rule or standard imposed, presumably, the greater is the discretion that can be exercised by resource managers in achieving that standard at the lowest cost. Further, the more acceptable the constraint, the less onerous it is, the more it harnesses self-interest, the greater the likelihood that compliance with it will be achieved.

Both goals (greater discretionary scope for cost effectiveness and greater compliance) are emphasized in debate about free market environmentalism versus command and control, and in consideration of other options.

In public policy with respect to forests, however, what is new is not just recognition of individual demands for multiple uses in serving current consumers at minimum opportunity cost in terms of valued inputs, but also the addition of system-wide or collective goals of sustainability, the preservation of biodiversity, and the recognition of obligations of stewardship.

Even without these goals, the problems of utilizing resources from a complex natural system, and mobilizing inputs with varying and uncertain prices to generate products flowing through complex global markets to supply demands of varying specifications at varying market prices are intricate and uncertain enough in their pay-off. Now, with the addition of collective preferences expressed in terms of system-wide objectives with respect to aggregate harvesting rates or preservation of a range of other characteristics of the resource system in addition to its future value for production purposes (and with a wide range of new users or claimants to be accommodated as well), the management challenge is substantially increased.

In addressing this challenge, the necessary process of decentralization might be achieved by any one (or a mix) of several actions: utilization of market mechanisms based on price signals; delegation of property rights to individual owners or firms; devolution of rights of access to the resource to other groups or combinations of groups; or decentralization of responsibilities within or among government agencies. In some cases, it is useful to distinguish more community-based organizations such as cooperatives based on traditions of voluntary cooperation from more contractually oriented, constitutionally based self-governing institutions, and to distinguish self-generating from socially constructed structures.[22]

This section now turns to examine these alternative channels of linkage from general social objectives to individual behaviour, emphasizing that collective decisions or social policies are only fully defined as they are implemented through the exercise of individual discretion in carrying out individual acts. The incentive structures and reward systems inherent in the institutions through which behaviour is influenced, decisions are transmitted, and action organized, thus become key concerns.

The growing diversity of structures and interests is an important feature of the human institutional systems emerging in the new global order. In the previous section, it was noted that in contemporary society, humans find themselves involved in market activities and formal economic institutions; in democratic institutions and formal government hierarchies; in organizations of civil society (non-government organizations or NGOs) and the voluntary sector; and in a variety of informal networks or associations. Through markets, consumers express their preferences regarding the products desired from resource harvesting or exploitation. Through the institutions of parliamentary democracy and through extra-parliamentary channels, citizens express their views on stewardship requirements or sustainability constraints (such as those discussed earlier), and look to the apparatus of public administration for the articulation of policies and the implementation of programs that will see those goals realized. And, indeed, through international organizations and global civil society, people everywhere may formulate frameworks or codes of conduct to assure appropriate stewardship.

In the context of a particular resource system, then, one can think of the task as finding or designing some institutional arrangement that can link an overall social or system objective with the discretionary individual acts, such as harvesting, conservation, and enhancement, which have impacts on that resource system. The key determining features, therefore, are the physical features and dynamic characteristics of the resource system concerned, the institutional structure shaping individual human activities impinging on it, and the preferences and goals of the populations served by those activities.

The point of this section is that there is a wide range of choice of institutional designs by which this linkage might be achieved, or by which, in other words, the task of attaining the overall social goal might be "decentralized." These structures call into play a variety of decision-making mechanisms or processes of governance within the differing institutional settings, all, in a sense, representing a delegation of collective authority to manage individual access to ecological resources.[23]

Two elements, or linkages, extend beyond the boundaries of the particular institutional structures that are charged with organizing the harvesting and investment activities relating to a resource system. The first is a set of feedback or adjustment mechanisms that may impinge on factors (signals) entering the institutional decision process. In the case of government departmental or agency hierarchies, formal and informal accountability mechanisms provide an information flow from field operations to central organizations in such a way that operating rules and guidelines at various levels of generality may be adjusted in search of better compliance and more complete achievement of objectives. In the case of self-governing institutions or cooperatives, a nested structure or federated system provides a flow of reporting from more local to more central structures, again for purposes of adjusting policies and rules. And in the case of market mechanisms, individual enterprises see the adjustment of market prices reflecting, in principle, changing costs and relative scarcities that in turn call forth adjustments in decisions on harvesting scales and resource allocations.

In each case, it is not only the decisions of the resource "managers" (the institutions charged with decisions on harvesting or investment impinging directly on the resource) that might be influenced. In the case of governmental institutions, the judgments of voters or citizens as consumers of the product from the activity may be registered at local or central levels of the hierarchy. In the case of self-governing institutions or cooperatives, the voices of members will reflect their assessment of the degree to which individual decisions are well harnessed towards agreed general purposes. And in the case of market mechanisms, the willingness of consumers to pay for a product will rest on the information they have about the process of harvesting and the state of the underlying resource as well as on the characteristics of the product itself. This market readiness to pay for the product will presumably be a determining influence in management decisions about acceptable practices. And for this reason, any attempt to block the flow of full information to customers (and indeed the exercise by interested stakeholders of their persuasive powers upon them) should be met with great concern as an interference in the operation of informed markets.

The second major external influence on decisions that must be recognized in this discussion of institutional choice and governing instrument is the existence of both a collective ethical framework which may influence conduct through the articulation of explicit overarching principles (as mentioned), and personal moral codes which shape the way in which individual decisions and conduct will respond to the objective influences in a decision setting. In the case of forest policy, international discussions may lead to formulation of a code of forest conduct. Through a long process beginning with national action to comply with such commitments,[24] this framework will provide explicit guidelines intended to shape the actions of individuals within any of the institutional structures elected for management of human activities related to forests. At the same time, the ultimate determinants of the impacts of logging operations on the state of the forest resource are the moral codes which shape how far individual operators and workers are prepared to conform to the codes, regulations, policies, guidelines, operating manuals, and instructions which are intended to influence their behaviour (along with the informal incentive systems, which may or may not be designed consciously to influence anything), and how far consumers are prepared, if necessary, to pay a premium for goods produced according to what are deemed to be socially acceptable processes.

From this discussion it will be clear that the distinctions among the various instruments are not so much categorical as simply questions of organizational structure, information flow, and the credibility or immediacy of the perceived impacts on incentives and rewards. The apparent advantage of community-based organizations arises from the opportunities they offer for all those affected to participate in management decisions. This advantage may appeal to people who have lost confidence in their ability to influence outcomes through the larger-scale institutions of representative democracy, which, ironically, evolved precisely to provide for popular participation in decisions setting the framework for decentralized action. The argument that market mechanisms can shape incentives better than regulatory mechanisms may reflect at root simply a belief that the sanctions associated with non-compliance with rules are unlikely (by the time they are adjusted to reflect probabilities of detection, prosecution, conviction, and penalty) to be significant factors in management decisions. But a regulatory structure in which permits for continued access to the resource (or corporate charters themselves)[25] were truly contingent on compliance with agreed codes and policies would certainly create an incentive system at least comparable in its impacts to the influence of price signals.

What is important about this picture is that all these options ultimately are inter-organizational structures beginning and ending with non-market ethical frames or moral codes. Here, there is no market independent of an overall regulatory frame: the issue is simply how general that frame can be. In particular, it might be asked how far economic instruments can create incentive structures within which the exercise of independent discretion by economic agents pursuing their self-interest can serve the interests of the public owner.

These observations leave us with a number of questions and puzzles. Is an association of individual licensees who form themselves into a corporation for broader ecosystem-based management in harvesting and distribution significantly different from a corporation granted an allocation as a single enterprise, and different again from a community holding a single licence, which is then broken down into individual allocations among community (or cooperative) members for harvesting purposes? Apart from the questions of enforcement and compliance, what will be different, presumably, are the rule books (modes of governance) and incentive systems relevant in each case. The next two sections briefly review these issues.

Economic Instruments and Voluntary Compliance Mechanisms

These alternative institutional designs suggest an array of possible "governing instruments" by which individual behaviour (including institutional or corporate decisions) might be influenced for the purpose of ensuring that individual discretion is exercised in actions consistent with achievement of agreed targets for system performance. Those targets might relate to allowable levels of harvest at any one time; to the efficient allocation of resources to ensure that harvests are achieved at the lowest economic cost; or to an appropriate allocation of resources to investment in assuring the continued health of the resource in the future. Particularly important are targets or constraints relating to the impact of timber harvesting operations on other ecosystem features and values.

This array of governing instruments includes (as noted) the whole range of more responsive or interactive regulatory structures that bring "command and control" options into much more of a negotiated setting, where the quantities regulated may be general performance indicators rather than choices of technology or outputs. It includes the variety of alternative service delivery mechanisms by which regulatory responsibilities might be devolved in public-private partnerships or other contracting arrangements; a variety of community-based or spontaneously generated self-governing

institutions; and government interventions through both pricing mechanisms and granting or redefinition of tenures and rights.

The test of institutional effectiveness then is whether its system of incentives, rewards, and accountability can be expected to be successful in influencing decisions and shaping the exercise of discretion so as to assure
compliance with the broader guidelines and targets specified or delegated,
and thus with the overall sustainability goal.

Each of the institutional designs or organizational forms is managed
according to its own structures of information flow and rules of governance, and each may emphasize different goals or purposes. Each, therefore, generates different incentives and may arrive at different decisions
even in the face of the same systems data and market circumstances. This
outcome is true both for appropriation (resource allocation) decisions and
for investment decisions. Further, it is important to recognize that these
organizational forms are themselves mixed: within firms, decisions are
made and transmitted through hierarchical mechanisms, while within
governments, much is accomplished through contractual machinery (and
the newly fashionable public-private partnerships).[26]

In all these cases, we may view the selection of organizational form as
the selection of an agent engaged to act for the polity which lies behind
the system-wide social targets. The agent is expected to achieve these overall goals through the exercise of discretionary authority in the face of
diversified and changing local circumstances.

A substantial literature deals with the range of economic instruments
that might be deployed to "internalize the externalities"(discussed previously) and align the interests of individual operators with those of the public owner. Stanbury and Vertinsky, in Chapter 3, review a number of
reasons why, despite almost universal enthusiasm among resource economists, such instruments are not widely used. A recent review commissioned by the Canadian Council of Ministers of Environment finds similar
enthusiasm and similar barriers in relation to problems of pollution and
environmental quality.[27]

Many of the identified barriers have to do with the income and wealth
distribution consequences associated with the introduction of economic
instruments, and the tendencies towards concentration of economic activity to which they might give rise in the current context of global markets.
Such questions of concentration and inequality are not usually addressed
in the arguments for the efficiency of economic instruments. In a policy
setting, however, as Stanbury and Vertinsky emphasize, they are central.
And, with the many players and interests involved, transactions costs and
wealth effects cannot be assumed away or ignored as they must be if the
economic armament of spontaneous individual (Coasean) contracting is to
be brought to bear in support of market mechanisms.[28]

Nevertheless, arguments for economic instruments can still be strongly made simply on the basis of the greater discretion they offer to individual operators to respond to widely varying and substantially changing individual circumstances in individual settings. With such discretion, opportunities can be exploited to achieve agreed goals by more cost-effective means. If the destination can be agreed upon, the most efficient route to get there can be left to the judgment of those familiar with the facts and circumstances of the operation.

If the price system were not substantially incomplete with respect to ecological services and social capital, then much could be done through appeal to outcomes-based management relying on market signals about the value of resources and true costs of decisions.[29] But in the absence of the thoroughgoing ecological tax reform recommended in the references just cited, the degree of government intervention in tax and pricing mechanisms would be extreme.

Alternatives have therefore developed in the form of a wide range of negotiated deals. The attempt to achieve balance between system goals and opportunity costs leads to contemporary interest in voluntary compliance plans ("bureaucratic covenants"),[30] which emphasize the value of willing, uncoerced participation in setting standards and goals, and to results-oriented regulation or performance-oriented regulatory programs (PORPs),[31] which emphasize the value of the discretion on specific operational decisions that can be exercised within general agreement on outcomes to be achieved.[32]

In all these cases, however, we still face the well-known principal-agent problems[33] of balancing the costs of organizing and monitoring the actions of agents against the benefits of the improved decisions achieved by them. Issues of feedback mechanisms, information flow, and the identity and interests of the groups to whom accountability is to be rendered enter into the assessment of organizational effectiveness.

Of course, all such covenants presume, as does the use of economic instruments, that a government regulatory framework remains in the background and that failure to achieve agreed results ultimately will require appeal to much grosser mechanisms of detailed "command and control." Enforcement then takes one of two forms: formal monitoring and audit resulting in formal government action, or informal scrutiny and reporting by NGOs and others to inform employees, investors, and above all customers about the conditions violated by operators in the processes of production of goods to be marketed.

Enforcement creates, in effect, economic incentives for compliance, but the enforcement process itself must work its way through an extended sequence of events. Operational response to anticipated new regulations may (or may not) be seen during regulatory development or during a notice period for implementation. This response may begin only when inspections begin; in many cases, it may not be significant until investigations with serious prospect of penalty are launched. The value of publicity at various stages in this process is a subject of continuing discussion.

Unfortunately, one sad implication of this growing appeal to negotiated arrangements and enforcement through efforts of civil society is that there will be no end to the increasing need for talk, interaction, consensus-building, and dispute resolution. The possibilities for freeing up energies to go on to other things by delegating resource management responsibilities cleanly once and for all through generalized ownership or tenures (or other administrative means) seem more and more limited and remote.

Intense debate surrounds the degree of "voluntariness" that can be accepted in the stewardship of natural capital, and the degree of scrutiny essential to make it acceptable in the less deferential, more sceptical world of consultative mechanisms and participatory democracy.[34] Again, however, the conclusion is that only tests of pragmatism will apply, and only assurances of ongoing public scrutiny and reporting by non-government organizations can supplement the increasingly limited resources governments can put into monitoring and enforcement activities on the ground.

So in assessing the effectiveness of various governing instruments, the concern ultimately is with information flows and certification to shape individual perceptions and beliefs, and with processes of education, social learning, and communications to shape the moral codes individuals bring to bear in exercising their judgment about the achievement of general goals in specific, changing circumstances.

Adaptive Institutional Design: Constraints on Interpretations of Property Rights

Advice based on the experienced insight and common sense of Lord Keynes has been much called into question recently; but there may still be something to teach us here. Consider his observation that "the spectacle of modern investment markets has sometimes moved me towards the conclusion that to make the purchase of an investment permanent and indissoluble, like marriage, except by reason of death or other grave cause, might be a useful remedy for our contemporary evils. For this would force the investor to direct his mind to the long-term prospects and to those only." [35]

This reasoning has immediate relevance in the case of natural capital, and in the possibilities of adequate investment directed towards its maintenance in an integrated world where global companies can readily switch their capital and activities from region to region.

> The outstanding fact is the extreme precariousness of the basis of knowledge on which our estimates of prospective yield have to be made ...
>
> Decisions to invest in private business of the old-fashioned type were, however, decisions largely irrevocable, not only for the community as a whole, but also for the individual. With the separation between ownership

and management which prevails to-day and with the development of organized investment markets, a new factor of great importance has entered in, which sometimes facilitates investment but sometimes adds greatly to the instability of the system ... The Stock Exchange revalues many investments every day and the revaluations give a frequent opportunity to the individual (though not to the community as a whole) to revise his commitments ... Thus certain classes of investment are governed by the average expectation of those who deal on the Stock Exchange in the price of shares, rather than by the genuine expectations of the professional entrepreneur ...

If I may be allowed to appropriate the term speculation for the activity of forecasting the psychology of the market, and the term enterprise for the activity of forecasting the prospective yield of assets over their whole life, it is by no means always the case that speculation predominates over enterprise ... Speculators may do no harm as bubbles on a steady stream of enterprise. But the position is serious when enterprise becomes the bubble on a whirlpool of speculation. When the capital development of a country [the sustainability of its ecological base] becomes a by-product of the activities of a casino, the job is likely to be ill-done.[36]

For present purposes, the point to be drawn from this traditional knowledge is the observation that mobility is much more constrained for the resource-dependent community than it is for the resource-based industrial giant firm – or indeed for any owner of tradable market instruments representing claims on natural capital. Communities are locked in, and dependent on the long-term yield of the local natural capital base. For communities that suffer losses to this resource base, there are no safe havens; footloose firms, however, can revalue their prospects every hour, and decide on the exploitation of different stocks in different regions any time. For communities and society more generally, therefore, the risks may well be irreversible, and not well reflected in the decision rules of private owners.

What is of still more basic importance is that the property rights in the case of natural capital are fundamentally social. The real natural capital is the ecosystem itself – that is, a complex web of relationships, or a network, rather than a simple stock. That complex web of life is the common heritage of humankind. It cannot be privately owned. Even though the access rights must be efficiently allocated and effectively used, the increasing value of the natural capital itself ought not to be privately appropriated. This argument suggests that if there are tradable rights – as there might be for purposes of economically efficient allocation and effective harnessing of incentives and motives to individual action – there must be either full capture of the flow of scarcity rents or full taxation of the speculative gain due simply to increasing population and increasing scarcity value of natural capital. This fact poses an obvious difficulty because it raises the question of how to separate the returns to genuine investment in conservation and good management of the stocks from the increases in scarcity value that should accrue to the public owner. In practice, there will be large margins of uncertainty in setting royalties and other returns to the public owner. But the point is surely that the bulk of the increasing value of access rights to public resources (the windfall gains from enclosures measures) ought somehow to be captured for the community rather than become the object of speculative trading of ownership claims. This requirement to ensure full returns to the public owner limits the degree to which self-interest can be enlisted to assure achievement of the social goals of continuing investment in sustainable forest resources.[37]

Beyond institutional design and instrument choice, then, there are questions of ownership and distribution of endowments. Constraints arising from the need to recognize the interdependence of human activities with the rest of the biosphere lead to rules that may restrict the exercise of rights to claim access to the resources of the global commons, and, in particular, may limit the discretion to be exercised in managing forest resources on behalf of the public owner.

These rules become the implicit or explicit constraints on the interpretation of property rights or other rights of individuals to participate in the management of human activities with significant ecological impacts. Over the last few centuries, as institutions have evolved towards more formal economic mechanisms, there has developed what may be interpreted as a general struggle for primacy. This struggle pits those who would express the basic rules through market institutions with spontaneous contracting, and derive from these a variety of constraints on the more general exercise of governance, against those who would express the basic rules through social structures, and derive from these any necessary constraints on the allowable scope and exercise of commercial activities. These tensions and swings remain, but this chapter argues that in any case appeal to economic instruments must be set within more general social rules articulated to balance conflicting rights and competing claims on congested ecological space. And these rules must provide for the adaptation of institutional designs in the face of continuing change in resource systems and our understanding of them.

This discussion suggests that it is necessary to distinguish more finely among regulatory institutions and reforms of tenures or property rights. In the context of fisheries management, for example, a number of recent studies have proposed decomposing the notion of "property" as a bundle of rights into a long array of component rights, roles, and responsibilities, or rights and corresponding duties, and allocating each element in this array explicitly to one organization (enterprise or individual) in the chain extending from aggregate social purpose to individual discretionary action.[38] Each distinct assignment of authorities defines a distinct institutional arrangement. With experience and learning, institutions may then need to be adapted by reassignment of these authorities or rights, within the nested institutional structure.

Conclusion

This chapter approaches the question of economic instruments from a point of view based in public policy and public administration. It focuses on incentives as part of the context and information flow that shape management decisions, and on compliance in the exercise of individual discretion, within a framework of institutions that can be designed or redesigned. In this respect, it pursues questions opened up by Cohen, Scott, and Robinson in a recent article.[39]

Contemplating a range of governing instruments, the debate on economic instruments conventionally contrasts one approach (coercive "command and control") which prescribes behaviour in detail and leaves it to state hierarchies or the courts to enforce compliance, with an alternative approach (economic instruments) which offers incentives to economic agents – usually corporate managers – and leaves it to corporate hierarchies or markets to enforce compliance. In each case, it is the enforcement, more or less probabilistic, that actually creates the incentives – penalties or rewards – shaping individual behaviour.

The spectrum between these poles must be filled in and explored further. Rather than prescribe behaviour or impose detailed standards, one might negotiate agreement on more general outcomes or desired results, and let means to achieve such results (the required behaviour and intermediate outputs) be endogenously determined or discretionary. This practice is the VCM or PORPs or "reg-neg"[40] approach to decentralizing regulation towards a more interactive mode. Or one might use government powers to influence prices (or subsidies) if suitable markets do not exist to establish prices appropriately, and let the outcomes, standards, and requisite behaviour all be determined endogenously. This approach is the Pigovian pricing solution: the way of ecological tax reform. Or, finally, one might use government powers to create suitable markets, defining or redefining ownership, property rights or endowments, thus leaving prices themselves, as well as outputs, standards, and behaviour all to follow as endogenously determined.[41] This approach is the route of Coasean contracting, and it of course raises distributional issues of the most fundamental sort, opening up the whole debate over "new enclosures movements."

The conclusion of this chapter is that in any case, it is illusory to talk about realizing social goals solely by "leaving it to the market" and relying on economic instruments in decisions about human activities in timber harvesting and forest management. Ethical judgments and individual moral codes bracket the market. Overall objectives and ground rules on conduct flow from social consensus, possibly expressed explicitly in covenants in an international forum. The exercise of individual discretion in behaviour on the ground is governed by a personal sense of responsibility, moral obligation, and awareness of social scrutiny and peer pressures. Market mechanisms and economic instruments must therefore be judged ultimately according to how well they form an effective bridge from general social obligations of stewardship (and the covenants or codes of conduct to which they give rise) to individual commitment to conduct consistent with them. The contractual claims, tenures, and property rights that form the foundation of economic instruments thus are in no sense absolute; they must all be interpreted in light of the changing, increasingly stringent, obligations of stewardship in the context of changing institutions of government in a growing human society pressing increasingly on the limits of ecological space.

So we finish with the notion that forests, as habitat, as part of a complex web of life, and as a key element in the carbon cycle, are part of a common heritage of humankind and not simply bodies of resources up for claim. This awareness demands an ethical framework that entails a fundamental obligation of stewardship bounded by the constraints of sustainability. This obligation requires that we not compromise the value of that heritage to subsequent generations. Such an ethical frame leads to principles like those expressed in the "precautionary approach," and general codes of conduct such as those presently being negotiated, either as international agreements or as voluntary industry codes and certification programs.

To realize these obligations, nation states, corporations, and other institutions must be viewed as agents of humanity acting collectively. Indeed, for these purposes of access to a common human heritage, individuals cannot be viewed as sovereign entities with fixed preferences to be unquestioningly respected, but as subject to agreed rules of conduct designed to regulate conflicts of rights arising in the competition for access to common resources (some of which are presently unpriced but will later come to be valuable and marketable as the commercial exploitation of new species and new products works its way through former waste and weeds).

In practice, vast uncertainty surrounds this resource system even though, in principle, the forest inventory might be knowable. Knowledge of the system is drowned in records, observations, and facts, from which usable information can be extracted only with great difficulty. Monitoring of practice is lost in the wilderness. But, even if distant from any central control, ongoing operational decisions on practice build to cumulative impacts of vital importance to all citizens, including those outside the local community. We need sets of linking institutions, therefore, with a focus on information, instruments, and incentives (structured according to ideas of subsidiarity and local knowledge), all of which can be viewed in a sense as shaping individual decisions, exercising individual discretion.

The problem is to link individual discretion adequate to assure requisite variety with overall direction adequate to assure sustainability. This goal requires that we resolve the old puzzle about performance measurement and accountability in the face of uncertainty.

The ultimate economic instrument, then, is information. In a client-centred, customer-driven world, the most powerful market mechanism is the informed consumer – or more generally, feedback from informed stakeholders and citizens able to influence the behaviour of resource managers. But how can an effective market, in this sense, be assured?

The key requirement is easy to describe and hard to assure – good information accessible to all participants, with a forum for effective deliberation. This requirement demands good environmental reporting, and adequate attention to the stocks and health of resource systems. It requires widespread monitoring, audit, and the play of "intrusive sunshine."

In the end, perhaps we are not talking so much about preferred instruments to achieve right outcomes as about institutional designs for good process – for an evolutionary dynamics with survival value. These adjustment processes will involve scrutiny, audit, and ultimately opportunity to adjust or revoke tenures[42] and corporate charters in light of failure to fulfil commitments to social goals of sustainability.

In a complex and changing interdependent and congested world, there can be no certainty of outcome; the only certainty can be on the agreed rules for sharing the costs of adjustment among the various claimants to a share of the resource or returns from it. In this setting, there is no case for compensation simply because the rules change in response to new circumstances – though there is an obvious case for fair procedure in implementing such change.

In summary, this chapter recognizes that the existence of an ecological frame for decision-making poses a conundrum for corporations or communities involved in harvesting of forests. All those making decisions face ambiguous marching orders. As usual in market settings, they seek to do good (and do well) by serving customers. But, as discussed in the second section, sustainable forestry seems to involve objectives of stewardship of the resource and social equity as well as economic efficiency. Those making decisions must also serve the public owner as agent, acting as steward and guardian of the resource, and with a concern also for the impacts of decisions on income and wealth distribution.

The third section explored how these larger objectives might be reflected through alternative institutional arrangements focused on the achievement of compliance. The fourth section examined some reservations about the workings of economic instruments for this purpose and moved on to consider a variety of "voluntary compliance measures." Noting that the various structures tend to shade one into another, rather than posing a clearcut choice as between p-rules and q-rules,[43] the discussion concluded that it is unclear which approach might do better in particular circumstances; but certainly no single approach could be adopted universally with certainty. In the fifth section, therefore, an argument for an adaptive approach to institutional design itself was suggested, recognizing that this approach demands acceptance of a complex social and political setting. That setting involves continuing private scrutiny by non-government organizations as well as public scrutiny by public agencies, with the prospect of adaptation of rights, roles, and responsibilities as learning proceeds. In this setting, perfect tenures fixed for indefinitely long periods do not seem plausible. Agents working with a common heritage can only work within more general rules for sharing returns and adjustment costs in changing circumstances reflecting both natural variability and unknown consequences of human influence.

This procedural approach, or insistence on some kind of communicative rationality, takes us, interestingly, into the whole question of multi-stakeholder processes, and the difficulties of transition from consultation to implementation in such processes. Hence, it is not too extreme to argue that in practice, appeal to economic instruments in fact shades inevitably into consultative process. But that is a whole different story, for some other time.

Notes

1   H. Daly, "From Empty-World Economics to Full-World Economics: Recognizing an Historical Turning Point in Economic Development," in *Environmentally Sustainable Economic Development: Building on Brundtland,* ed. R. Goodland, H. Daly, Salah el Serafy, and Bern van Droste, 78-91 (Paris: UNESCO 1991), as reprinted in K. Ramakrishna and G.M. Woodwell, *World Forests for the Future* (New Haven: Yale University Press 1993).

2   See R. Dobell, "Complexity, connectedness and civil purpose: Public administration in the congested global village," *Canadian Public Administration* 40(2) (1997): 346-69.

3   The expressions "sunshine and scrutiny" or "intrusive sunshine" refer to processes of reporting and audit. Such processes have been suggested as means to avoid formal extra-territorial enforcement activities in the implementation of international agreements. See the North American Institute Workshop Report, "The North American Environment: Opportunities for Trinational Cooperation" (Santa Fe, NM, February 1993) on the design of the Commission for Environmental Cooperation.

4   I draw the idea for these labels, and later comments on designs to match institutional form to resource systems, from the work of my colleague Darcy Mitchell, specifically her 1997 study *Sustainable by Design: How to Build Better Institutions for Fisheries Management in British Columbia,* PhD dissertation, University of Victoria.

5   J. Habermas, *The Theory of Communicative Action* (Boston: Beacon Press 1984).

6   Daly, supra, note 1 at 79-80.

7   V.I. Vernadsky, "The Biosphere and the Noosphere," *American Scientist* January (1945): 33.

8   P.M. Vitousek, P.R. Ehrlich, A.H. Ehrlich, and P.A. Matson, "Human appropriation of the products of photosynthesis," *Bioscience* 36 (1986): 368.

9   J.N. Rosenau, "The relocation of authority in a shrinking world," *Comparative Politics* April (1992): 255.

10 P.M. Haas, "Environmental institutions and evolving international governance," paper presented at Harvard University, 1995; Dobell, supra, note 2.

11 K. Hanf and L.J. O'Toole, Jr., "Revisiting old friends: Networks, implementation structures and the management of inter-organizational relations,' *European Journal of Political Research* 21 (1992): 163-80.

12 N. Nevitte, *The Decline of Deference: Canadian Value Change in Cross-National Perspective* (Peterborough, ON: Broadview Press 1996).

13 The standard discussion of "public goods" focuses on the properties of "rivalry," "excludability," and "congestibility," which when absent (in the case of the first two) or present (in the case of the third) may lead to a degree of market failure and a possible case for government intervention in market mechanisms. See, for example, D.L. Weimer and A.R. Vining, *Policy Analysis: Concepts and Practice,* 2nd ed. (Englewood Cliffs, NJ: Prentice-Hall 1992), 45-57.

14 The size of that natural capital stock, and the value of the services it yields, are enormous. Robert Costanza and a large group of colleagues have developed rough estimates suggesting that the value of ecosystems services is two to three times the currently measured world total gross product (with about one-third of that value coming from terrestrial ecosystems, two-thirds from marine ecosystems). See R. Costanza, et al., " The value of the world's ecosystem services and natural capital," *Nature* 387 (1997): 253-60.

15 See, for example, E. Ostrom, *Governing the Commons: The Evolution of Institutions for Collective Action* (New York: Cambridge University Press 1990); R.D. Putnam, *Making Democracy Work* (Princeton: Princeton University Press 1993); D. Ronfeldt, *Tribes, Institutions, Markets, Networks: A Framework about Societal Evolution* (Santa Monica: RAND 1996).

16 Vernadsky, supra, note 7.

17 Vitousek, supra, note 8.

18 N. Elias, *The Civilizing Process: State Formation and Civilization* (Oxford: Basil Blackwell 1982).

19 See, for example, T. O'Riordan and J. Cameron, *Interpreting the Precautionary Principle* (London: Earthscan 1994); and J. Simon, *The Ultimate Resource 2* (Princeton: Princeton University Press 1996).

20 Weimer and Vining, supra, note 13.

21 Some authors make a strong distinction between government intervention through Pigovian pricing as contrasted with "free market environmentalism" based on Coasean contracting, using the argument that the latter may eliminate, in principle, the need for any government action (other than to define the initial distribution of property rights and enforce subsequent contracts). See, for example, T.L. Anderson, "The Market Process and Environmental Amenities," in *Economics and the Environment: A Reconciliation,* ed. W.E. Block (Vancouver: Fraser Institute 1990). Both approaches, however, fall in the category of non-coercive economic instruments.

22 See, for example, I. Ayres and J. Braithwaite, *Responsive Regulation: Transcending the Deregulation Debate* (Oxford: Oxford University Press 1992); or M.I. Lichbach, *The Cooperator's Dilemma* (Ann Arbor: University of Michigan Press 1996).

23 Three qualifications should be noted. First, general targets such as an overall allowable cut could conceivably be set locally through market mechanisms, by communities, or in decentralized and interactive regulatory structures – indeed, for many observers, the possibility of delegating such conservation responsibilities is the prime purpose of institutional networks with distributed powers and delegated authorities. Second, none of the linkages just discussed is really only one-way: "values and preferences" do not just move or find expression through institutions, they may well also be formed and shaped by those institutions and relationships within them. Third, individuals and societies have values and preferences with respect to choice of institutional form as well as other objectives or targets. Institutional arrangements are not purely instrumental. Indeed, the relative merit of alternative institutional forms in matters of procedure, fairness, or freedom is surely the basic question addressed in current controversy about the role of markets in governance. Here, however, the emphasis is on the basic task of achieving compliance in individual behaviour with rules of conduct established socially, and achieving coordination of individual action so as to realize aggregate targets for management.

24 See L.E. Susskind, *Environmental Diplomacy: Negotiating More Effective Global Agreements* (Oxford: Oxford University Press 1994); or P.M. Haas, "Why comply: Some hypotheses in search of an analyst" (unpublished 1996).

25 Interest appears to be growing in the idea that corporate charters conferring authority to carry on business as an enterprise with limited liability also carry social conditions on the exercise of that delegated authority, and that such charters can reasonably be revoked if those conditions are not adequately met. See, for example, R.L. Grossman and F.T. Adams, *Taking Care of Business: Citizenship and the Charter of Incorporation* (Cambridge: Program on Corporations, Law and Democracy 1993).

26 See R. Ford and D. Zussman, *Alternative Service Delivery: Sharing Governance in Canada* (Toronto: Institute of Public Administration of Canada and KPMG Centre for Government Foundation 1997).

27 See "A Fresh Look at Economic Instruments," a consulting report prepared for the Canadian Council of Ministers of Environment by Pembina Institute for Appropriate Development with Apogee Research International, Ltd., released November 1996.

28 D. Bromley, "Entitlements and Public Policy in Environmental Risks," in *The Social Response to Environmental Risk: Policy Formulation in an Age of Uncertainty,* ed. D.W. Bromley and K. Segerson (Boston: Kluwer Academic 1992).

29 P. Hawken, *The Ecology of Commerce* (New York: HarperCollins 1993); E.U. Von Weizsacker, *Earth Politics* (London: Zed Books 1994).

30 D.J. Kra'an and R.J. in't Veld, *Environmental Protection: Public or Private Choice* (Boston: Kluwer Academic 1991).

31 J.K. Martin, "Performance-oriented regulatory programs: A way to modernize regulatory systems," *Optimum* 26(3) (1995/1996): 3-10.

32 Extensive discussion of the extent to which "voluntary compliance" is a self-contradictory phrase has led to adoption of a working definition as "measures internally generated and voluntarily undertaken to improve compliance with externally imposed requirements" – thus discretionary instruments to fulfil commitments to meet non-discretionary requirements. See K. Welks, "Voluntary Compliance with Environmental Laws in North America," notes on a discussion paper prepared for the Commission on Environmental Cooperation, 1997. (Electronic document available at http: //www.cec.org.)

33 O. Williamson, *Mechanisms of Governance* (Oxford: Oxford University Press 1996).

34 Much concern has been expressed that the negotiation of "voluntary" arrangements will move from the "openly cordial" to the "covertly cosy," and will lead again to the arbitrary excesses of "sympathetic administration." Recent "streamlining" of the *Forest Practices Code* gives rise to much the same concerns.

35 J.M. Keynes, *The General Theory of Employment, Interest and Money* (New York: Harcourt, Brace and Co. 1935), Ch.12.

36 Keynes, supra, note 35.

37 It also limits the degree to which these returns flowing to the public owner can properly be considered earmarked for re-investment only in the particular region or sector from which they flow. More comprehensive goals of overall economic transition would be compromised by such restrictions of benefits to only one segment of the community adversely affected by ongoing economic structural adjustment.

38 D.A. Mitchell, supra, note 4; E. Pinkerton and M. Weinstein, in *Fisheries That Work: Sustainability Through Community-based Management* (Vancouver: David Suzuki Foundation 1995); E. Schlager, "Fishers' Institutional Responses to Common-Pool Resource Dilemmas," in *Rules, Games and Common Property Resources,* ed. E. Ostrom, R. Gardner, and J. Walker, 247-65 (Ann Arbor: University of Michigan Press 1994).

39 D. Cohen, A. Scott, and J. Robinson, "Institutions for Sustainable Development of Natural Resources in British Columbia," in *Managing Natural Resources in British Columbia,* ed. A. Scott, J. Robinson, and D. Cohen, 3-21 (Vancouver: UBC Press 1995).

40 The term "reg-neg" refers to processes of negotiating individual regulatory commitments appropriate to specific production circumstances. See, for example, the federal government's proposed Regulatory Efficiency Act, originally introduced in the House of Commons in December 1994.

41 One set of prices cannot be determined on markets in this manner: the terminal values of assets, especially natural capital. As noted in the quotations from Keynes, in economic decisions we generally fall back on the convention of taking current market valuations as representing long-term social value; but for natural capital, there is no reason to endorse that convention.

42 Insisting on the right to revoke tenures does not really make them uncertain, only contingent on compliance with agreed conditions. The rules are known and certain; rights are stable provided conditions are met.

43 L. Thurow, *The Zero-Sum Society* (Markham: Penguin Books 1980), 6.

5
Structural Instruments and Sustainable Forests: A Political Ecology Approach
Michael M'Gonigle

Introduction

This book addresses the complex problem of which instruments, or policy tools, might be used to sustain British Columbia's forest industry. The challenge poses a deep conundrum for those seriously concerned about the quest for sustainability not just for an industry, but for the forests that underlie it. Indeed, the evidence of a contrary proposition is widespread. That is, whether it is the need to ensure the survival of terrestrial forest ecosystems or wild fish populations, to shift our cities onto a less consumptive path, or to halt the erosion of the planet's atmosphere, a simple truth seems evident: our existing economic or regulatory instruments are not really leading us to a sustainable future. Understanding why this is so after almost three decades of official public "environmentalism" is a prerequisite to constructing any policy instruments that seek to achieve in the future what has failed in the past.

A few articles in this book address this situation by taking what I call a "structural" (as distinct from an incremental) perspective. Across a range of environmental sectors, an unconscious acceptance of the systemic assumptions on which existing corporate and bureaucratic institutions are founded continues to determine the solutions that are deemed acceptable. This situation has ensured that the status quo has been maintained throughout the past decade of "sustainable development." This is not to deny that there have been reforms, even sweeping reforms. Nevertheless, these reforms have occurred within a limited productivist paradigm which, whether on the Left or Right, whether in MacMillan Bloedel or the Ministry of Forests (or, indeed, the Ministry of Environment), continues to constrain the possibilities for achieving real sustainability. Indeed, any attempt to challenge the deep-seated assumptions embedded in our economic and political institutions risks dismissal for a multitude of reasons, not the least of which is the threat it poses to the existing configuration of production and authority. This is the post-modern condition – existing power determines acceptable knowledge. This is also the true challenge of sustainability.

As a result, a battle has been running for many years in British Columbia, a battle that has dominated the headlines and gained the province an international reputation as one of the world's great environmental hotspots. At the heart of the battle are two related issues that are the topic of this book: the ecological and economic sustainability of the BC forest industry, and the role of the state and other actors in achieving this sustainability. Despite a recent lull in the conflict, its ultimate resolution depends on changes that will sustain forests across the broad provincial landscape. Despite a plethora of specific initiatives in recent years, this challenge remains because the basic structure and dynamics of an unsustainable forest industry have not been addressed. Understanding this situation, and considering the alternatives to it, is the task of an ecological political economy, or what I will call "political ecology." In this chapter, I will set out the basic character of an approach that goes beyond traditional neoclassical economic (or even innovative ecological economic) analyses by examining the systemic dynamics of BC forestry.

Other commentators in this book suggest that they too take a basic, systemic approach. This is certainly how many neoclassical economists see themselves insofar as they argue that the state has failed in its objective of protecting the environment, and has actually made matters worse because it has intervened with rules when it should have been relying more on market mechanisms. These authors assert that a multitude of social problems, including environmental sustainability and economic competitiveness, are exacerbated by governments' attempts to "command and control" the world about them. In the process, governments impede more effective agents of change, especially the market, which could be harnessed through a broader use of economic instruments. As this book demonstrates, there are certainly insights to be gained from this perspective, and we will explore some of them below. At the same time, this debate often also reflects the opposing ideological preferences of the advocate: on the one side, the marketeer/deregulator (Right) taking the initiative, while on the other side is the regulator/interventionist (Left) in the defensive posture. In this debate about the choice of policy instruments, we find not just an issue of means, but of the ends for our forests – and whose interests should determine these ends.

In this debate, a third school should be added to that of the interventionist and the marketeer – that of the political ecologist. If the interventionist defines the ends as being those chosen by the political and bureaucratic apparatus of the *state*, and the marketeer defers to the *corporate* calculations of those with economic power, the ecologist begins with the needs of the *ecosystem*. By taking sustainability seriously as an overriding end in itself, the political ecologist does not defer, as an assumed point of departure, either to the interventionist state (with its sovereign authority to command and control) or to the business corporation (with its property-based market power). In this regard, this chapter posits a larger choice than one between the competing means of market-based versus regulatory-based instruments. Instead, it looks at a choice of ends between, on the one hand, incremental reforms within the context of established state and corporate hierarchies, and, on the other hand, structural change that reconstructs those hierarchies to whatever is dictated by the primary goal of maintaining ecosystem health. Today, these hierarchical structures stand astride natural ecosystems as the basis from which they draw their wealth – and to the needs of which, they are largely oblivious.

In this situation, to reconsider openly the ends of forest policy from a political ecology perspective is difficult, especially for those in positions of authority and credibility. Nevertheless, the perspective illuminates forest policy in new ways and, in the process, leads to the consideration of a novel set of ways to achieve the ends. This approach is the task of political ecology – to take responsibility for our social self-constitution, indeed, our re-constitution. As one such writer put it, our aim is "to achieve a consciously self-regulating society in the face of the ecological abyss, to climb off the roller-coaster of run-away social evolution and actively take responsibility for social organization into our own hands."[1] This task not only threatens existing institutions of authority and power, but challenges our most fundamental beliefs, particularly the "instrumental rationality" on which neoclassical economics and bureaucratic managerialism are both founded and, without which, they both become substantively indeterminate. In this larger critique of the structures of central power, the differences between the dominant schools of marketeer and interventionist fade in significance.[2]

Ends and Means in BC Forest Policy: A Retrospective View

To appreciate the political ecological perspective, we will first reconsider our forest history to assess how the values and structures that have evolved either support or undermine the quest for sustainability. This reconsideration is not the history of an authoritarian, "command and control" regime that dictates the marching orders for a corps of corporate supplicants. It is, instead, the history of a cooperative integration of the two hierarchical forces that dominate society: big corporations and big government – supported in large measure by organized labour, that equate their special perspectives and interests with those of the larger social and natural worlds.[3]

In the latter years of the nineteenth century, and the early days of the twentieth, the government's primary concern was to create a stable and growing forest economy in the face of a remote and unreliable market.[4] To attract capital, the province made outright grants of forest land to large railway and logging companies, and leases to less well-off private loggers.

In these leases (the antecedent of today's tenures), the Crown retained ownership of the land while the licensee got ownership of the timber, subject to a royalty fee or payment of stumpage upon cutting. Thus, noted the Forest Resources Commission in 1991, "the private sector and the government became partners in the management of BC's forests."[5] Although the government retained the power to regulate these tenants, it was reluctant to do so for fear of discouraging investment and slowing development. For most of the first half of this century, forest cut rates were unregulated, reforestation was not funded, and rules relating to wood product utilization were unwritten.[6] As to environmental regulation, no one gave it much thought. In this development history, public and private interests have not been adversaries, but cooperative.

The use of state instruments reflected this cooperative approach. The province's major policy instrument, the tenure system, was conceived as the primary instrument for a broad, state-directed strategy of economic development, but it was devised in a manner which, through a contractual property right, granted huge discretionary power to a handful of powerful, private actors with minimal need to resort to coercive regulatory actions. In its regulatory role, the primary concern of government was to facilitate the province's economic development, and many restrictions on the exercise of private tenure rights were directed to this end. Such requirements as the "use it or lose it" rule (that required a licensee to use the assigned quota or lose it),[7] and the "appurtenance" rule (that required a tenure holder to build and maintain a mill as a condition of receiving a tenure)[8] were aimed at ensuring economic growth and stability in the province's hinterland.

With government being ever concerned about industry profitability, stumpage rates to the government were meanwhile kept low, often lower than the government's own management costs. At the same time, allowable annual cut rates were historically set high, well above the so-called long run sustained yield, and they were raised as the evolving economies of scale demanded. Above all, in the early days, the provincial government eschewed interference in the economic affairs of the industry, allowing it to grow, consolidate, and merge, becoming ever more productive in accordance with the higher commands of market-based competitiveness. Ironically, since the Second World War, the policy on which all this growth was accomplished was "sustained yield management," a policy predicated on clearcutting stands of slow-growing, old-growth natural forests, and replacing them with faster-growing, even-aged stands of "normal forests." This liquidation forestry has always been seen as the best way of ensuring both a high production of forest fibre and forest industry competitiveness.[9] In short, economic "laws" directed public policy, and it is according to these laws that the natural world has been obliged to conform.

This strategy has had a distinct socioeconomic character to it. First, ecological values have been minimized throughout the history of the industry in British Columbia. Forest management science has long been enmeshed in an almost exclusively productivist paradigm so that concern for a forest's "ecological" character (its composition, structure, and function), let alone for the diverse values it embodies, has been minimal. Even today, this neglect is manifest in the variants of the sustained yield policy, such as multiple use, and integrated resource management. Second, the driving force of forest policy has been the need for corporate-based economic growth. On the one hand, labour-shedding and business consolidation are the rules of this game. On the other, to maintain provincial employment, cut levels have been constantly increased. With competitive growth assured in this manner, the forest industry has historically provided an engine for provincial economic development, an engine that has been marked in recent decades by increasing corporate concentration.[10] Today, with timber supplies running short, the appurtenance requirement in the Forest Act has been increasingly criticized as an obstacle to the consolidation of mills which is seen as essential for maintaining a profitable economy of scale. Third, with the Ministry of Forests appointed as the arbiter of the public interest in forest management, bureaucratic oversight has grown in partnership with the growth of the industry. Approval processes are extensive and time-consuming, even though the substantive requirements imposed by forestry regulations have historically been neither onerous (especially from an ecological viewpoint) nor economically constraining.

The regime which emerged can thus be characterized as that of a corporate/bureaucratic partnership, the *modus operandi* of which has been the liquidation of natural ecosystems to fuel corporate economic growth as the basis for provincial social well-being. As recently as the late 1980s, for example, annual expenditures to maintain the Ministry of Forests regularly exceeded the direct revenues (through stumpage and royalties) from the entire forest industry. With the partial exception of the new Forest Practices Code, this regime has not been an adversarial system in which the state has operated in a strict command and control mode. Quite the contrary, legislation has maximized the discretion of Ministry of Forest employees at all levels; violations have almost never resulted in prosecutions or penalties; and public review has been minimal with limited legal controls being afforded to the citizenry.[11] Instead, with critical voices historically excluded from a closed "policy community," the recourse of forest critics was to set up remote blockades of forestry operations and to organize downtown rallies before corporate headquarters.[12] As between the partners in development, however, everything is subject to in-house negotiation based on the tenure agreement.

To understand the operation of this partnership, recourse to the currently popular academic distinctions between public powers and private rights, and between legislative and market instruments, is not fully informative. A more important distinction is between big established institutions with authoritative power, and small entrepreneurial and local ones without. The critical point is that, in looking to future policy, once the resource base is embedded in a legally formalized institutional system dependent on high-volume resource production, it is extremely difficult for public bodies to extricate themselves. Existing obligations dictate future opportunities; to alter course is difficult regardless of the merits of doing so.[13] Strong sustainability becomes an elusive goal; instead, sustainability itself is redefined to accord with prevailing practices and institutions.[14]

The New Democratic Party's terms in office have demonstrated this inability to undertake the structural reforms that would alter historical power relationships. Tenure reform was quickly removed from the government's agenda early in both of its two terms of office. Its land use planning processes have led to the creation of a plethora of new parks that were all disproportionately skewed away from the best forest lands, and they have not changed industrial relationships on the rest of the land base. The land use planning process that led to the new parks did, however, lay the foundation for designating large forested valleys as "special resource development zones" suitable for more sensitive, non-industrial forestry ("eco-forestry"), but these zones have never been implemented. Quite the contrary, the Ministry of Forests actually delayed the introduction of stricter standards while approving industrial operations at an accelerated rate in such areas, effectively preventing such an alternative forestry from taking root.[15] With a corresponding expansion in the percentage of the forest land base being directed to more intensive forest management (i.e., super-industrial forestry) through such processes as the Vancouver Island Resource Targets (VIRT), zonations under the land use planning exercises are becoming a vehicle for the consolidation of even tighter industrial control of the forest land base.[16]

Meanwhile, the government's new Forest Practices Code was intended to increase the weight of environmental regulations on the forest industry, but instead such regulations have been rolled back time and again as a result of industry complaints. Never have these regulations challenged either the productivist economic structure of that industry or the monopolistic centralism of the Forest Service. Thus, the regulations fail to foster both sustainable alternatives to unsustainable industrial forestry and greater community-based management. They do impose greater paperwork on the industry, a criticism made all the more plausible by the hugely bureaucratic character of the Code. Meanwhile, the new Act (and Regulations, and Standards, and guidebooks) leaves large amounts of discretion available to ministry officials, and channels public interventions through a government-appointed board and commission. By superimposing them onto an unchanged structure of production, the government has put the cart before the horse – constraining the industrial mode with more regulations rather than first changing tenure and production systems, and then developing a new regulatory system appropriate to those new systems. Indeed, as part of the "political management" associated with implementing the Code, it was "downwritten" to minimize its impact on the level of cut (and jobs) out of deference to the government's supporters in the IWA who see their interests allied with those of the corporate sector. Indeed, in the year after the proclamation of the Code, the provincial cut actually increased. Thus, the implementation of new zoning systems subsequent to the CORE land use planning processes and the application of the Forest Practices Code have not resulted in significant changes on the ground, but are portrayed as constituting the full range of possible reforms. In light of this seemingly comprehensive approach, tenure reform becomes a non-issue.

And so, today, large integrated companies rule the forests. In 1995, the largest tenure types (area-based Tree Farm Licences, and volume-based Forest Licences) secured over 80 percent of the province's allowable annual cut, and virtually 100 percent of this cut was undertaken through traditional industrial forest methods, including large-scale clearcutting.[17] In contrast, small operators account for a small fraction of the total cut, about 13 percent. Even here, this cut is overwhelmingly dominated by freelance small business operators working under the Small Business Forest Enterprise Program (SBFEP) (about 12 percent of the provincial total), most of whom employ the traditional industrial logging techniques and supply their product directly to mills owned by the majors.[18] With the possible exception of some 500 individual public woodlot operators supplying just 0.7 percent of the cut, no tenures have been permitted on Crown land where an alternative approach to traditional forms of logging has been allowed.[19] Instead, the provincial forest land base is overwhelmingly dominated by high volume, timber commodity producers maximizing fibre production through industrial forestry techniques based on ecosystem liquidation.

The liquidation of BC's forest base throughout this century has produced the wealth on which the province has been built, and it continues to keep the province afloat. In 1994, the forest industry was, by the industry's own calculation, at 26.6 percent, the largest contributor to the goods-producing industry. This industry constituted some 52 percent of direct manufactured shipments, directly employing 92,000 people.[20] Highly productive in low-value commodity products, the centralized and inflexible forest industry complex that extracts this wealth from the forests is nevertheless sustainable neither economically nor environmentally. It is beyond the scope of this chapter to review these well-worn arguments,[21] but they underpin the central thesis of those chapters that point to the need for transformative new models of tenure and management.

Understanding Ecological Centralism
To a political ecologist, the history of industrial forestry in British Columbia can be understood as one more manifestation of an expansionist

social and economic structure that is inherently unsustainable. In this light, sustainability requires not just incremental reform *within* this structure, but reformation of the structure itself. This situation has been characterized thus:

> Looking back in history, we recognize that every age has, under changing circumstances, been faced with the unexpected, unwelcome challenge of radically adjusting the way it does things. In school, we all learn about the Scientific and Industrial Revolutions and, standing on the afterside of this history, we understand why these revolutions occurred ... Today, the dominant fact of global life is that we are catastrophically overshooting our resource base ... [To counter this momentum, we] must embark on an Ecological Revolution, a revolution equal in scale to the great revolutions of the past, only this time we stand not after the revolution, but before it.[22]

The revolution that awaits us in the BC forests, and worldwide, is a revolution against what might be called "ecological centralism." This centralism cannot be understood through the essentially *intra-systemic* tools of neoclassical economic analysis and bureaucratic regulation. Instead, by examining the dynamics of that centralist system as a whole structure, one can begin to appreciate the character of the instruments needed to create a new set of dynamics that inherently tend towards, not away from, sustainable outcomes. Geographically, our central institutions are supported by resource flows from the hinterland to the heartland, from the periphery to the core, from rural to urban. Politically, our social decision-making is dominated by an integrated hierarchical structure of power, whether this structure be corporate offices or Crown bureaucracies, where authority is concentrated at the top. Economically, these central/hierarchical structures are fuelled by energy and economic flows of materials, flows of wealth that are *linear*, moving from one place to a distant locale, down the freeway and into the cities. These linear flows displace more traditional processes that are *circular*, sustaining communities in place wherever they might be. This fact explains, for example, the importance of creating wealth through *trade-based* economies of scale, rather than creating social and community wealth through *multiplier-based* economies that recirculate smaller volumes of materials and capital in place. Socially and culturally, ecological centralism is dominated by the top-down intelligence of technocratic expertise and management, rather than the bottom-local experience of living with, and in, community.

In short, a political ecology perspective identifies the problematic character of a structure that is dominated by *centralized hierarchies of power* that are *sustained by distant resource flows* out of local environments and communities, and at great cost to them. As Rees and Wackernagel explain, our "ecological footprint" is far larger than the place where we live because we draw so many more resources than we could produce locally from afar – we "import sustainability."[23] This practice has indeed long been the history of the colonial expansion; it is, in fact, the history of the "rise and fall" of countless civilizations of the past.[24] This pattern and process characterize the past and present of the BC forest economy. Economic values and legal rules seeking to constrain the linear patterns of these hierarchies are inadequate, and doomed to failure, insofar as they remain internal to the maintenance of these hierarchies. Instead, the task of structural instruments is to juxtapose both spatial (i.e., *territorial*) relations and new forms of *noncentralist* social and economic organization against these hierarchies.

It is beyond the scope of this chapter to discuss this systemic analytical perspective in detail.[25] Nevertheless, recasting the histories by which we have come to our present juncture of widespread unsustainability by considering how centralized hierarchies have arisen and been maintained over space and time is a critical challenge for scholarship that will be relevant to the quest for sustainability. This recasting applies in all corners of the globe,[26] and it takes our historical enquiry beyond some of our most established landmarks. For example, the patterns of hierarchical social growth discussed earlier clearly mark a variety of civilizations over many millennia. Non-market forms of production, exchange, and hierarchical power long predated the liberation of market forces in the West (from Rome to the Anglo-Saxon kings), and did so in such ways as should move market-based economics from a primary to a subsidiary position in our historical understanding. To take the market or monetary values or regulatory instruments as the assumed foci of policy development is thus fatally limiting. Instead, policy discourse must venture beyond the boundaries of these *intra-systemic* techniques by drawing attention to, and dealing with, the profound thermodynamic costs inherent in the *materially based growth system itself*.

Ecological Sustainability Through Structural Change
This brief chapter highlights several approaches to the search for structural instruments for sustainable forestry, approaches that implicitly inform some of the chapters that follow in this book. To start, structural change means that "intra-systemic" economic values and regulatory approaches are of limited utility. In the current debate in forestry, this understanding implies a shift from the pursuit of increments of "sustainable development" to a broader quest to "develop sustainability," that is, to develop the tools that will promote a transformation in the dynamics of our economic and management institutions.[27] More specifically, it requires that our standard, "ends-means" policy tools be reconceived. For example, relying on existing economic values in the forest industry implicitly means the acceptance of a raft of problematic factors which are themselves embedded in these values – the existing distribution of legal entitlements (tenures) in the forests, the lack of monetary accounting for the liquidation of the very natural capital (old-growth forests) that constitutes the financial stuff of the industry, the

range of uninternalized and unpriced environmental externalities that further skew economic values, and so on. Instead, as Alexis de Tocqueville once remarked, a new science is needed for a new world.

Today, economic and legal "sciences" remain embedded in the thinking of the old world. In response, many today call for "a new political economy which combines the breadth of vision of the classical political economy of the 19th century with the analytical advances of twentieth century social science."[28] Indeed, the need for a re-inventing of scientific thinking applies to the natural and applied sciences as well. For example, industrial foresters still argue that clearcutting is a broadly acceptable prescription because it "mimics a natural disturbance," while equally well trained ecological foresters argue precisely the opposite – that forestry must move to adopt techniques that retain the "composition, structure, and function" of the forest ecosystem. What is one to make of this contradiction in our knowledge systems, especially where it is broadly pervasive? Again, this situation is but one aspect of the larger crisis in modern science in which competing paradigms have become the rule, rather than the exception, a situation that has given rise to a concern for what is often called "post-normal" science.[29] Neoclassical economists will make the mandatory genuflection in the direction of the recognized loss of objectivity and the inevitable presence of values in their science, but, once done, they quickly return to their posture of instrumental rationality, ignoring any further the consideration of its obviously ideologically bounded character.

The overall direction of the change contemplated by a political ecology critique is, instead, away from the imperatives associated with ecological centralism (that is, away from corporate and bureaucratic forms of rationalistic organization that dominate natural systems) towards the more stable processes associated with ecological communities (that is, forms of participatory organization that exist *within* natural systems). As one political ecologist put it, "the kind of reflexive institutional re-creation appropriate in a new ecological political economy involves reasoning constitutively rather than instrumentally about institutions."[30] This analytical perspective informs the growing interest in ecoforestry.[31] In looking to a future different from our past, this approach to forestry embodies a different set of starting principles from corporate-based industrial forestry, which follows the "laws" of the competitive market. Sustainability, in contrast, is rooted in what might be seen as even greater "natural laws."

This different starting point entails a host of alternative techniques for logging and forest product manufacture. In the late 1980s, this shift manifested itself among resource managers in a new school of "ecosystem-based management."[32] This school takes its impetus from the rise of ecological science, the growing alarm which this science has stimulated over the loss of biodiversity, and the recognition that existing regulatory systems have failed to stem the tide of decline. In their prescriptions, it is interesting to note that ecosystem scientists are as concerned with the need to reform agency mandates, alter jurisdictional boundaries, and restructure decision-making processes as they are with positing strict scientific formulas for protecting biodiversity and ecosystem integrity.[33] Aware of the institutional context for their science, these authors make a telling distinction between "ecosystem management" and "ecosystem-*based* management." The former often entails an extension of established institutional mandates to improve their management of ecosystems through the dedication of greater financial resources, improved information, better monitoring and enforcement, and so on. The latter takes a more precautionary approach by seeking to constrain human activity within the limits of ecosystem functioning. This approach identifies the issue not as *forest* management but as *forestry* management, controlling human activities and impacts within the limits of ecosystem sustainability, rather than attempting to manage the consequences of this activity after the damage is done.[34]

This approach is also inherently based in a respect for the territorial "community," because, when fully defined, the maintenance of such community is the most encompassing foundation of sustainability. It embraces complete continuity with both the spatial (including human and non-human elements) and the temporal dimension; including past, present, and future generations. An ecosystem-based approach thus seeks to situate economic processes with the carrying capacity of local ecosystems. The task of an ecological politics is to help communities move – economically, politically, and culturally – to this state.

British Columbia's forest conflicts are aptly characterized from these differing perspectives. Take, for example, the struggle over Clayoquot Sound. On the one side, are the industrial foresters, MacMillan Bloedel and International Forest Products, with their Tree Farm Licence tenures. From the viewpoint of tenure rights, the Ministry of Forests stands on this side as well. While it has tried to establish new and better forest practices in the Sound, it has assiduously avoided challenging the contractual property rights of these companies by reconsidering their tenure rights even when opportunities to do so have arisen.[35] On the other side are the environmental protesters, Nuu-Chah-Nulth First Nations, and ecological analysts. To the protesters, if ecosystem-based alternatives are ever to have a chance, the dominant issue is tenure. So, too, First Nations seek self-government and, with it, a greater control over the forest base. And both seek to change forest practices to those that are in keeping with the community-oriented planning recommendations of the Clayoquot Sound Scientific Panel.[36] The paradigmatic contrast is clear: on the one side, economically driven forestry, based on established corporate property rights and, on the other side, ecosystem-based forestry that is enshrined in, and determinative of, community rights and duties.[37]

Ecosystem-based management clearly speaks to a more locally bounded relationship with the natural world, but, as the Clayoquot Sound example indicates, it speaks to such a relationship with the social world as well. Across the planet, the expansion of the Western market/state has historically been fuelled by the colonization of local resources *and* self-regulating communities and cultures. Whether one is speaking of British Columbia, Sumatra, or Amazonia, the pattern is the same – privatization (and, now, increasingly, the corporatization) of hitherto communal resources. *The Ecologist* magazine refers to this as "development as enclosure."[38] In this light, although the "policy community" in BC might attempt to technocratize the forestry debate, the issue is not really a managerial one. It is deeply political; at root, it is all about power. Hence the need for a new perspective – political ecology – to make sense of the economic and regulatory instruments associated with these power relations.

If one is to address clearly the options for the state, one must reckon with this basic historical and structural fact. As a noted economic theorist has observed, the "real tragedy of the commons is the process whereby indigenous property rights have been undermined and delegitimized. This destruction of local-level authority systems [is] the principal cause of natural degradation."[39] From this historical perspective, it will only extend past trends of unsustainability if we disempower the state further by consolidating existing corporate property rights against regulatory changes. Difficult as it may be, the lessons of history point to the need to muster new forms of authority to create a counter-balance to the one-sided reign of hierarchical power itself, whether this reign is manifest through economic or regulatory instruments. Only in this way might we set in motion a different evolutionary dynamic that will take us beyond the centralist weaknesses of both bureaucratic regulation and corporate-based incentives. This reorientation is, for example, the thrust of the literature on "common property" resources.[40] At stake in the forestry debate is thus not a superficial debate about economic versus legal instruments as the most efficient regulatory approach, but about using all such means to transform the character of industrial forestry and, with it, the state itself.[41]

Conclusion

In looking at new economic instruments and new approaches to regulation, subsequent chapters discuss a fundamentally different model of "economic development" and a fundamentally different role for the state. Central to these proposals is a basic principle applicable to regulatory and economic instruments, state and corporate structures alike – the need to move our social systems from linear to circular processes. To be sustainable, ecological and community processes

> must maintain themselves, living on the stock of natural and social capital
> with which they have been endowed, so that they can return long-term

stability to the forest and long-term value to the community ... Just as healthy ecosystems recirculate nutrients and water locally, so too a healthy community is internally dynamic, recirculating wealth locally, in a diversified economy that is only partly dependent on outside employers and outside markets.[42]

To this concern for the local recirculation of natural and human capital might be added the parallel need for local recirculation of political power and authority, i.e., enhanced community decision-making. With the appropriate democratic design, new opportunities arise both for the use of a range of economic and regulatory instruments, and for their redefinition.[43] In so doing, the potential of both the democratic state and the ecological economy will begin to be realized.

Many of the proposed changes are singularly appropriate to the age of government downsizing, though in a very different way than is envisioned by proponents of privatization and economic instruments. Instead of securing the resources of remote territories for centralist growth, the overall thrust of a political ecology perspective means the reverse, redesigning the institutions of central power for the purposes of protecting territorial integrity. This reorientation is certainly the unnoticed significance of the much-debated need to shift from economically driven to "ecosystem-based" management. Even more, for progressive forces, community power should not be seen as a threat to political and bureaucratic authority, but as an alternative set of values and strategies to the privatization and corporatization that is advocated by neoclassicists as the desired avenue for cash-strapped governments seeking greater "efficiency" in their regulatory functions. At stake is a revitalized, historically appropriate role for the interventionist state as a central steward of the territory – that is, the ecologically sustainable territories and socially coherent communities within it.

It is a great leap of imagination and will for progressive thinkers to embrace this potential. To do so, they must be able to break free from long-held ideologies of neutral markets and instrumental regulation, and deal in a fully rational manner with the institutions of power and authority that these ideologies serve.

Acknowledgment

The author would like to thank the Eco-Research Secretariat, Ottawa, for its support of this research.

Notes

1 A. Atkinson, *Principles of Political Ecology* (London: Bellhaven 1991), 180. This necessity is reminiscent, in the Critical Legal Studies tradition, of Roberto Unger's call for the revision of our "formative contexts" through a move to a "radical democracy." See his three-volume opus, *Politics, A Work in Constructive Theory* (Cambridge: Cambridge University Press 1987).
2 This lack of concern for contextual questions is manifest in the very terms by which these

structures make decisions. There is, for example, a complex of arguments concerning the inherent biases in using economic values, including problems with assigning monetary values to biospheric functions and accounting for future generations. One area of special relevance to the topic of this book is the growing recognition of the disparity that exists between the market value one will assign to a particular commodity depending upon whether one is buying or selling it. This debate points, in particular, to the importance of the prior allocation of legal, or property, entitlements as determinative of economic values. This allocation is exactly the issue in question with forest tenure allocations, but neither corporatists nor state interventionists address the fundamental indeterminacy for economic calculations which is inherent in this recognition, preferring instead to focus on economic production within the existing structure of legal allocation. Thus is neoclassical "science" based on transparently ideological foundations. For an empirical review of this issue from within the discipline, see D. Kahneman, J.L. Knetsch, and R.H. Thaler, "Anomalies: The endowment effect, loss aversion, and status quo bias," *Journal of Economic Perspectives* 5(1) (1991): 193-206; and D. Kahneman and J.L. Knetsch, "Valuing public goods: The purchase of moral satisfaction," *Journal of Environmental Economics* 22 (1992): 57-70.

3   This analytical critique is developed in greater detail in my book (with B. Parfitt), *Forestopia: A Practical Strategy for A New Forest Economy* (Madeira Park, BC: Harbour Publishing 1994).

4   BC Ministry of Forests, *1994 Forest, Range and Recreation Resource Analysis* (Victoria: Queen's Printer 1994), 267.

5   BC Forest Resources Commission, *The Future of Our Forests* (Victoria: Queen's Printer 1991), 268.

6   Ibid., 269.

7   *Forest Act*, R.S.B.C. 1979, c. 140, ss. 55, 55.2, 55.3, 55.4, 55.5, 56, 56.01, 59, 61, and 61.1. In this regard, s. 63 provides that there is no compensation for deletion or reduction of AAC. Sections 28(k), 55.2(10), and 55.3(4) allow a discretionary grant of relief from the provisions.

8   Ibid., ss. 12(g), 27(3)(b), and 35(c).

9   This orientation is written into the Ministry of Forests' mandate which is required to "encourage maximum productivity of the forest and range resources in the province" [*Ministry of Forests Act*, s.4(a)]. Indeed, all the functions of the Ministry set out in its enabling legislation are oriented to productivist goals, including encouraging a "world-competitive" timber-processing industry.

10 For example, the share of cutting rights held by the ten biggest companies increased from 37 percent in 1954 to 59 percent in 1975 to 69 percent in 1990 [*Report of the Forest Resources Commission* (Victoria: Queen's Printer 1991), 37].

11 A series of cases initiated by the Sierra Legal Defence Fund attests to the constrained right of public oversight. See, for example, *Western Canada Wilderness Committee* v. *British Columbia (Attorney General)* (24 May 1991), unreported (B.C.S.C.), dismissing an application to require the district manager to complete a new management and working plan before allowing further logging in the Nahmint Valley. Instead, the company was operating under an outdated plan that had been extended three times. In rejecting the application, the Court stated that the Forest Act was "a living thing capable of responding to the varied and competing rights and interests of the forest ... [The Minister] has the power to amend. He is not locked into the four corners of any document including the Management and Working Plan."

12 On the concept of the "policy community," see the detailed treatment by B. Cashore, "Explaining forest practice and land use policy network divergence in British Columbia and the U.S. Pacific Northwest," address to the Annual Meeting of the Pacific Northwest Political Science Association, Bellingham WA, 1995; and B. Cashore, "Comparing the forest policy communities of British Columbia and the U.S. Pacific Northwest," address to the 1995 Association for Canadian Studies in the United States Biennial Conference, Seattle WA, 1995. The author suggests that state autonomy is extremely high in BC because of three factors. First, the Westminster model of government gives the cabinet a great degree of autonomous policy-making power. The incumbent political party thus has a direct influence on policy change. In contrast, the more fragmented macro-institutional

system of the United States "weakens" the state which allows societal interests (the forest industry in Washington state and environment groups with respect to US federal lands) to initiate new policy processes more readily through legislation and litigation. Second, the large amount of state-owned land limits the common law protections and economic power of private landowners. Third, a discretionary environmental statutory regime provides little legal constraint to the provincial cabinet or the Ministry of Forests from autonomously initiating new policy. Thus, in BC, environmental statutes with ad hoc discretionary protection allow the state (government or government agency) to initiate policy change with little fear of legal challenge. In this situation, the informal links between corporate and bureaucratic interests are highly influential. Other writers have made a similar distinction, although they have used the term policy "network" to refer to the insiders, and the term policy "community" to refer to the larger group with an interest in forest policy. See M. Howlett and J. Rayner, "Do ideas matter? Policy network configurations and resistance to policy change in the Canadian forest sector," *Canadian Public Administration* 38 (1995): 3.

13 One example of this problem is the litigation initiated by MacMillan Bloedel and Timberwest challenging the provincial government's right to increase royalty rates on Timber Licences to a level equal to the stumpage assessed on other tenures. The case, initiated in December 1995, has yet to go to trial.

14 This redefinition has happened generally with the distinction between "strong" and "weak" sustainability. The former refers to sustainability based on the maintenance of natural systems, while the latter refers to sustainability based on the economic substitution of "human" for "natural" capital. The effect of even making such a distinction is to obscure the concept of sustainability in a fog of prevarication. In this way, even-aged timber plantations are equated with old-growth ecosystems as a basis for future "sustainability."

15 See the report by the Sierra Legal Defence Fund entitled *Business As Usual: The Failure to Implement the Cariboo-Chilcotin Land Use Plan* (Vancouver: Sierra Legal Defence Fund 1996). In this report, data from approved logging plans show that logging in these zones will occur at higher rates than outside the zones. Although they account for only 18 percent of the productive forest land base, forest companies are planning to extract over 30 percent of their cut from them, using almost exclusively traditional clearcutting methods with no virtually consideration for non-forest values.

16 See, for example, the author's editorial on this subject, "Backroom plans threaten Vancouver Island's forests," *Victoria Times-Colonist* (18 August 1996): C2.

17 "Since [1976] the 20 largest forest companies have increased their control of harvesting rights from 74 to 86 percent. The share of cutting rights held by the ten biggest companies has increased from 37 percent in 1954, to 59 percent in 1975, to 69 percent in 1990" [K. Drushka, "Forest Tenures," in *Touch Wood, B.C. Forests at the Crossroads,* ed. K. Drushka, B. Nixon, and R. Travers (Madeira Park, BC: Harbour Publishing 1993), 11]. Peter Pearse, Commissioner of the Royal Commission on Forest Resources, warned about the problems of corporate concentration in 1976: "In my opinion the continuing consolidation of the industry, and especially the rights to Crown timber, into a handful of large corporations is a matter of urgent public concern ... Finally, the trend toward concentration of cutting rights in the hands of a diminishing number of large firms has been accommodated, if not stimulated, by the tenure policies in the [TSAs]" [*Timber Rights and Forest Policy In British Columbia: Report of the Royal Commission on Forest Resources,* Vol. 1 (Victoria: Queen's Printer 1976), 62, 78].

18 Some of these operators are invisibly backed by major companies, in a process called surrogate bidding. Because majors are effectively subsidized by below-market stumpage on their huge volume of AAC, they can bid high on "incremental" wood from these sources, still maintaining an average cost for wood below the true market value. In this way, the majors regularly subvert the SBFEP program, outbidding legitimate candidates who have fewer resources. D.W. Gillespie's *A Review of the Small Business Forest Enterprise Program and Woodlot Program: Report to the Minister of Forests* (Victoria: Ministry of Forests 1991) states incorrectly that there is no effective way to prevent surrogate bidding because large companies can always find a way to circumvent any restriction. A simple way to prevent this practice would be a government-imposed requirement that no SBFEP wood could come directly into the possession of a major producer at any stage, except through an open,

publicly permitted, bidding process. With strict enough sanctions, and a modicum of enforcement, the process could be quickly changed.

19 It is difficult to consider even public woodlots as an "alternative" tenure because they too must comply with high, productivist cut levels dictated by the Ministry of Forests. With the exception of a small forest owned by the Municipality of Cowichan, all tenures held by municipalities (such as Mission and Revelstoke) and Native bands (such as Tanizil) are in the form of Tree Farm Licences or are joint ventures with existing industrial forestry companies.

20 *BC Forest Industry Factbook* (Vancouver: Council of Forest Industries 1995), 17, 18, and 24.

21 For a comprehensive recent review, see the author's study, *Forestopia*, supra, note 3.

22 Ibid., 14-15.

23 W. Rees and M. Wackernagel, "Ecological Footprints and Appropriated Carrying Capacity: Measuring the Natural Capital Requirements of the Human Economy," in *Investing in Natural Capital: The Ecological Economics Approach to Sustainability*, ed. A.M. Jansson, M. Hammer, C. Folke, and R. Costanza, Ch. 20 (Washington: Island Press 1993).

24 See, for example, A.G. Frank and B.K. Gills, "World System Economic Cycles and Hegemonial Shifts in Europe: 100 BC to 1500 AD," *Journal of European Economic History* 22(1) (spring 1993): 155-83. This analysis is part of the "dependency" school of political economy. A similar perspective informs the work of geographers writing in the area of regional development. See especially, D.R. Matthews, *The Creation of Regional Dependency* (Toronto: University of Toronto Press 1983). For a specific application of dependency theory in the context of forestry, see C. Bailey, J.C. Bliss, G. Howze, and L. Teeter, "Dependency theory and timber dependency," paper presented at the 56th Annual Meeting of the Rur. Soc. Society (1993), cited in T.M. Beckley, "Pluralism by default: Community power in a paper mill town," *Forest Science* 42 (1996): 1.

25 For a more detailed, and recent, presentation of this analytical framework, see the author's paper "Between centre and territory: Toward a political ecology of ecological economics," presented to the Fourth Biennial Conference of the International Association of Ecological Economics, Boston MA, 1996. This paper is available as Discussion Paper 95-2, Eco-Research Chair of Environmental Law and Policy, University of Victoria. See also the general introduction to ecological economics, H. Daly and J. Cobb, *For the Common Good: Redirecting the Economy toward Community, the Environment and a Sustainable Future* (Boston: Beacon Press 1989). See also the journal *Ecological Economics*.

26 For example, with reference to forest degradation in Southeast Asia, one scholar notes: "One of the most efficient ways of 'legitimizing' their acquisitions of forest products was by expropriating them from indigenous populations. In this, they were aided by prevailing conceptions of western (a.k.a. Roman) law that had come to dominate contemporary European jurisprudence. According to this doctrine, all land, and hence all natural resources, belonged to the acknowledged sovereign and were his to use, abuse or parcel out as he best saw fit" [O.J. Lynch, "The Road to Baguio: Keynote Address," in *Common Problems, Uncommon Solutions: Proceedings from the NGO Policy Workshop on Strategies for Effectively Promoting Community-Based Management of Tropical Forest Resources, Lessons from Asia and Other Regions*, ed. M.S. Berdan and J.P.A. Pasimi (Baguio, Philippines: World Resources Institute 1994), 2].

27 For an early exposition of this distinction, see the author's paper "Developing sustainability: A Native/environmentalist prescription for third level government," *BC Studies* 84 (winter 1989/1990): 65.

28 A. Gamble, A. Payne, A. Hoogvelt, M. Dietrich, and M. Kenny, "Editorial: New political economy," *New Political Economy* 1(1) (1996): 5. As they note, "The methodology of the new political economy rejects the old dichotomy between agency and structure, and states and markets, which fragmented classical political economy into separate disciplines. It seeks instead to build on those approaches in social science which have tried to develop an integrated analysis, by combining parsimonious theories which analyze agency in terms of a conception of rationality with contextual theories which analyze structures institutionally, and historically" (5-6).

29 See, for example, S. Funtowicz and J. Ravetz, "Science for the post-normal age," *Futures* 25(7) (1993): 739-56. For a detailed treatment of this approach in light of the philosophical foundations of political ecology, see the author's paper *The New Naturalism: Is There a*

(Radical) "Truth" Beyond the (Post-Modern) Abyss? Discussion Paper 96-1 (Victoria: Eco-Research Chair of Environmental Law and Policy, University of Victoria 1996).

30   J.S. Dryzek, "Foundations for environmental political economy: The search for homo eco-logicus?" *New Political Economy* 1(1) (1996): 36.

31   M. M'Gonigle, K. Stratford, and F. Gale, eds., *The Business of Good Forestry: A Symposium of Practitioners*, Report Series 96-1 (Victoria: Eco-Research Chair of Environmental Law and Policy, University of Victoria 1996).

32   R.B. Keiter, "Beyond the boundary line: Constructing a law of ecosystem management," *University of Colorado Law Review* 65 (1994): 293; R.E. Grumbine, "What is ecosystem management?" *Conservation Biology* 8 (1994): 27.

33   Ibid. Grumbine, 29.

34   Ibid. Grumbine, 32: "In the academic and popular literature there is general agreement that maintaining ecosystem integrity should take precedence over any other management goal ... This may be due partially to the fact that, given the rate and scale of environmental deterioration along with our profound scientific ignorance of ecological patterns and processes, we are in no position to make judgments about what ecosystem elements to favor in our management efforts."

35   The then-NDP Forests Minister Dan Miller approved the sale of two Fletcher Challenge sawmills with associated logging rights in Clayoquot Sound to International Forest Products Ltd. in December 1991. Although the government attached twenty-one conditions to the sale including maintaining jobs, prompt reforestation, and good forestry practices, they did not take the opportunity to convert the licences to alternative tenures despite intense public pressure to do so [B. Parfitt, "Interfor awarded cutting rights to huge tracts on Vancouver Island," *Vancouver Sun* (4 December 1991); "Fletcher Challenge to be concentrating on pulp and paper," *Victoria Times-Colonist* (5 December 1991)]. The issue is thoroughly discussed by the former editor of *Forest Planning Canada*, Bob Nixon: "The minister consciously decided he could ignore the *ongoing* efforts of a myriad of citizen groups who had already invested the previous three years in an attempt to build consensus on how to manage the forest resources of Clayoquot Sound to ensure long-term sustainability" [B. Nixon, "Public Participation," in *Touch Wood, B.C. Forests at the Crossroads*, ed. K. Drushka, B. Nixon, and R. Travers (Madeira Park, BC: Harbour Publishing 1993), 29].

36   The Scientific Panel for Sustainable Practices in Clayoquot Sound, *Sustainable Ecosystem Management in Clayoquot Sound: Planning and Practices*, Report 5 (Victoria: The Scientific Panel for Sustainable Forest Practices in Clayoquot Sound 1995). The primary planning objective of the panel is to sustain the productivity and natural diversity of the Clayoquot Sound region and the stability of local communities. The panel recommends replacing the AAC as a harvest determinant with a method in which cuts are determined through an analysis of the geographical distribution of harvesting and percent of territory cut per unit time in a watershed area. Other suggestions include: replacing administrative planning units with physiographic or ecological planning units; combining scientific with traditional knowledge and engaging local people in all phases of planning and management; developing internally and externally consistent planning at the subregional, watershed, and local site levels; basing planning for large areas on a long-term perspective (100 years) to incorporate natural cycles; making an inventory of resources, values, and activities at an early stage of planning to assess ecological responses to change; monitoring ecological effects of activities and employing adaptive management procedures. The panel explicitly bases planning on the "precautionary principle."

37   Many economic critics of state regulation assume that the only choice is between public and private property rights. In fact, a third form of rights – communal rights – long predates both, and is demonstrably successful (as discussed in my article below despite ideological, neoclassical preconceptions to the contrary. These preconceptions arise to a great degree from Garrett Hardin's famous article that first coined the concept of the "tragedy of the commons," which advocates of privatization often point to. In addition to its historical inaccuracy, Hardin's analysis mistakenly equated the self-defeating over-exploitation associated with "open access" resource regimes (where there are no property rights) to "common property" regimes that have, contrary to Hardin, long proven to be capable of successful self-regulation under communal rules.

38 See the special issue of *The Ecologist* magazine, published in book form as *Whose Common Future? Reclaiming the Commons* (Philadelphia and Gabriola Island, BC: New Society Publishers 1993). For example, the Contents lists these sub-heads for the last chapter: Communities around the world are determined to safeguard, to revive, or to recreate their commons; their land and environment; their vernacular knowledge; their community-controlled markets; their arenas for decision-making; their commons in everyday life.

39 D.W. Bromley, *Environment and Economy: Property Rights and Public Policy* (Cambridge: Blackwell 1991), 104. Bromley was specifically referring to the tropics, but this same analysis has been applied to Canada, and other places with an established indigenous culture.

40 For a good example of this perspective, see E. Ostrom, "Designing Complexity to Govern Complexity," in *Property Rights and the Environment,* ed. S. Hanna and M. Munasinghe (Washington, DC: Beijer International Institute for Ecological Economics and The World Bank 1995), 33.

41 At the risk of extending the corroborative argument further than might be familiar to forest policy analysts, the argument advanced here is also in keeping with an ecofeminist critique of patriarchal structures of dominance and exploitation. This orientation was very much in evidence in the 1993 "peace camps" during the summer of protest at Clayoquot Sound. More generally, this argument might be situated in the broad debate on democratizing state administration. See G. Albo, D. Langille, and L. Panitch, eds., *A Different Kind of State?: Popular Power and Democratic Administration* (Toronto: Oxford University Press 1993).

42 *Forestopia,* cited supra, note 3 at 54 and 55.

43 Although it is outside the scope of this chapter to look at non-forestry models, a recent example of such an innovative design that seems to be working very well in managing a diversity of conflicting fisheries interests can be found in the Skeena Watershed Committee. On this point, see E. Pinkerton, "The contribution of watershed-based multi-party co-management agreements to dispute resolution: The Skeena watershed committee," *Environments* 23(2) (1996): 51-68.

Part 3:
Sustainable Forestry Policy:
Lessons from British Columbia

Over the last decade, Canadian forest policy has been in a state of unprecedented flux. Nowhere has this phenomenon been more pronounced, in terms of both public debate and policy innovation, than in the Province of British Columbia.

What lessons emerge, for sustainable forest policy, from the British Columbia experience? How much closer, if at all, have recent policy developments brought this province towards achieving sustainable forestry? In Part 3, we confront these questions by examining areas of forest policy critical to the transition to sustainable forestry.

The current form of large-scale industrial tenure, prevalent in British Columbia and throughout most of Canada, is considered by many to be one of the most significant impediments to achieving sustainable forestry. Yet in British Columbia, and elsewhere in Canada, existing tenure arrangements have been highly resistant to change. Chapters 6 and 7 – authored by Haley and Luckert, and M'Gonigle respectively – reflect on the prospects for tenure reform as part of a broader sustainable forestry agenda.

In contrast, in other policy areas, British Columbia has embarked on significant policy innovations, most notably timber pricing, forest practice regulation, and land zonation. A critical assessment of recent timber pricing reform initiatives, and their relationship to ongoing trade disputes with the United States, is offered by Scarfe in Chapter 8. In Chapter 9, Cook evaluates BC's unique and controversial new Forest Practices Code. And in chapter 10, Rayner considers the promise and potential of evolving land zonation proposals.

Another area of undisputed importance to the achievement of sustainable forestry concerns regulation of the rate of cut. In this area, the subject of ongoing and heated political controversy, the BC government has initiated modest reforms. In Chapter 11, Dellert argues that these reforms, like earlier attempts to regulate harvest rates, are unlikely to promote, let alone achieve, sustainability.

In the final chapter in this part, Gale and Burda consider the potential role of eco-certification – a highly touted, but as yet largely untested, new market-based policy instrument – in the quest for sustainable forestry.

6
Tenures as Economic Instruments for Achieving Objectives of Public Forest Policy in British Columbia
David Haley and Martin K. Luckert

Introduction

Social, political, and economic environments, within which public forests in British Columbia are managed, have gone through major transitions in recent years and continue to change at an accelerating pace. Although governments have reacted to these changes with arrays of new regulations, the underlying framework of institutions under which forests are managed remains essentially intact. There is a pressing need to re-examine forest policy in British Columbia to address its efficacy in furthering social objectives. Key elements in this process of restructuring are the economic instruments embodied in the forest tenure system.

In investigating the economic incentives provided by forest tenures, we will first examine the rationale for public regulation in a market economy with particular emphasis on the forestry sector. Next, we will briefly review the nature and role of forest tenures as property rights that may be used as economic instruments of public forest policy. Our attention will then turn to British Columbia. Specific market failures in the forestry sector will be discussed and the ways in with which they are addressed by current tenure policies will be critically examined from a social perspective. Finally, modifications to the current forest tenure system to meet changing social imperatives in British Columbia will be explored and some considerations in designing a more effective portfolio of forest property rights will be presented.

Market Failures as the Rationale for Public Regulation of the Forestry Sector

Public regulation of market-based economies is predicated on the assumption that private markets, and the unfettered private property rights on which they are based, fail to adequately promote public objectives. Three main areas in which private markets may fail to meet the public purpose can be identified (Boyd and Hyde 1989): allocation, distribution, and stabilization. All are important in the forestry sector and may justify varying types and degrees of regulation.

Allocation problems arise when a market system fails to arrive at mixes of resource inputs that reflect society's true values. This market failure may be the consequence of industrial concentration and resulting market control in product (output) or resource (input) markets. For example, one company may control regional stumpage or log supplies as a monopolist, or the number of pulpchip buyers in a region may be small in number, indicating oligopsony.

However, a more ubiquitous source of allocational inefficiencies is externalities. Externalities arise when decision-making units within the economy – individuals, households, or private companies – make production or consumption choices that do not take account of all the socially relevant benefits and costs associated with their actions. Externalities may exist because property rights and, therefore, market forces for certain goods and services are absent. For example, in the cases of air, water, wilderness, and biodiversity, difficulties in establishing exclusive property rights can prevent markets from recognizing the values of these resources, thereby causing inefficient allocations of resource inputs. Another common source of externalities is when producers do not have access to all the markets relevant to their actions and, therefore, ignore some of the consequences of their activities.[1] For example, a company building logging roads will provide benefits in the form of recreational access, but, unless it can charge recreationists for use of the roads, the company will have no incentive to take the needs of recreationists into account when designing its road system. This situation describes a positive externality of production. On the other hand, stream sedimentation resulting from road construction may reduce the catch and revenues of downstream fishers. The logging company, unless it is also involved in the fishing industry or has access to a market in which it can sell clean water, will not factor these forgone revenues into the costs of log production, thereby creating a negative externality. From a social standpoint, the external costs cause logs to be overproduced and underpriced while fish are underproduced and overpriced. In effect, consumers of fish end up subsidizing consumers of logs.

Distribution becomes a justification for public regulation when the market system fails to allot income in a manner that conforms to a society's notions of equity. Distributional concerns fall into three categories: interpersonal, interregional, and intergenerational. An example of concerns for interpersonal equity is embodied in log export controls. While such measures favour labour in the wood-processing industry, the owners of timber resources suffer as log prices are depressed domestically. Social concerns over interregional distribution may arise from the fact that firms in the private sector locate capital in those areas where it will generate the highest rents, enriching certain regions and impoverishing others. Concerns over interregional equity may be alleviated by public policies that provide subsidies to encourage industrial activity in disadvantaged regions or relegate economic efficiency to a subordinate role in the spatial allocation of public expenditures on infrastructural developments or activities such as silviculture. Intergenerational equity is manifest in concerns that private firms fail to give adequate weight to the welfare of future generations in their allocative decisions. This failure may be expressed as a divergence between private and social discount rates. These differences may result from imperfections in capital markets, or a more fundamental difference in time preferences between private firms and society at large.

Many public forest policies are directed towards promoting economic stability, particularly in forest-dependent regions or communities. Decisions made by profit-seeking private firms may fail to consider social imperatives such as community survival and stability of employment and revenue. Sustained yield policies, which remain a major driving force of forest policy in many jurisdictions, including British Columbia, are predicated on the assumption that a sustained flow of timber will dampen the economic impact on economies of cyclical activity in timber markets and maintain forest industry activity on a regional basis in perpetuity.

The Role of Property Rights as Instruments of Public Forest Policy
In reacting to market failures, which stand between private actions and the attainment of social objectives, governments are faced with a wide variety of alternative strategies. Some options take the form of policy instruments or tools that can be introduced within the existing institutional framework of fundamental political, social, and legal arrangements which comprise the ground rules for production, exchange, and distribution within a society (North 1991). For example, a government may introduce modified procedures for assessing taxable property values or subsidies designed to encourage tree planting. Other options may require that more profound changes be made in the institutional framework itself, such as the expropriation and nationalization of a hitherto privately owned resource. Public intervention is not without its problems, however. Wolf (1988) asserts that market failures have their counterpart in the public sector as "government failures." Public intervention can result in sub-optimal solutions from a social perspective and public agencies may have goals that discourage management in the best social interest. For example, Clawson (1976) and others have argued that the US Forest Service has not acted in the public interest.

Among all the ways governments can influence the behaviour of individuals and firms in the marketplace, perhaps none is more important than changes in the arrangements affecting the distribution and specification of rights to property. In fact, property rights provide a comprehensive conceptual framework within which a broad array of public regulatory tools can be examined.

Concepts of Property Rights

Definitions of property rights have been provided by scholars from many disciplines including law, political science, sociology and economics (see, for example, MacPherson 1978; Hallowell 1943; Scott 1983; Dahlman 1980; Barzel 1989; Furbotn and Pejovich 1972; and Bromley 1991). Drawing on the plethora of available definitions, property can be regarded as a physical asset or service of value to human beings – individually or collectively. A property right is a socially sanctioned and enforceable claim of an individual or group to the benefits (pecuniary or non-pecuniary) flowing from property subject to the conditions society places on the use of an asset or service.

As implied in this definition, the use of property is rarely absolute or unfettered. Instead, property rights are frequently attenuated, or limited, in various ways. Attenuations may take the form of legal (customary, statutory, or contractual) prohibitions and responsibilities that impose costs on property right holders and thereby reduce the value of their rights. For example, the owner of a right to harvest timber may be constrained by provisions that proscribe certain harvesting practices and that limit the volume cut annually or periodically – requirements which may reduce considerably the potential stream of benefits from timber harvesting.[2] Conversely, conditions placed by society on the use of property can include subsidies that may be thought of as "negative attenuations," which augment the flow of benefits to which a property right holder has a legal claim.

Given these concepts, property rights may be specified by the following components: a description of the benefit stream to which the right is granted; recognition of the claimant(s) to the right to the exclusion of society at large;[3] the relationship between the property right holder and society at large as defined by attenuations and subsidies; and the extent to which the property right is recognized and enforceable.[4] Note that these concepts are consistent with the simultaneous existence of multiple property rights in a physical asset. For example, forests have several valuable resources including timber, water, wildlife fodder, recreational amenity, and biodiversity, the rights to which may be vested in individuals, corporations, groups such as cooperatives and communities, or governments.

Public policy analysts, particularly economists and jurists, frequently describe property rights as bundles of attributes that can be disaggregated for taxonomic and analytical purposes. Many such schema exist (see, for example, Alchian and Demsetz 1973; or Milgrom and Roberts 1992); however, in our opinion, the most useful for evaluating economic behaviour was proposed by Scott and Johnson (1983), who identified the characteristics of property rights as

- *comprehensiveness*, which describes the number of asset attributes to which a single property right holder has claim;

- *duration*, which refers to the period over which rights may be exercised;
- *transferability*, which describes the extent to which holders are allowed to sell, or otherwise transfer, their property rights;
- *right to economic benefits*, which describes the extent to which right holders can capture and retain the stream of benefits flowing from the property;
- *exclusiveness*, which refers to the degree to which a holder can exclude others from enjoying the stream of benefits to which they have rights; and
- *security*, which is a measure of the confidence tenure holders have in the sanctity of their rights.

These attributes of property rights have been augmented by Haley and Luckert (1990), and used to describe and analyze forest tenures in Canada.

Property Rights as Instruments of Public Policy: The Potential for Government Failures

Property right arrangements embody systems of incentives that affect the ways individuals and groups behave and, thereby, influence the extent to which social objectives are achieved. From an economic perspective, property right holders may be characterized as having incentives to maximize the net values of the benefit streams to which they have claims. Within the context of the market system, this tendency may provide a powerful stimulus for the efficient allocation of resources. For example, a person holding the rights to an area of land will have an incentive to put the land into that use, or combination of uses, which produces the stream of benefits with the highest net value. A logging company holding rights to harvest timber will seek that combination of logging techniques and markets that will maximize the net value of the timber resource.

If private and social interests coincide – that is, if there are no market failures – property holders, in exercising their rights, will serve the social purpose. If, on the other hand, private and social values are disparate, steps may be taken by governments to rectify the situation through the placement of conditions on the exercising of property rights. These conditions may take many forms and are frequently characterized by taxonomies such as that presented above. For example, they may include fiscal measures which impose taxes, user fees, and other levies on holders of property rights; limits to the period over which the property rights can be exercised; restrictions on the transferability of property rights through sale, gift, or bequest; constraints on how a property can be used; and requirements that must be met in order to enjoy the stream of benefits flowing from a property.

Property rights' attenuations modify the incentives and thus the behaviour of tenure holders in ways which may, or may not, align social and private values. Poorly conceived conditions on the ability of property holders to exercise their rights may lead to unexpected and, from a social standpoint, perverse results. Requirements, which are frequently difficult and expensive to administer, may be met at minimum cost, evaded, or only partially accomplished, despite prospective penalties. In some cases, cost-minimization behaviour may be designed to serve the social purpose and will produce positive outcomes. However, in other situations, cost minimization incentives may result in negative social effects.[5]

In forestry, and other natural resource sectors, changing societal perceptions about environmental values, driven by intensifying environmental activism, have given rise to public regulatory frameworks that increasingly encroach on the values of property rights awarded to tenure holders. Indeed, in some cases, encroachments are sufficiently severe to cause tenure holders to argue that a "taking" of property has occurred. Such concerns have given rise to the question of compensation. Particularly, what is a compensable taking and, if a taking is compensable, how should the level of indemnity be calculated?[6] Despite the difficulties in addressing these questions, they are germane to the issue of the security of property rights, which influences investment, resource allocation, and economic and social stability, and will be discussed further in a later section.

Market Failures, Government Failures, and Forest Policy in the British Columbia Forestry Sector

To facilitate discussion, market failures in the British Columbia forestry sector and the policies currently used to address them will be divided here into several categories. These categories are neither mutually exclusive nor all-inclusive, but they are intended to focus attention on what we consider to be the most important questions. First, we will examine negative externalities associated with harvesting trees. Second, we will investigate positive externalities related to tree husbandry. Third, we will examine market failures that have specific impacts on multiple use of forest lands. Fourth, we will direct our attention to distributional problems – interpersonal, intergenerational, and interregional – and, finally, problems perceived to arise from the current structure of the province's wood-processing industry.

We will assume that the reader has some knowledge of the forest tenure system in British Columbia, because, beyond some brief introductory remarks, this involved topic is beyond the scope of this chapter. The historical development and current structure of the province's forest tenures is encapsulated in a number of sources including the BC Forest Resources Commission (1991) and Haley and Luckert (1990, 1995).

Public Forest Policy in British Columbia

In British Columbia, title to 96 percent of forest land is held by the provincial government. Of the remainder less than 1 percent is held federally and

3.4 percent privately (Forestry Canada 1993). As in other parts of Canada, partial usufructory rights to Crown forest lands have been transferred to the private sector to create a diverse system of Crown forest tenures. These arrangements can be viewed as a complex system of temporary rights attenuated by regulations and fiscal obligations. This system, which has evolved over about 130 years, has been, and remains, the principal instrument used by successive governments to pursue the goals of public forest policy.

Historically, the overriding concern of British Columbia forest policy was the utilization and management of timber resources to provide wealth – in the form of direct public revenue, but, more importantly, as value added in timber processing – and a continuing, stable source of employment on a regional basis. Policies were developed that relied heavily on the Crown tenure system to encourage the establishment of timber-processing facilities and create an institutional environment in which manufacturing companies could flourish and become world competitive. Forest conservation and management policies, until quite recently, have been almost entirely directed at the provision of timber supplies over the long term (sustained yield) and the full utilization, in the sense of minimizing physical waste, of timber harvests. Reforestation – viewed since the 1960s as an essential component of sustained yield – is delegated to Crown tenure holders who, since 1987, have been required to meet all the costs of their contractual silvicultural obligations.

Awareness of forest values, other than timber and range, developed during the 1970s. Outdoor recreation, water, fish, and wildlife were recognized in the 1978 Ministry of Forests Act[7] as legitimate uses of provincial forests. This piece of legislation was the first statute in British Columbia to provide the Ministry of Forests with legally mandated goals.

Since the passage of the Ministry of Forests Act, two major forces have influenced the goals of public forest policy. First, there has been a growing conviction that the increasingly concentrated structure of the forest industry discourages efficiency, innovation, and diversification into higher-value manufactured products. Second, the rise of the environmental movement, and since the late 1980s, the philosophy of sustainable development, have had a considerable impact on the objectives of forest policy. Policy initiatives arising from these latter concerns include the maintenance of biodiversity, the preservation of wilderness values, a heightened awareness of the need to protect the integrity of forest ecosystems, a greater commitment to integrated resources management, and increased public involvement in decision-making. The cornerstone of recent policy changes is the Forest Practices Code of British Columbia Act[8] which was proclaimed in April 1995. This Act provides a comprehensive legal framework within which the Crown forest lands of British Columbia are to be managed sustainably to achieve a balance among "productive, spiritual, ecological and recreational values – to meet the economic and cultural needs of peoples and communities including First Nations." The Forest Practices Code itself consists of the Act, Regulations pursuant to the Act, Chief Forester's Standards, and a comprehensive series of guidebooks.

While public forest policies in British Columbia have undergone major modifications in recent times, the essential components of forest tenure have remained virtually intact for close to fifty years. Changes in policy objectives have mainly been addressed by an increasing number of attenuations on an existing set of property rights. These attenuations have principally taken the form of regulations of ever-increasing scope and stringency.

Policies Designed to Address Negative Externalities Associated with Harvesting Trees

External costs of harvesting activities are widespread and affect a number of different forest values. One set of costs occurs because harvesting practices may adversely affect the ability of forests to provide adequately for other (non-timber) goods and services. These costs are borne by many groups in society, which suffer from reduced water quality for downstream users, impacts on future timber production through soil loss and fertility reduction, and the loss of recreational and visual amenity values. A second set of costs associated with harvesting, which may occur no matter how carefully harvesting is done, is due to the loss of pristine standing forests, particularly old growth, which may have values associated with wilderness and biodiversity.

Solutions to externalities of the first type have been sought in British Columbia through regulations. Historically, flexible guidelines were established, leaving much to the discretion of Ministry staff and tenure holders. However, with the introduction of the Forest Practices Code, regulations have statutory backing and are much more comprehensive, rigorous, and inflexible. Monitoring and enforcement procedures have been strengthened and penalties for non-compliance have become more severe.

The second type of external cost associated with timber harvesting has largely been addressed by zoning for the protection of social and environmental values embodied in undisturbed forests. Amendments to the Forest Act in 1987 recognized that land within provincial forests might be "maintained as wilderness" if this single use, in the opinion of the Chief Forester, "provides the greatest contribution to the social and economic welfare of the Province." This legislation was given administrative force in 1993 with the introduction of the Protected Areas Strategy (Province of British Columbia 1993), which has as a goal the protection from commercial exploitation of 12 percent of the provincial land base by the year 2000.

Although these policy solutions adopted by the British Columbia government may internalize external costs, the costs of code regulations and land withdrawals are considerable. For example, the annual cost of implementing the Forest Practices Code has recently been estimated at over \$2 billion (Haley 1996) including over \$600 million in private compliance costs, \$1.4 billion in economic activity forgone as result of reduced timber harvests, and \$50 million for public monitoring and enforcement costs. KPMG Ltd. (1997), in a study commissioned by the British Columbia Ministry of Forests, estimated Code compliance costs at \$750 million annually; that is, approximately \$11 per cubic metre of Crown timber harvested.

Policies Designed to Address Positive Externalities Associated with Growing Trees

In the case of private forest land, many of the benefits of timber growing are realized by the owners. Nevertheless, external benefits of tree husbandry do exist and are increasingly recognized. These benefits include the ability of forest crops to sequester carbon and modify microclimates. Positive externalities may also be claimed to exist in terms of regional growth and stability, and the provision of employment in rural areas.[9] In recognition of these positive externalities, many jurisdictions provide subsidies to forest owners for silvicultural activities, including reforestation and stand management. For example, in British Columbia subsidies have been granted to small-scale forest owners through the 1991-1996 Canada-British Columbia Forest Resource Development Agreement (FRDA II).[10]

In addition to the positive externalities that may be caused by failure of private markets, the Crown forest tenure system in British Columbia also externalizes, from the tenure holder's viewpoint, most, if not all, the benefits of silviculture associated with enhanced timber production and value. This practice is because the tenure system provides rights to harvest timber but reserves for the Crown the rights to future timber crops. Rent collection arrangements on Crown forest tenures are designed to capture the values generated when stocks of old-growth timber are harvested. When these same levies and regulations are applied to second-growth crops, which must bear the costs of reforestation, stand tending, and capital opportunity costs, little or no returns to tenure holders remain (Luckert and Haley 1989, 1993). Furthermore, institutional insecurity associated with British Columbia's current forest tenure arrangements (Luckert 1991) may further discourage long-term investments in timber production.

In the absence of market incentives for tenure holders to invest in timber growing, provincial governments in British Columbia have, historically, relied on regulations, cost disbursements, and penalties to promote silviculture operations. However, since 1988, the costs of meeting reforestations regulations have been borne entirely by tenure holders. Under these arrangements, firms may carry out reforestation on public land in order to avoid penalties and, thus, they have a strong incentive to simply achieve minimum standards at the lowest cost. Performance prescriptions and standards which are supported by increasingly stringent and costly auditing procedures are difficult to set, and, in the interests of administrative ease, are generally fairly uniform. The result is least-cost, basic crop replacement to relatively uniform standards which precludes decisions about how to optimally manage denuded lands for future forests based on specific site characteristics and economic and social values.

Regulations only require reforestation on Crown land to certain "free to grow" standards. More intensive silviculture is voluntary and principally takes place on private land where benefits are mainly internal (Luckert and Haley 1990; Zhang 1994). To encourage private investments in silviculture on Crown forests beyond contractual requirements, the main instrument used by Canadian provinces, including British Columbia, is the allowable cut effect (ACE). Under these arrangements, tenure holders are allowed to increase the volume of timber they are allowed to harvest annually, if they can demonstrate that their silvicultural activities have increased the productive capacity of the management unit within which they are operating. However, such policies have been generally ineffective.[11] Reasons for their failure have been analyzed by Haley and Luckert (1995), who conclude that if private firms are to contribute significantly to the management of Canada's timber resources, over and above contractual requirements, then alternative policy instruments designed to provide the necessary incentives must be developed.

Recently, the provincial government embarked on a policy of more direct involvement in silvicultural activities. Under the 1994 British Columbia Forest Renewal Act,[12] a Crown agency – Forest Renewal BC – was established to plan and undertake major investments in stand management beyond the free to grow stage. This agency is financed by major increases in Crown stumpage charges, which provide it with an annual budget of about $600 million, of which $450 million will be allocated to silviculture. While it is too early to assess the success of this initiative, available information suggests that budget allocations will be driven by distributional rather than allocational concerns. For example, the major political justification for the program, which has been made quite explicit, is to absorb labour displaced from the forest industry as a consequence of such programs as the Commission on Resources and Environment (CORE) land use plans,[13] the Protected Areas Strategy (Province of British Columbia 1993), the Forest Practices Code,[14] and the Timber Supply Review[15] process.

Policies Designed to Address the Multiple Use of Forest Land
In most jurisdictions, even those with a substantial private forest land base, pervasive externalities surrounding the production of many non-timber forest products mean that private forest managers cannot be relied on to manage optimally for multiple land use from a social viewpoint.[16]

In British Columbia, all Crown forest tenures provide their holders with exclusive rights to harvest timber but no opportunities to benefit from the production of other forest products (Haley and Luckert 1990). Rights to non-timber forest products are generally retained by the Crown. Consequently, all values, other than commercial timber harvesting, are external to the decision-making processes of Crown forest tenure holders.

Instead of granting rights to non-timber resources, the general approach to multiple use in British Columbia is regulatory. Regulations addressing these problems include contractual and statutory requirements, constraints on timber harvesting practices, and zoning.[17] For example, under the Forest Act,[18] a tenure holder may be required to establish and maintain a recreation site or trail. Many of the provisions of the Forest Practices Code are designed to enhance, and integrate with timber management, the production of non-timber values. Areas set aside under the Protected Areas Strategy provide for those values which are incompatible with extractive forest uses. In 1992, CORE was mandated to "develop, implement and monitor regional planning processes." CORE concentrated its attention on four regions: Vancouver Island, Cariboo-Chilcotin, West Kootenay-Boundary, and East Kootenay. Subsequently, the provincial cabinet approved planning frameworks for these regions. These frameworks recognize integrated resource management areas where all relevant values are considered and jointly managed; enhanced resource development zones where intensive resource development, including timber management, is a priority; and special resource management zones where particularly sensitive non-timber values receive priority. The focus has now turned from broadly based regional plans to Land and Resource Management Plans (LRMPs). These plans are on a smaller scale than the regional plans, typically embracing one forest district or a small group of districts. Like regional plans, they involve local stakeholders' participation and subsequent cabinet approval.

In summary, externalities arising from the multiple use of forest land are dealt with mainly by means of command and control instruments. Private firms occupying Crown land have no rights providing them with a stake in the values they are required to create. Consequently, they have no incentive, beyond meeting minimum requirements, to plan for multiple use in an integrative, optimum manner. Instead, the government, through its regulations, and with little or no access to market signals, must attempt to arrive at socially optimal mixes of goods and services and enforce management strategies through increasingly complex and costly monitoring and enforcement procedures.

Policies Designed to Address Distributional Concerns

Interpersonal Distribution
Examples of concerns over interpersonal distribution are common in the forestry sector. For instance, there is the perception that market forces will lead to the export of products of low labour content which will reduce

labour's share of the net wealth generated by the forest industry sector. In essence, this issue is a concern for the distribution of value added, particularly that portion made up of wages and profits.

One policy addressing these concerns are log export controls, which represent important attenuations of tenure holders' rights. Export restrictions reduce benefit streams to log production operations, hence reducing the value of harvesting rights. Accordingly, such restrictions have widespread implications, generally unacknowledged at the public level, for industry structure, investment in silviculture, and allocational efficiency in timber processing.

Increasingly, concern about the socially optimum product mix extends beyond the issue of raw material exports into the realm of commodity production (e.g., dimension lumber, market pulp) versus the production of further manufactured wood and pulp products – such as furniture components, doors, and windows – and paper and paper products.[19] It is argued that the increasingly capital-intensive, labour-saving technologies employed by manufacturers of producer-oriented products entering world commodity markets results in increased returns to capital and entrepreneurship in these industries at the expense of wage-generating employment. A more socially desirable balance can be restored, it is said, by encouraging the manufacture of consumer-oriented, specialized products in smaller-scale, more labour-intensive plants. Policies in pursuit of this objective in British Columbia include a special category of timber sales, or "bid proposals," under the Small Business Forest Enterprise Program (*Forest Act*, s. 16.1). Such sales are awarded, not on the basis of the highest bid, but to the competitor presenting the most desirable business proposal in terms of value added to the raw material used and the number of jobs created per unit volume of wood input. Also, Forest Renewal BC, which has as part of its mandate the encouragement of further manufacturing activities, has recently announced BC's Value Added Strategy (Forest Renewal BC 1996). This plan includes initiatives to improve timber utilization standards, to encourage the use of, hitherto, low-value hardwood species, to create electronic log-marketing networks and regional log markets, to establish training programs in secondary wood products manufacturing, and to promote BC's further manufactured products in overseas markets.

The BC government is now embarking on a more direct program of job creation in the forest sector. In June 1997, Premier Clark announced the British Columbia Jobs Accord, involving government, the forest industry, and organized labour. The goal of the Accord is to create 21,000 new direct jobs in the forest sector by the year 2001.

Intergenerational Distribution
The limited terms of Crown forest tenures in British Columbia, combined with attenuated transferability of rights and institutional and political

uncertainty, externalize future costs and benefits from a tenure holder's perspective. However, solving tenure problems in this regard (i.e., by creating perpetual, fully transferable, and secure rights) would still leave the matter of disparate private and social discount rates, which is the central issue of concern for intergenerational equity.

The primary forest policies addressing this issue in British Columbia are those that regulate the rate of harvest and replacement of timber crops over time to ensure a sustained supply of timber in perpetuity. Unlike the United States, British Columbia has never pursued a policy of "non-declining even flow." Rather, an indicated falldown in timber harvests below current levels is accepted, provided future harvest levels do not fall below the estimated long run-sustainable yield. Since sustained yield polices were introduced fifty years ago, controversy has surrounded allowable annual cuts. As mentioned above, the Ministry of Forests has just completed a province-wide, detailed Timber Supply Review. However, this review has done little to allay the fears of those who maintain that current levels of harvest are unsustainable and that the province will eventually "run out of wood."

British Columbia has yet to come to grips with the real questions of intergenerational equity embodied in the concept of sustainable development. The volumetric flow of timber remains a proxy for social welfare but addresses neither the value nor the intertemporal welfare implications of planned yields (Haley and Luckert 1995).

Interregional Distribution
Concern for interregional equity is the driving force underlying many of British Columbia's forest policies including the spatial distribution of Timber Supply Areas, land use planning initiatives, broadly based provincial forest practices and reforestation standards, and the allocation of silvicultural budgets. Guidelines published by Forest Renewal BC give high priority to regional equity in the allocation of funds for silvicultural and other projects.

Policies Designed to Address Industrial Concentration
The public welfare implications of integration and concentration in the British Columbia forest industry have been of concern to policy analysts for many years (Pearse 1976). Most recently, the BC Forest Resources Commission (1991) recommended radical reforms, including major changes in the tenure system designed to restructure the forest industry to reduce vertical integration and reverse the trend towards greater concentration.

Public interest in the structure of the forest industry in British Columbia is focused on the hypothesis that industrial concentration in control over rights to harvest public timber results in the inefficient allocation of the timber resource. Ironically, today's forest industry structure owes much to a tenure system that has severely limited competitive stumpage and log markets and, through mill appurtenancy conditions, has made vertical integration a requirement for holders of the major tenure types.

Policy initiatives to influence industry structure in recent years include the Small Business Forest Enterprise Program, which currently reserves about 15 percent of the annual timber harvest for competitive sales to various categories of small business; statutory control over the transfer of Crown forest tenures and a requirement that when a tenure is transferred, 5 percent of its allowable cut reverts to the province for reallocation; and an increase in the number of smaller tenures available as woodlot licences.

In Search of Forest Tenure Solutions to Potential Market Failures

The above discussion reveals that, to date, perceived market failures and corrections for government failures in British Columbia's forest sector have largely been approached with "command and control" regulations layered over rights to harvest timber. The last five years, particularly, have seen a significant increase in measures that attenuate the rights of Crown forest tenure holders. Regulations erode the potential stream of economic benefits enjoyed by tenure holders and, in so doing, may weaken the incentive structure the tenures were designed to create. Furthermore, as the costs of meeting regulations increase, so do the incentives to avoid them. As a result, as demonstrated by the Forest Practices Code, proliferating regulations must be supported by increasingly stringent monitoring and enforcement procedures, resulting in mounting public costs, and increasingly punitive penalties. Requirements, at best, are met at minimum cost, which may result in perverse consequences from a social perspective. While social cost-benefit analyses rarely, if ever, precede the introduction of new regulations, some studies suggest (van Kooten 1994) that the social costs of some of the major regulatory initiatives may exceed the value of the benefits they are intended to protect and/or create. Moreover, regulations have proliferated with seemingly little analysis of their interactions and cumulative impacts.

These problems arise, in part, from the fact that British Columbia's forest tenure system is anachronistic. It evolved to its present form during a period when the provincial government's primary goal was to attract private capital, both domestic and foreign, to liquidate a stock of old-growth timber in an orderly manner, and establish a world-competitive timber-processing industry. The system is ill-equipped to deal with the problems that beset the province's forest sector today, particularly the transition from old-growth timber harvesting to second-growth forest management, and the production of an increasingly complex array of non-timber forest products and environmental services.

The question we will now examine is whether forest tenure arrangements can be reformed in ways that will more effectively achieve social objectives while reducing reliance on regulations. It is contended that such a policy shift will increase efficiency in government by attaining social goals at lower total cost – that is, private cost plus public cost – than at present. First, we will discuss the opportunities for increasing the rights granted to tenure holders beyond timber harvesting. The possibility of conferring partial rights based on the principles of sharecropping will receive particular attention. Second, we will address the arguments for reallocating rights among the provincial government, communities, the industrial private sector, and other private groups and individuals.

Opportunities for Augmenting Rights to Forest Resources

Important sources of market failure can be dealt with by internalizing, to the decision-making unit concerned, the external costs and benefits that distort resource allocation from a public perspective. One approach is to specify property rights in ways that allow their holders to enjoy a stream of benefits hitherto beyond their control. The holders of these newly granted rights will be motivated to maximize the net value of the stream of benefits they produce and, provided no serious residual externalities remain, the social purpose will be served with no further cause for intrusive and potentially costly public intervention. This approach will only work if the benefits concerned can be realized through market transactions; that is, they are not public goods, and the relevant markets are relatively competitive and accessible to rights holders.

Forest attributes to which rights could, potentially, be granted are the soil's capability to produce timber crops in managed stands and certain recreational values such as hiking, camping, sports fishing in forest lakes and streams, hunting, and wildlife viewing.

Providing Rights to Grow Timber

Luckert and Haley (1989) suggest that rights to grow trees would require tenures which were granted to specific geographical areas – that is, they would have to be area- rather than volume-based tenures.[20] They would have to be granted for periods at least as long as the forest rotation and be transferable. Rent collection arrangements would have to allow tenure holders a share in the timber values resulting from their silvicultural activities, possibly replacing stumpage payments with an annual rental charge based on the natural site productivity. Finally, tenure insecurity may have to be reduced by statutory procedures for awarding compensation in the event of rights being attenuated or extinguished.

Such right-based incentives will only work if silvicultural opportunities present profitable investments to the private sector. However, if there are positive externalities associated with growing trees, of the types described above, or if market interest rates do not reflect society's intergenerational preferences (Markandya and Pearce 1991), then private investments in silviculture may fall short of the social optimum.

Providing Rights to Non-Timber Values

Pearse (1988) examines the possibility of extending rights granted over forest land to include other values – particularly wildlife and fish. Citing European and Canadian examples, he argues that internalizing these values would result in multiple product management that would more closely approach a social optimum than under present arrangements.

While market-based solutions to multiple product management are appealing in many ways and may have a much greater role to play in British Columbia, they are limited in their application. Values of an increasing number of forest-based services are too diffuse to be captured through market prices. For example, many values associated with wilderness, biodiversity, and old-growth retention are the result of "passive use." That is, they accrue to individuals who do not use the resource directly but derive benefit from the knowledge that it exists. Some products display characteristics of both market and public goods. For example, wildlife as a source of recreational hunting might be efficiently allocated and integrated with other resource values through conventional market channels. However, wildlife also has value to those who derive utility from wildlife populations as part of an intact ecosystem. Granting transferable rights to recreational hunting may fail to preserve other wildlife values and may create an environment that is incompatible with their production. For example, a firm managing land jointly for timber and hunting may have an incentive to produce the maximum number of large ungulates and, as part of this strategy, reduce the population of natural predators which are valued by many as essential components of balanced, diverse ecosystems.

In some cases, the specialized skills of individual firms, or distributional concerns, may suggest that, instead of granting rights to one tenure holder across a spectrum of forest values, rights to various products produced from the same area of land should be separated and granted to different individuals or companies. For example, some Canadian provinces grant to separate firms the harvesting rights to different commercial timber species growing on the same area of land (Luckert 1993). Likewise, rights to extract oil and gas are not held by firms that harvest trees growing above energy deposits (Haley and Luckert 1990). Under this approach to joint management, following the ideas of Coase (1960), firms have incentives to negotiate agreements which result in the socially efficient coordination of interrelated activities. However, problems of indivisibility, imbalances in negotiating power, and transactions costs may prevent the realization of such solutions (Luckert 1992, 1993).

There is a further complication of granting private rights to many recreational values associated with public land: relatively unrestricted access to public resources for a variety of outdoor activities is so firmly entrenched in Canadian culture that any attempt to control admission and charge for use would be met by fierce opposition. User-pay solutions might also have socially unacceptable distributional implications.

Dealing with Insecurity

A crucial question surrounding any initiative to transfer property rights in public forest land to the private sector is that of uncertainty. From a private perspective, incentives to invest in and manage forest resources in an optimal manner will be greatly affected by security of property rights. From a public perspective, providing the necessary level of security to private right holders involves sacrificing much of the flexibility considered necessary to deal with inevitable changes in social values and aspirations.

Some would argue that adequate security can only be accomplished through privatization (Haley 1985). Although private property rights enjoy no constitutional protection in Canada and the rights to regulate all aspects of property remain with the Crown, rules for the expropriation of private property in the public interest are well established; private, or Crown-granted, rights, are generally considered to be the most secure form of land tenure. It is doubtful whether the transfer to the private sector of exclusive, fully comprehensive, and transferable rights to forest land in perpetuity, as described by Luckert and Haley (1989), would be currently politically acceptable in British Columbia. An alternative might be to sell exclusive, transferable rights to manage for the production of timber while retaining public title to the land itself. This procedure has been used successfully to "privatize" timberlands in New Zealand where rights are limited to a period of seventy-five years, which is two rotations in the New Zealand context. The success of such arrangements depends on the land concerned being zoned primarily for timber production and for rights of sufficient duration to allow returns to investments in timber management to be realized.

Procedures for awarding compensation in the event of property rights being attenuated or extinguished is a topic that is receiving much attention in many jurisdictions (see, for example, Epstein 1985; Schwindt and Globerman 1996; Innes 1995). Blume, Rubinfield, and Shapiro (1984) have shown how complete security, provided by full compensation for publicly induced changes in values of property rights, isolates firms from the risks of changing social values, thereby causing them to over-invest in resource development from a social perspective. In the absence of compensation, it is argued, private firms will appropriately hold back on some of their investments given the possibility that changing social values will result in public intervention affecting property right values. A counter-argument to this view is that compensation is necessary to keep the behaviour of governments in check. In the absence of compensation, it is argued, governments will fail to correctly consider the private costs of takings.

The insecurity facing forest tenure holders may be alleviated by creating institutional frameworks flexible enough to accommodate inevitable changes in social values (Luckert 1993). As social objectives change, property rights may be adjusted to changing conditions through renegotiation or respecification. If property rights are in private hands and transferable, private negotiations may facilitate adjustment to changing values as markets reallocate resources in response to changing demands. Alternatively, if property rights have a larger degree of public control, negotiations may have to occur between tenure holders and the government. Flexibility here depends on the ability of government to respond to changing values with new policies. In essence, insecurity may be thought of as having two sources: changing public values, and the inability of inflexible institutions to adjust to changing public values. While the first source of insecurity is inevitable, the second source may be alleviated through the creation of institutions that efficiently accommodate change.

Partial Private Rights: The Potential For Sharecropping
The above discussion indicates that many forest products provide a combination of social and private benefits and their production gives rise to both social and private costs. Granting property rights to private firms will encourage them to invest resources in ways that maximize net private benefits. However, this practice will not result in optimal allocation from a social perspective unless social values are also taken into account. These considerations suggest it might be fruitful to investigate tenure arrangements that allow holders partial rights to the benefits they produce while recognizing that certain social values should be reserved for the benefit of society.

A common means of granting rights to part of the proceeds of a property holder's activities is through sharecropping. In general, under such arrangements, the lessee and lessor share the costs of inputs and the benefits of outputs. In silvicultural investments, for example, the Crown could reimburse tenure holders for a portion of their expenditures in return for a portion of the output created. One advantage of such an approach is that, unlike requirements enforced by regulations, incentives created by sharecropping may be neutral in their effect on resource allocations.

Luckert (1994), following Cheung (1969), has shown that if the proportion of costs is shared between two parties according to the proportion of benefits each party will receive, then the lessee will have incentives to invest in those projects that yield the greatest economic benefits.[21] A further advantage of sharecropping is that, unlike the granting of complete rights, the Crown can retain an element of control over investment decisions, thereby correcting for market failures. For example, suppose that positive externalities from planting trees exist in the form of carbon sequestering. If complete rights are granted to private firms, the benefits of planting trees to sequester carbon will be external to the benefits of the

Table 6.1

Examples of sharecropping to correct market failure

Example	Crown	Tenure holder	Total
A. Timber only			
Shares (%)	25.0	75.0	100.0
Value of benefits in year 60 ($)	2,630.0	7,890.0	10,520.0
Present value of benefits at 4% ($)	250.0	750.0	1,000.0
Present value of costs ($)	-375.0	-1,125.0	-1,500.0
Net present value ($)	-125.0	-375.0	-500.0
B. Timber and reduced carbon			
Shares (%)	62.5	37.5	100.0
Present value of benefits at 4% (S)	1,250.0	750.0	2,000.0
Present value of costs ($)	-937.5	-562.5	-1,500.0
Net present value ($)	312.5	187.5	500.0
C. Timber and wildlife			
Shares (%)	60.0	40.0	100.0
Present value of benefits at 4% ($)	1,500.0	1,000.0	2,500.0
Present value of costs ($)	-1,200.0	-800.0	-2,000.0
Net present value ($)	300.0	200.0	500.0
D. Timber at social discount rates			
Shares (%)	25.0	75.0	100.0
Present value of benefits at 2% ($)	801.5	2,404.5	3,206.0
Present values of costs ($)	-375.0	-1125.0	-1,500.0
Net present value ($)	426.5	1,279.5	1,706.0

firm, and investors will allocate fewer resources to reforestation than is socially optimal. Under these circumstances, the Crown could consider a portion of its benefits to be the increased sequestration of carbon. With these benefits entering into the calculation of output shares, input shares for the Crown would increase accordingly and incentives to invest optimally could be maintained. An example of this concept follows.

In Table 6.1, Example A, the Crown and a tenure holder each hold, respectively, 25 percent and 75 percent of the rights to the benefits derived from the harvested timber of a future forest stand, predicted to be $10,520 in sixty years. Accordingly, silvicultural investments of $1,500 are split 25/75 percent between the Crown and tenure holder, respectively. If the discount rate used by the Crown and the tenure holder is 4 percent, neither party would be tempted to undertake this investment, because the present value of $10,520 at 4 percent is $1,000, which is lower than the costs. However, assume that, over the life of the forest stand, $1,000 of carbon sequestration benefits will be generated in present value terms at 4 percent. Further assume that the Crown, on behalf of society, considers these benefits as part of its return on investment. This case is presented in Table 6.1, Example B. The total amount of benefits available, in present value terms, is now $2,000, with the Crown receiving $1,250, or 62.5 percent of the total benefits. The project is now feasible for both parties. In return for its $1,250, the Crown's share of costs would increase to 62.5 percent, or $937.50. The tenure holder would receive the remaining $750 in benefits for its 37.5 percent, or $562.50, share of the expenditure.

This example describes a situation in which it could be possible to establish incentives for silvicultural investments that would internalize positive externalities. Theoretically, the same principle could be applied to other external values associated with growing trees. For example, assume that a silvicultural operation under consideration is thought to be harmful to some species of wildlife while beneficial to others. As in Example A, the silvicultural investment would cost $1,500 and would yield a present value of $1,000 in increased timber values, making the investment not feasible for either the Crown and the tenure holder. However, if the silvicultural activity is conducted with consideration for the values of the wildlife species (i.e., considering the species that would be both negatively and positively affected), then assume the cost would be $2,000 with a total present value of timber and wildlife habitat of $2,500. This scenario is presented in Example C. If the tenure holder adjusts the silvicultural operation to accommodate these non-timber values, the Crown's portion of the $2,500 is $1,500, causing the Crown's share to increase from 25 percent (as in Example A) to 60 percent. Accordingly, the costs of the tenure holder drop to 40 percent of $2,000, or $800, making this option clearly preferable to both the tenure holder and the Crown. In short, under such a scenario, it is in the tenure holder's interest to provide for non-timber values, because if the tenure holder can increase the Crown's share of the benefits, its costs will be reduced accordingly.

One problem with this procedure is its assumption that tenure holders are certain that in sixty years they will receive their share of the timber harvest. Tenure insecurity might create uncertainty which in turn could result in a reduced investment (Luckert 1991). To avoid uncertainty for the tenure holder, and allow greater flexibility for governments to respond to changing social values, another variation on the above scheme is for the Crown to pay the tenure holder upfront, in present value terms, based on expected future returns to investments. In essence, the tenure holder would be reimbursed for expenditures according to the present value of social benefits their investments are expected to create. In Example B, the tenure holder would receive $750 upfront for an expenditure of $562.50. In Example C, the tenure holder would receive $1,000 upfront for an expenditure of $800.

Sharecropping also provides a simple means of correcting for disparate social and private rates of time preference. For the examples presented in Table 6.1, assume that there is no positive externality from sequestering carbon, but there is a divergence between the social and private discount rates. This scenario is presented in Example D. Let the 4 percent discount rate, used in Examples A, B, and C, represent a private discount rate. Therefore, the $1,000 return to silvicultural activities in Example A represents a present value in terms of private sector time preferences. Assume a social discount rate of 2 percent. Accordingly, the $10,520 of timber revenue generated in sixty years is now worth $3,206 in present value terms from a social perspective (Example D). According to the 25/75 Crown/tenure holder split, the Crown would receive $801.50 for its expenditures of $375, while the tenure holder would receive an immediate payment, or reimbursement, of $2,404.50 in return for its expenditure of $1,125. The tenure holder in Example D would therefore have an incentive, lacking in Example A, to undertake the investment.

Despite the potential for sharecropping agreements to align social and private values, there would likely be many difficulties in implementing these arrangements. To begin with, values would have to be derived for many things that are not currently valued. Although this task is formidable, it is endemic to any management exercise dealing with competing resource values. To the extent that non-timber values can be expressed, sharecropping arrangements could be used to correct for market failures. To the extent that non-timber values cannot be expressed, we will be less able to meet social objectives through any forest policy.

An additional concern involves the costs associated with negotiating and monitoring sharecropping agreements. The Crown and tenure holders would likely have to negotiate shares to be held by each party on a project-by-project basis, as non-timber benefits would vary among projects.

Although tenure holders would have incentives to propose investments that further the public interest, after agreements are negotiated, monitoring would likely still be necessary to ensure that benefits, as outlined in the agreements, were actually produced and distributed. In cases where large shares are granted to the Crown, costs associated with negotiating and monitoring could be such that it may be more efficient for the Crown to conduct the management activities directly, rather than attempting to align private and public values.

Reallocating Forest Property

Crown forest tenures in British Columbia are mainly held by industrial forest companies and are concentrated in the hands of a relatively small number of large integrated forest product manufacturers (Haley and Luckert 1995). The transition to second-growth management and increasing demands on forests for non-timber products and environmental services raises questions, not only about the types of rights granted and how they are granted, but also to whom they are granted. There is a growing body of opinion that society might be well served by reallocating some rights to forest values to individuals, groups, and public agencies whose goals may be more closely aligned with social objectives than those of industrial corporations. The public inquiry into British Columbia's forest resources conducted in the mid-1940s (Sloan 1945) recognized the value of diverse forest ownerships. The lack of diversity in ownership rights to the province's forests was cause for alarm to the Pearse Royal Commission (Pearse 1976) and, fifteen years later, to the BC Forest Resources Commission (1991), which made a number of major recommendations in this regard.

Of particular interest are the possibilities for concentrating more rights in the hands of the Crown or transferring some rights to non-integrated, smaller tenure holders including individuals, small firms, Native bands, and communities. We will focus our discussion on community forests, although many of the observations we make are more broadly applicable.

Management by Crown Agencies

In many countries, including Australia, France, Germany, and the United States, public forest lands are managed directly by public agencies. The logic of such an approach is because public and private objectives are so different, policy instruments cannot be used to satisfactorily merge them, and, consequently, public agency management best serves public interests. In British Columbia, over the last twenty years, failures of the Crown tenure system have become more evident as differences between public and private forest management objectives have become more pronounced. The result has been increasingly restrictive and costly attenuations of the rights of forest tenure holders in order to promote the public interest. It seems likely that, in some cases, it would be more efficient to phase out the current forest tenure system and vest all rights to public forest resources in the provincial government. Under such an arrangement, the high public administrative costs and private compliance costs associated with increasingly complex forest tenure arrangements would be avoided and public resources, including the highly trained foresters and other professionals employed by the Ministry of Forests, could be devoted directly to forest management in the pursuit of social objectives.

Public management of forest lands might be undertaken directly by an agency such as the BC Forest Service or, indirectly, through a Crown corporation. Under the former arrangement, provincial forests would be managed by the Forest Service to produce the mix and amounts of goods and services demanded by society. While the problems of balancing the production of market and public forest values would remain, serious externalities would be eliminated. Timber would be sold to the private sector either standing, as is the practice in the United States, and/or in log form, as in a number of European countries.

A Crown corporation would take control of public forest resources and manage them to maximize their net worth subject to constraints designed to safeguard non-market values. Such a scheme was proposed by the BC Forest Resources Commission (1991) and has been tried elsewhere – most recently, but unsuccessfully, in New Zealand. Because Crown corporations are intended to encourage the management of forest land according to recognized commercial practices and distance the process from political influence, they are only feasible where commercial values, particularly timber values, dominate.[22] They can be regarded, in fact, as a form of quasi-privatization.

While Crown corporations are, in theory, arm's length from government, they are, in fact, rarely free of political influence. Given the many non-pecuniary goals governments expect forests to fulfil, political interference is likely to be exceptionally pervasive in the case of a forest corporation. The political consequences of allowing such a corporation to founder would be unacceptable and, if it became insolvent, a public subsidy would likely be used to ensure its survival. A Crown forest corporation would not be exposed to market competition and would have little incentive to be efficient. Rather, management may have incentives to increase the size of the corporation's administrative structure.

Community Forests
There is increasing enthusiasm in British Columbia for the establishment of community forests. Some examples of this type of arrangement already exist in the province, the earliest – the Mission Tree Farm – dating back to the 1950s.[23] Property rights in community forests may be vested in municipal or regional governments or directly in the people of the community. This latter arrangement – sometimes described as common property – does not exist in British Columbia but is common in Scandinavia.

The rationale upon which the concept of community forests is predicated is that those who are most directly affected by forests should be given responsibility for their management. It is reasoned that, if people in the direct vicinity of forests reap the benefits and bear the costs of their use, then community values become internal to the management process and optimum management from a community perspective is more likely to take place. Underlying these arguments seems to be a perception that community values are closely aligned with the values of society at large. If this perception is true, then the arguments favouring community forests are compelling. However, several considerations suggest that community and broader societal values may differ considerably. For example, recent events in British Columbia and elsewhere seem to indicate that the values of rural communities may be quite different from values held by the majority of the population located in urban centres.[24] Although rural community residents are important primary users of forest resources, the majority of the population are urban dwellers who increasingly enjoy existence, or passive, use values. This divergence of values suggests that the management of forests according to community objectives may exacerbate rural/urban conflicts.

Transferring rights in public forests to communities has important distributional and political implications. Do residents of local communities have more right to benefit from adjacent forest resources than other provincial residents? Community forests might be viewed as a means of promoting regional equity. But are resource transfers, which are inflexible by their very nature even over the long term, a better means of achieving this goal than transfer payments and the allocation of provincial resource development budgets? From a political viewpoint, defining communities, and, therefore, the beneficiaries of community forest benefits may be a difficult and potentially divisive task. Also, the devolution of provincial government powers to local interests, which the establishment of community forests entails, is likely to be politically unacceptable except on a very minor scale.

If community forests are to replace, in part, existing industrial tenures, an important question to address is whether interests of local communities coincide more closely with broader social values than do the interests of industrial corporations. Insofar as community and societal values diverge, are these differences easier to redress with economic instruments than divergences between the goals of industrial corporations and society? Industrial corporations are frequently characterized, and successfully modelled, as profit maximizers. On the other hand, the literature on non-industrial forest landowners indicates that income is not a sufficient determinant of behaviour and must be supplemented with non-market values.[25] This observation suggests that non-industrial groups, including communities, may have a wider set of management objectives, reflecting more closely the diverse nature of social welfare than the more unidimensional objective of profit maximization.

Another potential benefit of community forests is that they may help alleviate problems associated with industrial concentration. Community and other non-industrial forests could provide a more diversified timber supply source, thereby increasing domestic competition for standing timber and logs. It is suggested that non-industrial forest managers, if provided with the type of rights to grow timber described earlier, would undertake more intensive silvicultural operations and achieve higher levels of productivity than their industrial counterparts. However, empirical evidence to support this hypothesis is unavailable. While it might be argued that timber growing is a labour-intensive activity which, relying on intimate knowledge of site, soil, and climatic conditions, may lend itself to efficient small-scale production, little is actually known about economies of scale in forest management and smaller-scale, non-industrial tenure holders may find themselves at a competitive disadvantage.

Finally, if timber production incentives for non-industrial forest managers, including communities, are to be created, there must be markets in which they can obtain competitive prices for their products. Creation of adequate markets may require considerable restructuring of the industrial forest sector and the relaxation of log export restrictions which, by depressing domestic timber prices, favour integrated forest products manufacturers but discourage the production and sale of stumpage and logs.

Towards a Portfolio of Property Rights

The increasing scope and complexity of public objectives for the management of forest resources in British Columbia are resulting in more serious and pervasive market and government failures. The reaction of public policymakers to this situation is the use of increasingly intrusive command and control policies which are accompanied by high transactions costs, have eroded rights granted under the province's Crown forest tenure system, may encourage socially undesirable behaviour on the part of private forest users, and are frequently inflexible with respect to specific locational conditions. There is some scope to improve this situation and achieve the public purpose more efficiently by taking advantage of the economic incentives that can be provided through the redefinition and reallocation of property rights to forest resources.

The Crown tenure system, designed to attract capital to liquidate a stock of old-growth timber and establish an efficient timber-processing industry, is ill-equipped to meet today's challenges. Rights granted under the tenure system are limited to timber harvesting and are concentrated in the hands of a relatively small number of integrated forest products manufacturing companies whose management objectives are to produce the optimum timber supply profile for their manufacturing plants at minimum cost.

In redesigning British Columbia's forest policy, it must be recognized that one form of tenure will not serve today's varied and frequently conflicting public objectives. A system of diverse but complementary tenure arrangements is necessary. The form a particular tenure should take, and to whom it should be granted, will depend on the mix of values the land in question is expected to produce.

Property rights can be used to provide market-based, economic incentives for timber production and harvesting and, to a more limited extent, for the production of certain non-timber products. Regulations will still be necessary to protect the public's broader interests from market failures, but these regulations may be reduced by means of sharecropping agreements. In some circumstances, social and private values may be more closely aligned by granting forest property rights to communities and other groups and individuals. In any case, a more diverse population of forest tenure holders may be desirable for the resilience provided through the diversification of management objectives, organizations, and techniques. If the prevalence of non-market values dictates that public objectives cannot be achieved without resorting to sweeping command and control measures, a strong case can be made that efficiency is best served by direct public management of the forest resource. This point is already recognized in the case of areas zoned for wilderness values or environmental protection, but it should also be considered for multiple use situations where the production of non-market benefits requires major constraints on commercial timber production practices. Clayoquot is a good example of such a situation.

Choosing the most appropriate set of policy instruments to manage each area of land according to its physical capabilities and planned product mix clearly poses complex administrative problems. The costs of initiating and administering such procedures would likely negate all the benefits to be gained. These problems could be alleviated through land use zoning, which, in an area as large and spatially diverse as British Columbia, is a necessary prerequisite to efficient forest land management. Lands would be zoned according to the product or product mix for which it is planned to manage them. Then, lands producing similar types of values would be assigned similar property rights which would be subject to comparable attenuations. Such a process would create a mosaic of policy solutions. Instead of the present situation where, over large areas, uniform tenure arrangements are used to regulate private behaviour, combinations of economic instruments would create a diverse portfolio of property rights designed to efficiently produce varying combinations of forest goods and services.[26]

Notes
1 Boyd and Hyde (1989) refer to this issue as the problem of interdependent markets.
2 Uhler and Morrison (1986), Luckert (1992), van Kooten (1994), and Haley (1996) have estimated the costs of various attenuations on Crown forest tenures in British Columbia.
3 In some cases, property rights to resources are characterized as "open access," implying a complete lack of exclusion. One interpretation of this situation is "Everybody's property is nobody's property."

4 A right that is challenged by a faction within society and/or is nor offered complete protection is weaker and of less value – all other things being equal – than a right that is fully
sanctioned by society and fully enforceable.

5 The potential for improving regulatory instruments is discussed in a later section of this
chapter.

6 See Cohen and Radnoff, infra Ch. 13. For a review of compensation issues associated with
takings of private rights to public natural resources, see Schwindt and Globerman (1996).

7 R.S.B.C. 1979, c. 272.

8 S.B.C. 1995, c. 4. See Cook, infra Ch. 9.

9 Claims that the provision of employment is a positive externality of timber growing must
be approached cautiously. If new jobs are taken by previously employed labour, there may
be no new wealth created but simply a transfer of wealth from one region to another or
from one sector to another.

10 There may be positive externalities in ensuring a supply of timber to future generations
over and above that provided through the marketplace. However, these questions can be
categorized under the issue of intergenerational equity and will be dealt with in a later
section of this chapter.

11 A notable exception in British Columbia is the case of Tree Farm Licence No. 35. Here,
Weyerhaeuser Canada Ltd. was awarded an increase of almost 60 percent in its allowable
cut on the condition that the company maintain an approved reforestation and stand
management program.

12 S.B.C. 1994, c. 3.

13 In 1992, the Commission on Resources and Environment was mandated by the BC legislature under the *Commission on Resources and the Environment Act*, S.B.C. 1992, c. 34. It
aimed to "develop, implement and monitor regional planning processes."

14 Implementation of the Forest Practices Code is significantly reducing allowable annual
harvests. A 1996 analysis by the BC Ministry of Forests estimated these reductions at 6
percent of provincial allowable timber harvests in the short term and as high as 7.7 percent over the longer term (BC Ministry of Forests, 1996).

15 The Timber Supply Review, authorized under the *Forest Act*, s. 7, updated timber supply
assessments for all Timber Supply Areas and Tree Farm Licences. It commenced in 1991
and was completed in 1996. While, for the whole province, the process resulted in a
downward adjustment in the allowable annual cut (AAC) of only 0.5 percent, for certain
regions it was much greater. For example, the Vancouver Forest Region experienced a 6.3
percent reduction, and the Nelson Forest Region a 9.3 percent reduction.

16 To a certain extent, public concern has internalized some non-timber values, in the sense
that firms have an incentive to produce or protect them in order to enjoy the benefits of good
public relations and avoid the possibility of boycotts of their products in export markets.

17 Limited non-timber rights have been granted to private individuals and firms, usually
outside the timber-processing sector.

18 R.S.B.C. 1979, c. 140.

19 Such products are popularly referred to as "value-added products."

20 This reallocation may be thought of as arising out of Coasian-type negotiations (Coase
1960). Luckert (1993) reviews several reasons such transactions may, or may not, allocate
resources to their highest uses.

21 Indeed, Luckert (1994) has shown that even if costs are not shared between the tenure
holder and the Crown, allowing tenure holders rights to even small proportions of the
future benefits their expenditures create can greatly improve efficiency.

22 Under the BC Forest Resources Commission's proposals, a Crown corporation would manage those lands zoned primarily for timber production while a public agency would manage
multiple use areas.

23 For a review of community forests in Canada, see Duinker, Matakala, and Bouthillier (1994).

24 Consider, for example, the controversy over Clayoquot Sound in British Columbia or the
rural/urban splits over the spotted owl issue in the northwestern United States.

25 For a review of literature on non-industrial forest land-owner behaviour, see Jamnick and
Clements (1988).

26 Evidence suggests that zoning would not only reduce the transactions costs associated

with the public administration of a diverse forest resource for multiple products, but also enhance the production of many of the values concerned. For example, research in the Revelstoke Forest District (Sahajananthan, Haley, and Nelson 1996) showed that by zoning for intensive timber production, the sustainable timber yield of the whole unit, currently managed uniformly for multiple products, could be produced on only 34 percent of the area. If similar results are obtainable elsewhere in the province, the implications for the ability of forest resources to meet social demands are profound.

References
Alchian, A.A., and H. Demsetz. 1973. "The property rights paradigm." *Journal of Economic History* 33(1): 16-27
Barzel, Y. 1989. *Economic Analysis of Property Rights*. Cambridge: Cambridge University Press
Blume, L., D. Rubinfield, and P. Shapiro. 1984. "The taking of land: When should compensation be paid?" *Quarterly Journal of Economics* 100: 71-92
Boyd, R.G., and W.F. Hyde. 1989. *Forest Sector Intervention: The Impacts of Regulation on Social Welfare*. Ames: Iowa State University Press
British Columbia Forest Resources Commission. 1991. *The Future of Our Forests*. Victoria: Queen's Printer
British Columbia Ministry of Forests. 1996. *Forest Practices Code: Timber Supply Analysis*. Victoria: BC Ministry of Forests and BC Ministry of Environment, Lands and Parks
Bromley, D. 1991. *Environment and Economy: Property Rights and Public Policy*. Cambridge MA: Blackwell
Cheung, S.N. 1969. *Theory of Share Tenancy*. Chicago: University of Chicago Press
Clawson, M. 1976. *The Economics of National Forest Management*. Working Paper EN-6. Baltimore: Johns Hopkins Press
Coase, R.H. 1960. "The problem of social cost." *Journal of Law and Economics* 100: 71-92
Dahlman, C.J. 1980. *The Open Field System and Beyond*. Cambridge: Cambridge University Press
Duinker, P.N., P.W. Matakala, and L. Bouthillier. 1994. "Community forests in Canada: An overview." *Forestry Chronicle* 70(6): 711-20
Epstein, R.A. 1985. *Takings: Private Property and the Power of Eminent Domain*. Cambridge: Harvard University Press
Forest Renewal BC. 1996. *B.C.'s Value Added Strategy*. Victoria: Ministry of Forests
Forestry Canada. 1993. *Selected Forestry Statistics Canada, 1992*. Policy and Economics Directorate, Information Report E-X-46. Ottawa: Forestry Canada
Furubotn, E., and S. Pejovich. 1972. "Property rights and economic theory: A survey of recent literature." *Journal of Economic Literature* 10: 1137-62
Haley, D. 1985. "The forest tenure system as a constraint on efficient timber management: Problems and solutions." *Canadian Journal of Public Policy* XI: 315-20
–. 1996. "Paying the piper: The cost of the British Columbia *Forest Practices Code*." Paper prepared for the conference Working with the Forest Practices Code, Vancouver
–, and M.K. Luckert. 1990. *Forest Tenures in Canada – A Framework for Policy Analysis*. Information Report E-X-43. Ottawa: Forestry Canada, Economics Directorate
–, and M.K. Luckert. 1995. "Policy Instruments for Sustainable Development in the British Columbia Forest Sector." In *Managing Natural Resources in British Columbia: Markets, Regulations and Sustainable Development,* ed. Anthony Scott, John Robinson, and David Cohen, 54-79. Vancouver: UBC Press
Hallowell, A.I. 1943. "The nature and functions of property as a social institution." *Journal of Legal and Political Sociology* 1: 115-38
Innes, R. 1995. "An essay on takings: Concepts and issues." *Choices* 1: 4-7, 42-44
Jamnick, M.S., and S.E. Clements. 1988. *Modeling NIPF Wood Supply in the Maritimes: Description and Analysis*. Fredericton: Canadian Forestry Service
KPMG Ltd. 1997. *The Financial State of the Forest Industry and Delivered Wood Cost*. Prepared for the Economics and Trade Branch, British Columbia Ministry of Forests
Luckert, M.K. 1991. "The perceived security of institutional investment environments of some of British Columbia's forest tenures." *Canadian Journal of Forest Research* 21: 318-25
–. 1992. "Changing Resource Values and the Evolution of Property Rights: The Case of

Wilderness." In *Growing Demands on a Shrinking Heritage: Managing Resource Use Conflicts,* ed. M. Ross and J.O. Saunders, 85-89. Calgary: Canadian Institute of Resources Law

–. 1993. "Property rights for changing forest values: A study of mixed wood management in Canada." *Canadian Journal of Forest Resources* 23(4): 688-99

–. 1994. *Efficiency Implications of Sivicultural Expenditures from Separating Ownership from Management on Canadian Forest Lands.* Working Paper 208. Vancouver: Forest Economics and Policy Analysis Research Unit, UBC

–, and D. Haley. 1989. *Funding Mechanisms for Silviculture on Crown Land: Status, Problems and Recommendations for British Columbia.* Working Paper 131. Vancouver: Forest Economics and Policy Analysis Project, UBC

–, and D. Haley. 1990. "The implications of various funding mechanisms for privately managed public forest lands in Canada." *New Forests* January: 1-12

–, and D. Haley. 1993. "Canadian forest tenures and the silvicultural investment behavior of rational firms." *Canadian Journal of Forest Research* 23: 1060-4

–, and D. Haley. 1994. "The Allowable Cut Effect (ACE) as a policy instrument in Canadian forestry." *Canadian Journal of Forest Research* 25: 1821-9.

MacPherson, C.B., ed. 1978. *Property: Mainstream and Critical Positions.* Toronto: University of Toronto Press

Markandya, A., and D.W. Pearce. 1991. "Development, the environment and the social rate of discount." *World Bank Research Observer* 6: 137-52

Milgrom, P., and J. Roberts. 1992. *Economics, Organization and Management.* Englewood Cliffs, NJ: Prentice-Hall

North, D.C. 1991. "Institutions." *Journal Of Economic Perspectives* 5: 97-112

Pearse, P.H. 1976. *Timber Rights and Forest Policy in British Columbia.* Report of the Royal Commission on Forest Resources. Victoria: Queen's Printer

–. 1988. "Property rights and the development of natural resource policies in Canada." *Canadian Journal of Public Policy* XIV(1): 307-20

Province of British Columbia. 1993. *A Protected Areas Strategy for British Columbia.* Victoria: Queen's Printer

Sahajananthan, S., D. Haley, and J. Nelson. 1996. *Planning for Sustainability of Forests in British Columbia Through Land Use Zonation.* Working Paper 96.08. Victoria: Canadian Forest Service, Natural Resources Canada

Schwindt, R., and S. Globerman. 1996. "Takings of private rights to public natural resources: A policy analysis." *Canadian Journal of Public Policy* XXII(3): 205-24

Scott, A.D. 1983. "Property rights and property wrongs." *Canadian Journal of Economics* 16(4): 555-73

–, and J. Johnson. 1983. *Property Rights: Developing the Characteristics of Interests in Natural Resources.* Resource Paper 88. Vancouver: Department of Economics, UBC

Sloan, G.M. 1945. *Report of the Commissioner Relating to the Forest Resources of British Columbia.* Victoria: King's Printer

Uhler, R.S., and P.D. Morrison. 1986. *Utilization Standards and Economic Efficiency in British Columbia Forests.* Information Report 86-1. Vancouver: Forest Economics and Policy Analysis Project, UBC

van Kooten, G.C. 1994. *Cost Benefit Analysis of B.C.'s Proposed Forest Practices Code.* Victoria: BC Council of Forest Industries

Wolf, C., Jr. 1988. *Markets or Governments: Choosing Between Imperfect Alternatives.* Cambridge: MIT Press

Zhang, D. 1994. *Implications of Tenure for Forest Land Value and Management in British Columbia.* PhD dissertation. UBC

7
Living Communities in a Living Forest: Towards an Ecosystem-Based Structure of Local Tenure and Management
Michael M'Gonigle

Introduction
Whether in a policy thinktank in Washington DC or a First Nations' forum in northern BC, one of the most potent new buzzwords in public policy today is "community." From education to resource development, from health services to aboriginal rights, community-based alternatives are widely espoused for their potential to create jobs, sustain local environments, and reinvigorate democratic participation. As discussed in my earlier chapter in this book, there are fundamental reasons why this new emphasis on community has both theoretical and practical merit, and this fact has led to a growing literature on community-based alternatives in forestry worldwide.[1] In this chapter, I will examine a range of innovative but practical tenure alternatives available to those concerned to effect a sustainable forest strategy, with particular emphasis on the potential of, and obstacles to, a community-based approach.

The rationales for taking a community-based approach are numerous. As noted earlier, one of the primary rationales is theoretical; that is, it is based in an understanding of the structural character of sustainable systems. This understanding emphasizes the circular, as compared to linear, nature of their "metabolic" processes. To be sustainable, both environmental and human systems

> must maintain themselves, living on the stock of natural and social capital with which they have been endowed, so that they can return long-term stability to the forest and long-term value to the community ...
>
> Just as healthy ecosystems recirculate nutrients and water locally, so to a healthy community is internally dynamic, recirculating wealth locally, in a diversified economy that is only partly dependent on outside employers and outside markets.[2]

In addition, an analysis of the political economy of environmental erosion will point to the extractive relationship of centralist institutions over local territories, a relationship that privileges bureaucratic (including corporate) forms of organization over communal ones. This pattern characterizes the history of the forest industry in Canada, from the first large-scale cutting of the forests of Eastern Canada by a distant British Crown some 200 years ago, to the extraction of logs from the traditional lands of the Nuu-Chah-Nulth First Nations by multinational corporations today. Historically, the development of these linear, centralist patterns has been at the expense of a range of community-based values and skills of territorial self-maintenance and mutual social cooperation. To reverse the undermining of this "social capital" requires the creation of new instruments of economic and regulatory control that both create and maintain circular patterns of community action. This goal points to an explicitly structural, as compared to incremental, policy orientation.

Despite the generally adverse reactions that many state interventionists (whether of a social-democratic or environmentalist persuasion) have at the mention of deregulation or the greater reliance on market instruments, much is still to be learned from this debate. Many advocates of economic instruments may be criticized for their reliance on the theoretical advantages of their approach, as compared to their demonstrated practical successes, but their criticisms of the bluntness and inefficiency of regulatory instruments are persuasive.[3] Moreover, as a rule, the carrots of economic incentives and self-interest through some form of ownership are more persuasive than the sticks of externally imposed legal rules. Nevertheless, at least as it pertains to forestry in BC, the debate is ultimately not about the merits of employing economic incentives or about seeking economic efficiencies. Instead, if one takes sustainability (and social equity) seriously, the real debate is about finding cutting-edge, innovative ways to get the best of both the regulatory and the economic worlds in pursuit of sustainability.

To achieve this innovation, one must reconsider the nature of Crown and private property rights over the forest land base in Canada. In my earlier chapter, the concept of ecosystem-based management as a basic organizing framework for forest policy was set out. In this chapter, I will briefly canvass the existing forms of tenure in the province which all share a productivist orientation in contrast to "alternative" forms of tenures that are rooted in an ecosystem and community-based approach. Not only in British Columbia, but across Canada, such tenures are rare to the point of non-existence. Yet property is a flexible bundle of rights which, with imagination and commitment, can be combined – and recombined – in an array of options to reflect a range of objectives. Instead, a simplistic choice is often posed between privatization and public ownership, with little consideration given to new ways of combining private, public, and *communal* rights to fulfil the larger complex of social interests associated with long-term, sustainable management and production.

It is possible to characterize both existing and alternative forest tenures along a spectrum.[4] This spectrum is constructed around differing visions of what constitutes the foundations for economic sustainability. At one end is the traditional economic model that takes growth in production and its associated neoclassical calculations as the assumed points of reference. At the other end is the ecological perspective that challenges these patterns of growth as no longer physically or socially desirable but as actually erosive of future prospects – that is, they entail a net social cost. Instead, concerns for maintaining "linear" economic flows must give way to fostering "circular" patterns that maintain existing stocks of environmental and social capital. In the process of making this shift, industrial structures must be reconfigured.[5] Discussions of how to achieve economic "efficiency" do not address this systemic issue because they continue to take place within the operational context of the economic "flow" system, rather than moving beyond it.

In the present analysis, one end of the spectrum is a form of tenure in which the values of private production and profit are inherently sought to be maximized. This form corresponds most closely with a corporate tenure model. At the other end of the spectrum is a form of tenure rooted in the maintenance of ecosystem values, with the provision only for such economic production as does not undermine the integrity of the functioning ecosystem. To the extent that one defines territorial community in its fullest sense – embracing both the natural and human elements, and present and future generations of these elements – this ecosystem-based approach is naturally compatible with a variety of community tenures. In other words, corporate-based tenures are so inherently in conflict with the objectives of ecosystem-based management that one can approach the tenure issue through a spectrum of values from the corporate to the ecosystem. The closer one moves to one approach, the farther one necessarily moves from the other.

Options for a Sustainable Tenure System

Corporate Contracts
Posing such an oppositional corporate-ecosystem spectrum for characterizing different tenure arrangements reflects the fact that public corporations are driven first and foremost to manage for profits and business growth. Indeed, they are legally mandated to do so.[6] In this light, when commentators refer to the improvements that occur with increased tenure security, they are referring to improvements that generate enhanced economic returns to the company. These improvements essentially mean returns from "fibre" production, not general environmental or social benefits to

the community through improved forest ecosystem management. Even if plantation forestry may be "productive" in this narrow economic sense, it is certainly not what most people have in mind when they consider long-term sustainability. Indeed, not only does ecosystem-based management offer few, if any, economic incentives to the company, it actually embodies severe disincentives insofar as it leads to reduced cutting levels to protect other "non-economic" values. Neither is it possible (given the huge trans-action costs) nor desirable (given the mutuality and integrity of an ecosystem) to disaggregate the various features of a forest into a multiplicity of separate rights holders, as Haley and Luckert consider in Chapter 6.

On the more limited basis of timber production, the argument concerning the benefits of enhanced privatization is driven more by theoretical considerations than it is by practical evidence. In British Columbia, the growth rate of timber is generally so slow that re-investing profits in timber productivity (let alone in forest health) would generate returns below the level of the discount rate. If profits re-invested in the bank would outperform those put back in the ground, then, without public subsidies or regulations, corporations cannot justify such re-investments on economic grounds.[7] Instead, the rational economic strategy is one of old-growth liquidation, with re-investments limited to local facilities that are coordinated ("rationalized") with this planned liquidation, while other investments increasingly move to more productive climates that will produce new fibre supplies at a profitable rate.[8]

In this light, a heavy burden of proof exists for the proposition that further privatization of corporate tenures will lead to enhanced economic, let alone environmental, benefits. In British Columbia, examples of destructive forest practices on private land (such as in the San Juan River area) do not lend support to this contention.[9] A systematic analysis of the practical effects of different forms of tenure on a range of ecosystem (and not just silvicultural) values is clearly necessary. However, as one moves the tenure being considered away from the purely corporate model, some productivist pressures decrease. In this regard, privately held corporations (i.e., not publicly traded and thus not concerned with profit-driven share values) have greater leeway to incorporate ecoforestry principles. Examples of family-held corporations that have embraced ecoforestry, such as Collins Pine, exist in the United States.[10] The BC forest company widely regarded as most open to innovative practices is also a family-held company, Lignum Ltd. Unlike Collins Pine, however, Lignum's ability to adapt is constrained by its established high-volume, commodity-product orientation.[11]

It is (again) theoretically possible for the Ministry of Forests to redesign corporate tenure contracts to mandate a greater ecoforestry orientation. Tenure contracts are, however, largely procedural documents that stipulate the required planning processes, without much substantive content. In this regard, the opportunity exists to mandate a more ecosystem-based planning procedure within corporate tenure contracts. One such opportunity was rejected by the government for Tree Forest Licence No. 44 in Clayoquot Sound where the infrastructure for such planning already exists.[12] An interesting feature of the proposed Clayoquot Sound amendment was a provision that would have given the local planning body an entitlement to a proportion of the locally generated stumpage sufficient to cover the costs of planning. This provision was not well received by the Ministry of Forests, because it would have reduced its monopoly on public planning, and could have established a precedent for an innovative, community-based economic entitlement. A later amendment to the TFL would mandate the licensee to implement the report of the Scientific Panel by "planning for sustainable ecosystem management [including] the adoption of an ecosystem based approach to planning to sustain productivity and natural diversity."[13] This obligation is potentially far-reaching, especially as the new licence also commits the licensee to work with the local management body in the area, the Central Region Board (see below), which has demonstrated a strong commitment to a community-based ecosystem approach. The proposed new licence also mandates the licensee to support locally developed initiatives to increase community economic benefits and stability.[14] Again, however, this proposal has not yet been accepted. Moreover, how these new commitments will work in practice will hinge on the assertiveness of the Central Region Board, and on extensive financial support from the province.

In short, while reforms to corporate tenures are possible, corporatist and bureaucratic imperatives render the inclusion of ecosystem-based values difficult, especially for less high-profile areas than Clayoquot Sound. Within the present context, there is no reliable basis to conclude that privatization and deregulation will lead to greater sustainability. Indeed, to the extent that broad ecological goals constrain purely economic returns, the likelihood is that such privatization will harm broader public values, even if it might marginally improve silvicultural investments in plantation forestry. Nevertheless, as the Clayoquot Sound example demonstrates, if enough social pressure can be brought to bear (and this is admittedly in limited supply), innovation in corporate tenure arrangements is possible.

Tenures Rooted in Individual Property Rights
Moving along the corporate-ecosystem spectrum, the next stop is the fee simple individual tenure which confers virtually unlimited tree-cutting powers, but without the same productivist pressures on corporate forests. In the province, there are over 20,000 private forest landowners, many of which are fee simple corporate owners, but the vast majority of which are family-owned woodlands.[15] Most of these woodlands in BC are not yet regulated under the Forest Practices Code.[16] Only when forest landowners seek to achieve tax benefits for their private woodlands do the spe-

cial obligations for "managed forest lands" kick in under the Assessment Act.[17]

The obligations imposed on such private landowners have historically been minimal, requiring largely that they file with the BC Assessment Authority a management plan "for the production and harvesting of timber."[18] As a result, private ownership of such "managed forest land" has allowed for the worst – and the best – of practices, because individuals are, on the one hand, less subject to Ministry oversight that might protect the land, *and*, on the other hand, free of the Ministry's production quotas that often force a licensee to strip the ecosystem of its standing timber.[19] At the same time, under the Act, tax benefits from these managed forest lands accrue only for activities that improve timber production, not for those that enhance ecosystem health or support habitat restoration. The latter approach would require re-orientation as part of a larger reform of economic instruments to ensure that private woodlots can be managed on an ecosystem basis.[20]

The focus of woodlot policy in British Columbia has, however, been neither on fee simple private land nor on privately held managed forest lands. Instead, the major thrust of policy has been on *public* woodlots located on Crown lands. Even here, the role accorded to woodlots is minimal. Compared with the some 300,000 (private) woodlots that provide the backbone of timber production in Scandinavia, and comparatively high numbers in the Maritime provinces in Canada, there were only 480 public woodlots in BC in 1995, contributing less than 1 percent of the provincial allowable annual cut. Public woodlots are based on time-limited, renewable, area-based agreements, with the tenure holder responsible for a host of management activities under the close direction of the Ministry of Forests.

In looking to the future, experience with the existing woodlot system points to four important lessons. First, the issue is not to choose between private and public property, but to devise a mixture of property rights and obligations that will ensure ecologically sustainable forest management on all woodlot tenures. Freedom from regulation on private woodlands has led to ecological innovation – and to rapacious clearcutting. On the other hand, close regulation of public woodlots has not produced progressive forestry. Quite the contrary, it has led to widespread criticism from woodlot licensees who charge that they have been explicitly prohibited from practising ecoforestry because of the demands of the Ministry for high-volume timber production.[21] For all manner of such tenures, the first issue is thus the choice of objectives between industrial production narrowly conceived and community production broadly conceived. Only with the objective clearly defined can one then undertake the design of specific systems to give effect to that objective (something beyond the scope of this chapter).[22]

Second, with the objective clear, property rights can be unbundled and repackaged in myriad ways. For example, in Scandinavia where private woodlots predominate, extensive regulations ensure the protection of public environmental values and uses on private forest lands. Even though privately owned, "these forests can be called community forests because they provide local citizens with forest products (e.g., mushrooms and berries) and services (e.g., recreation). The term community forest in this case does not mean that the forest is owned or managed by the community."[23] As one basis for community forestry, it is at least theoretically possible in British Columbia to achieve such a mix of values by creating a diversity of woodlot tenures ranging from regulated private lands, to conditional licences granted in perpetuity on public lands, to public forest tenures with exclusive long-term rights of individual forest stewardship.

Third, a woodlot system cannot provide a basis for community forestry without a shift in scale, and at several levels. Individual woodlots are in many cases now so small that they can only be run as hobby lots and at a loss. In any sustainable system, the individual sizes need to be large enough to be financially viable for the licensee. As important as individual size is achieving a sufficient *number* and *proximity* of woodlots in the aggregate at both regional and provincial levels. Only with sufficient scale might licensees organize effectively to create the necessary infrastructure – extension services, professional consultants, equipment exchange, cooperative marketing – that will allow them to achieve a functional woodlot economy and culture across the landscape.

Fourth, the demand for woodlots is certainly there today, with dozens of applicants on file in every forest district. But this demand meets a bottleneck: over-regulation. Although the woodlot licensee is often a single individual, he or she must still undertake the same timber cruising and layout planning for development, and submit the same detailed management and logging plans, as a major corporate tenure holder. And, despite the small amount of timber involved, the Ministry must process these plans at great public cost. This hugely burdensome system meets the goals of neither the individual licensee nor the government official. For the Ministry, it is a stretch to undertake even the expansion in the program from 500 to perhaps 1,000 woodlots that is presently being developed. One woodlot advocate, Ken Drushka, in fact alleges that the Ministry of Forests has intentionally sabotaged the program with bureaucratic red tape – has "dragged its feet" and "dug in its heels" so that the allocation of new licences is but "a trickle."[24] Indeed, the paperwork and inappropriate industrial standards of the Forest Practices Code have so weighed down existing licensees that many of them are actually forfeiting their licences rather than comply. If a sustainable economy of scale could best be achieved with several thousand woodlots throughout the province, this change would first require a reform of the regulatory system in which they would be situated.

In addition, therefore, to the need for a change in management objectives and for greater innovation in the treatment of property rights, a successful community woodlot system demands a new approach to regulation, one that escapes the pitfalls of too much individual autonomy or too much state intervention. As we shall discuss below, the answer here is to turn neither to privatization nor to deregulation, but to new forms of associational and community-based management. This shift in the regulatory system is a prerequisite to any hope of realizing the potential that this form of tenure holds for community forestry.

Tenures Rooted in the Community
Next along the corporate-ecosystem spectrum is the designated community tenure. In combination with a large, well-designed woodlot system, community tenure is the most promising source of tenure reform. Traditional community forests have a demonstrated history of successful management throughout the world. As Lynch and Talbot write about such forests,

> contrary to enduring stereotypes, sustainable community-based management systems are operated neither by ecological "noble savages" living in symbiotic harmony with nature, nor by self-centred exploiters seeking to maximize short-term gain. Like participants in other sustainable systems, most successful community-based managers are rational strategic-minded individuals who assess existing conditions and act in their own best interests. The more they depend on the surrounding resource base, the more incentive they have to protect it.
>
> ... Community-based tenurial rights are not the equivalent of "open access" regimes. They include individual and group rights, and typically derive from long-term relationships established between local peoples and the natural resources that sustain them ... Community-based rights often derive from the precept that the present generation holds the natural resource base, including forests, in trust for future generations. The privileges of the individual are thus generally subservient to the rights of the greater community.[25]

In their review of "community forests" in Canada, Duinker, Matakala, Chege, and Bouthillier refer broadly to a "community forest" as "a tree-dominated ecosystem managed for multiple community values and benefits by the community."[26] While this broad definition is useful, a "true" community forest must have at least three characteristics to merit the name. First, it must truly include and respect the needs and *integrity of the whole "community,"* which means both the natural environment of which the human community is a part, and future generations of the existing population. As discussed, this "true" community forest is inherently situated within an ecosystemic perspective. Second, although the central state should continue to exercise an important facilitative role (see below), there must be a high degree of actual *local control of political decision-making*. Third, the local community must both derive the benefits and pay the costs of the exercise of these decisions. That is, there must be *local control of economic institutions* of production and marketing.

Given the larger constraints to creating such forests in British Columbia (in particular the existing tenure system, and the lack of bureaucratic support for such innovation), we will not attempt to set out a detailed structure for a community forest designation. If one were free to allocate and reallocate property rights in new ways, this exercise would not be difficult, although the specific components would vary from place to place. Instead, we will briefly examine the degree to which the above three ingredients exist in current tenure systems at present; that is, a commitment to the practice of ecoforestry, a high degree of local control, and an operational structure that ensures economic and social benefits are realized locally. As these ingredients are the basis of any transition to sustainable forestry, they are critical criteria to bear in mind in evaluating any existing and proposed future community forest tenures.

In British Columbia, a small number of what many people call "community forests" exist, but none of these meets the criteria set out above. For example, two often-cited examples in Mission and Revelstoke are actually Tree Farm Licences which, although held by communities, are subject to the productivist demands and industrial standards of TFLs everywhere in the province. In Mission, for example, although the municipality's professional foresters might manifest some concern for a range of values (by utilizing smaller clearcuts, hiring local contractors, and allowing a recreational trail system throughout the land base), there is no separate public oversight, and they have been criticized for running the TFL with an "explicitly commercial orientation, [not as] a community resource management partnership."[27] This point applies as well to the Revelstoke tree farm where a limited land base is being used to support an industrial-level cut. This new tree farm licence has been very profitable, is run through a municipal corporation that includes broader community representation, and has been useful in supporting local employment and supplying local mills. It is, however, still largely run in the industrial mode, its primary objective being to maintain a steady flow of revenues.[28]

Likely the best example is the North Cowichan Municipal Forest where the land base is owned outright by the municipality, and is thus exempt from the pressures of the tree farm licence.[29] The main objective for the forest is to provide jobs and income from forestry for the community while offering recreational programs and facilities. Clearcuts are the primary forest harvest method for the production of sawlogs and firewood products, although variable "patch cuts" are used to maintain biodiversity, and large, coarse debris is left in the clearcut areas for soil protection. Horse logging is even employed as one of the logging methods. Three ecological reserves have been set aside, and the scope exists to incorporate other community values and needs.[30] This community forest does, however, primarily serve a productivist function. An AAC calculation determines cut levels (set at 20,000 cubic metres per year in a new Five-Year Development Plan in 1996), and the municipality's Forestry Department's stated objective is to practise intensive silviculture with the goal of obtaining increased yields from well-managed second- and third-growth forests. In 1995, the forest produced record net profits to the community of over $1.2 million. In that year, intensive silviculture (cone collection, tree planting, brushing and weeding, juvenile spacing, pruning, fertilization, and slashburning) created 2,768 person days of employment.[31]

In British Columbia, therefore, while there are interesting examples of community-based forestry operations, there are neither specially designated community tenures, nor community forests where non-industrial forestry is either mandated or being extensively practised. In recent years, under provincial funding for community development (through Forest Renewal BC), several municipalities have looked at developing their own forestry base, and have undertaken imaginative investigations in the process. As above, these studies invariably mention three "essential features [that] define a community forest; the community makes the management decisions; the community benefits; and the forest is managed for multiple values."[32] Inevitably, however, the absence of any new tenure possibilities has led these investigations into the choice between a Tree Farm Licence (on the Mission or Revelstoke model) and an even less desirable volume-based Forest Licence. One of the most innovative proposals is the Malcolm Island Community Forest Agreement that would replace the traditional AAC with a set of "performance standards for overall ecosystem health and sustainability," and would potentially use a community cooperative to manage the forest. The feasibility study notes, however, "the very real obstacles to establishing a community forest." It proposes, therefore, that the community "get organized, exercise leadership, and work towards establishing a community forest," in particular, by collaborating "with other communities to work for reform of the tenure system to allow for a community forest tenure under the *Forest Act*."[33] In response to widespread pressure, the provincial government announced the creation of an advisory committee "to recommend new tenure models that will give communities a greater role in forest management." The eventual goal of the government is to establish three pilot projects in those few areas where excess timber exists; it is not to challenge existing timber tenures.[34]

In Canada, community tenures are also rare, and none exists that combines these three components. In Ontario, for example, a pilot program of community forests was established in 1991, but it was premised on the need to "respect existing resource commitments on the land base."[35]

Although three projects were initiated (Geraldton, Kapuskasing, and Elk Lake/Temagami), no tenure transfers were involved in any project, relegating the participants to such marginal activities as tree-planting, pre-commercial thinning, and some training. For the Wendaban Stewardship Authority (governing the Temagami area), the intent was to grant authority to "plan, decide, implement, enforce, regulate and monitor all uses of and activities on the land," but that authority has not materialized.[36] Given these constraints, experience with the program produced great frustration that new ways of doing forestry could not be undertaken. One commentator noted that, with decision-making power lying beyond the community boundaries, the program offered little more than enhanced input into the status quo system of timber management.[37]

Interestingly, a little-considered statute has long been in place in Ontario that provides a general guide as to how such tenures could be created, the 1946 Conservation Authorities Act.[38] This Act authorizes local areas to enter into agreements with the province to create conservation authorities, which have the capacity to hold land and other property. Although not community forests, these authorities actually own a large amount of land, and there is general recognition of community conservation interests in the legislation; indeed, the authorities are organized on an ecological basis: the watershed. Six principles were considered in establishing these authorities: use of the watershed as the management unit; reliance on local initiative; establishment of provincial-municipal partnerships; integration of environment and economy; a comprehensive approach; and coordination and cooperation.[39] By 1995, there were thirty-eight conservation authorities in Ontario, the jurisdictions of which covered over 100,000 square kilometres. Although there is development potential in conservation authorities, these represented the largest blocks of relatively protected green space (after provincial parks) in the province.

A conservation authority represents the municipalities that constitute it, with the municipal councils and the province appointing representatives to the authority. In 1946, for example, the Town of Port Hope and five other townships entered into an agreement with the province to create the "Ganaraska River Conservation Authority," one of several in the province. The authority bought land for reforestation; the province plants, maintains, harvests (by selective logging), and takes profits from fast-growing red pine, mixed maple, and other plantations, using local mills. Now extending west to adjacent watersheds and including some 15,000 acres, the "Ganaraska Region Conservation Authority" has created recreation areas, protected recreational interests, developed flood control, halted erosion and restored parts of the watertable. By 2002 AD, when it will have recovered its investment, the conservation authority has the choice to continue in partnership with the province sharing revenue and expenses, sell out to the province at net investment, or buy out the province at its net investment.[40] Considering when they were created, the "conservation" component of these authorities was set as only one goal among many conflicting objectives (including development), and has often not been given a high priority in practice. Nevertheless, the model offers an interesting precedent of integrated planning, local "tenure," revenue re-investment, citizen involvement, and provincial-municipal cooperation including cost-sharing and joint management. All these practices could be adapted to British Columbia's forested Crown lands. However, despite their success for over four decades, under the present Conservative government in Ontario, provincial support for these authorities is being withdrawn, and their futures are very much in doubt.[41]

Perhaps the closest experimentation with non-traditional community forests has occurred in Quebec where a system of forest cooperatives exists.[42] Pressure there to increase the number of locally managed forests comes from a variety of municipal governments in Quebec.[43] In particular, as a result of a 1993 amendment to the provincial Forestry Act, municipalities became entitled to assume control of local tenures. One innovative vehicle for this, suggested by Bouthillier and Dionne, is the "inhabited forest."[44] This designation could include everything from private to public woodlots, from municipally based to corporate-controlled forests, provided the forest was managed subject to a "territorial contract that balanced economic, ecological and community objectives." This concept is, however, still largely at the developmental stage. As Duinker et al. conclude regarding Canada generally, "community property regimes are almost absent, are still 'table issues', or are only beginning to emerge. Institutional and policy reforms will thus be required to encourage alternative community-based tenure strategies that confer community property rights and assure local control."[45]

Applicable examples can be found in the United States where the use of the land trust is an especially popular instrument for achieving social objectives – usually land preservation – on both private and public lands. For example, organizations such as the New Mexico-based Forest Trust and the Tennessee-based Woodland Trust, act as trustees for large tracts of forest land on which ecoforestry is required, the proceeds from which are channelled into the community. In addition to overseeing land stewardship activities, the Forest Trust runs a wood products brokerage to assist in marketing forest products, and has assisted in the creation of "flexible manufacturing networks" across the state to coordinate value-added production.[46] Along these lines, a non-profit organization in British Columbia, Turtle Island Earth Stewards, has proposed its own land trust model that allows private owners to place their forest land in a land stewardship trust to be managed by the community according to ecosystem principles.[47]

Finally, many precedents for community tenures exist internationally, although the specific conditions of other countries limit the applicability of their precise form of tenure to British Columbia. In Japan, for example, a richly textured community forestry system has existed for over a century, rooted in a complex of rural cooperatives and village-based management that demands strict adherence to the cooperative ethos among members.[48] In India, the practice of "social forestry" offers lessons in providing for rural community needs, although the system is primarily oriented to meeting social, not market needs.[49] Other examples exist in a variety of developing countries such as Bolivia, Mexico, Peru, the Philippines, and Thailand,[50] but the conditions of each country and forms of tenure vary greatly, so that the specific policy relevance of their projects to British Columbia is complex.

Nevertheless, a few common lessons stand out. One is the key criterion for the success of community-based sustainable forest management, a lesson that is applicable to British Columbia: the need to shift power from central to territorial-based communities. In Mexico, for example, the movement to utilize forests as a large-scale developmental resource for the benefit of the communities themselves necessarily had to utilize strategies that

> were essentially political in nature since local-level users first needed to regain control of their forest resources [from commercial concessionaires]. This was accomplished primarily by holding meetings and demonstrations, and committing various acts of civil disobedience, such as refusing to turn harvested timber over to the concessionaires.[51]

Another lesson, evident throughout the Third World, is the existence of "demonstrable connections between community-based tenurial rights and effective resource management ... Anecdotal and historical evidence suggests that in many instances national resources are best managed locally."[52] Finally, however, perhaps the most common lesson of all is that

> few national governments in developing countries recognize forest-dependent peoples' locally-based natural resource rights for their contributions to sustainable forest management. Nor do most countries give local resource users any meaningful say in decisions on national forest laws and policies. Instead, many adhere to colonially inspired and centralized systems of forest land ownership that legally disenfranchises many rural citizens.
>
> National legal systems that benefit political and economic elites ... undermine local incentives for sustainable development and contribute to the still-accelerating rate of tropical deforestation.[53]

With these precedents, it is possible to envision a range of vehicles for creating a province-wide public program of community forest tenures on Crown lands. Indeed, envisioning a workable design for a community tenure system is not difficult; putting one into practice is. As with an expanded woodlot program, the critical need is to free up an adequate land base to provide for a community forest system of sufficient size that it can be both ecologically self-sustaining and economically community-sustaining. As Duinker et al. note, any such land base should not just include second-growth forests for silvicultural activities but should include areas with "a balanced forest age-class structure, substantial amounts of good quality timber, and good site quality."[54] To facilitate local management, these authors also envision a "transfer of revenue-gathering power to the local level [so as to] put some community forest management agencies in a strong position to finance the needed transformation."[55] The best vehicle for doing this would be one that shifted authority to the local level, but under strong conditions for sustainability, democracy, and equity set by the provincial government.

As seems to be envisioned by the government with its pilot projects, a new community forest tenure could be created under the existing Forest Act without affecting any established interests on the ground. Ideally, however, a new legislative vehicle would go beyond the provision of merely a new form of tenure in such a way as to change the nature of forest practices, management, and production to achieve both environmental and community sustainability. One proposal would do this for a larger, regional area of Crown land, by redistributing authority over Crown land to local level governments under a cooperatively managed trust.[56] With the appropriate legislative design (and supportive economic mechanisms), a Community Forest Trust Act could embed the values discussed in this chapter in a long-term, cooperative fashion and provide a vehicle for a profound, but evolutionary, transition to a truly sustainable forest economy. Giving communities the opportunity to opt into the Act, the legislation would set out a comprehensive process for local tenure replacement, community-based management, and economic development. This legislation would be a landmark achievement for a social democratic government confronting the realities of global ecological change.

Unfortunately, despite the potential merits, such a large-scale regional transition is not on the governmental agenda. Quite the contrary, community tenures are expected to fit within the existing tenure framework. In contrast, in late October 1996, the provincial government made a watershed decision by overriding a vociferous local movement that proposed an "ecosystem-based plan" in the Slocan Valley,[57] and instead issued cutting permits in community watersheds to the large industrial company, Slocan Forest Products. These permits were issued under the rubric of implementing the local "land use plan," a characterization of the decision that highlights the larger nature of the regulatory model at stake.[58] Neither was this just any valley. It is home to the longest-running, and most sophisticated, ecoforestry movement in probably the entire industrialized world. The issuance of these cutting permits signals a rejection of the potential for a gradual experimentation with, and adaptive transition to, an alternative form of community-based ecoforestry. It may also signify the beginning of a new stage in the battle in the BC forests, one that moves beyond the concern for wilderness protection to the organization of the entire forest land base. The message here is clear: even if the economic and social potential of the community tenure could be accepted in theory, implementation in practice confronts the overwhelming inertia of corporate, bureaucratic, and organized labour power.

First Nations[59]

British Columbia today possesses a most potent vehicle for structural reform of the dominant institutions that affect forested land: the ongoing treaty process. At one level, this process offers the potential to create a unique type of community tenure for First Nations. Traditionally, an ethos of sustainability has infused Native cultures. Historically, these communities have suffered where their lands have been cut over, and their traditional hunting, gathering, fishing, and ceremonial practices destroyed. With self-government in the offing, theirs is a future where a level of community control could merge with an ecoforestry practice to create a new model of forestry. At a higher level, however, the ecosystem-based context in which First Nations have been traditionally rooted also presents a "territorially-based" form of legitimacy (of the character discussed in my earlier chapter) which non-Native society could build on to reform its own centralist structures. Given the profound historical assumption of Crown-based authority underlying these structures, such an opportunity for structural reform is a one-time opportunity which, if taken, could provide a key to ecological transition. Unfortunately, the nature and importance of this opportunity is almost nowhere understood and, certainly, nowhere being acted upon.

Instead, again, the future potential and the current reality diverge fundamentally. To start, historically there has been no recognized aboriginal right either to manage forest resources or to control non-aboriginal uses on traditional lands, although recent court decisions may change this situation.[60] Through recently created associations, however, First Nations are attempting to achieve this on a cooperative basis. For example, at a national level, the National Aboriginal Forestry Association (NAFA), formed in 1989, has developed a model First Nations Forest Resources Management Act that, with provincial acquiescence, would provide the basis for large-scale management of forest lands.[61] While the Act stipulates neither a particular tenure nor a model of forestry to be implemented, NAFA has developed a set of "community guidelines" that stress a "place-based" approach to forestry. The guidelines include extensive provisions to protect biodiversity and to facilitate a planning approach whereby "those who are closest to the land and experience the direct consequences of land use decisions should be the ones who are consulted first and last."[62] Meanwhile, in 1995, the First

Nations Forestry Council in British Columbia (FNFC) developed a "strategic plan" aimed at increasing the participation of aboriginal ventures in the forest industry.[63] Unlike the NAFA document, this plan was a consensus document developed in conjunction with industry and government officials, and it neither proposes changes to the tenure system nor addresses ecosystem integrity or traditional values.

Legislation and practice in the United States again provide some indication of the changes that are possible. In 1990, the federal government passed the National Indian Forest Management Act[64] which recognizes tribes as the primary decision makers for the future of their forests. Here, too, the Act applies only to reservation lands, although these lands are far more extensive than in Canada. As a result of this legislation, the Makah Tribe of Washington State runs its own tribal forestry operation on almost 13,000 hectares of forest land, and does so through a community-based, integrated resource approach that attempts to ensure harmony among economic, ecological, and cultural values. In Wisconsin, the Menominee Tribe runs a similar, much-heralded forestry operation on its 105,000 hectare reservation. Active for over a century, this operation's inventories now show that there is more volume in the forest than when logging started. The forest is still used for a large variety of traditional purposes, yet produces 12 million board feet annually, with an industry that employs about 300 people, 10 percent of the reservation's population. Management involves the professional preparation of plans, which are then ratified by the tribal legislature. The operation is so successful that it has been "certified" as sustainable by two non-governmental, eco-certifiers, Green Cross and Smartwood. As commentators have concluded, the Menominee have successfully developed "a market-oriented forestry enterprise that is ecologically and socially sustainable as well as economically feasible."[65] A major reason for success is the tribe's control over the forest land base, and its decision to keep the forest in collective community ownership.[66]

In Canada, such novel tenure arrangements do not yet exist. For the most part, reliance is on existing corporate tenures, with First Nations involved by way of joint ventures, or, occasionally, as the holder of a traditional tenure. In Saskatchewan, the Meadow Lake Tribal Council, for example, entered into a joint venture with NorSask, an arrangement that provoked one of the longest-running blockades in Canadian history by band members who were concerned about the impact of industrial forestry by *their own tribal members* on hunting, trapping, and berry gathering. The arrangements that have evolved from this conflict, while interesting, are still in the traditional mode of corporate tenure and state-based management.[67] In British Columbia, the Tl'azt'en Nation was awarded a Tree Farm Licence to supply a local sawmill that created a significant amount of local employment. Again, however, conflicts have been endemic with Elders and others in the community over the obviously negative impact of the industrial forestry practices on the region's wildlife. Several other bands in British Columbia have also been awarded traditional forest licences and timber sale licences, but none has established its own special tenure designations.

A range of possibilities for such designations exist, the design of which is not difficult to envision. Especially in light of the special character of First Nations, their landholding and management arrangements, including specific tenures, should provide a land base that is community controlled and directed so as to create cooperative decision-making structures that empower the whole community. These arrangements could draw on traditional, experiential knowledge as well as scientific expertise in such as way as to ensure the protection of ecological diversity while fostering a diversity of forest uses, including non-economic ones, as the Menominee have done. In this way, the arrangements would return both economic and non-economic values to the community. As Lynch and Talbot point out, however, the critical aspect of indigenous claims to such arrangements is their different place of origin:

> Government-sponsored community forestry programs based on public grants that can be cancelled don't provide adequate incentives for sustainable community-based forest resource management. Wherever local people are striving to protect and sustainably manage forests, the best way to establish and secure these incentives is to get appropriate government agencies and officials to recognize existing community-based rights ...
>
> Functionally, community-based management systems and the property rights that they establish and support draw their fundamental legitimacy from the community in which they operate rather than from the nation-state in which they are located ... Externally initiated activities with varying degrees of community participation should not be referred to as community-based, at least not until the community exercises primary decision-making authority.[68]

While such community-based tenures do not presently exist, some First Nations are working towards these in British Columbia. For example, the Gitksan propose tenure arrangements that recognize their communal clan "House" system, but the province does not recognize the legitimacy of this House system, let alone a specially-designated tenure that would take away from the existing forestry commitments in the territory.[69] In fact, the provincial treaty process is explicitly based on maintaining the status quo to the greatest degree possible. Thus, for example, the province's guiding policy document states:

> The objective of treaty negotiations is to replace the broad-based sustenance rights recognized by the courts – and currently covering much of British Columbia – with clearly-defined contemporary rights ...

A. Treaty negotiations will exchange these relatively undefined aboriginal rights with clearly-defined rights to land and resources in a manner that fits with contemporary realities of economic, law and property rights in British Columbia.[70]

In ·the pursuit of this objective, the province follows a "land selection model," whereby the total land to be held by First Nations in fee simple would be no more than 5 percent of the provincial land base, outside of which lands "will continue to be owned and managed by the Crown. Treaties may include some clearly-defined rights that apply on Crown lands, such as rights to practice certain traditional activities, a role in some planning and development decisions or involvement in the management of resources."

The impact of this approach is clear with the Nisga'a Agreement-in-Principle, in which, in addition to following the above prescriptions with regard to land selection, stringent restrictions are imposed on the Nisga'a to ensure that their forestry operations deviate as little as possible from the status quo. Much of the land granted to the Nisga'a is already cut over, and future cutting rates are mandated to remain near to the established provincial levels. The Nisga'a are prevented from establishing their own primary sawmilling facility for a decade, although they can enter into joint ventures with existing non-Native facilities. No new tenure forms are contemplated outside the settlement lands, where, on the contrary, the Ministry would entertain only a standard Forest Act tenure that contains terms which address "regional timber supply needs." And, certainly, no broader regional management changes are envisioned.[71] As a result, estimates of the impact of the treaty on the forest industry are expected to be very modest.[72] This modest settlement has provoked some public opposition, although it is possible that this opposition is also a function of the competitive conception of the treaty process which the government has embraced. This issue is complicated[73] but the implication is clear: the treaty process is not to be a vehicle for the larger changes that it could be.

Yet again, therefore, the chance for innovative change is not embraced. Instead, it is only in high-profile areas such as Clayoquot Sound where anything approaching innovation in tenure design has been entertained. Instead, the hegemony of the state and the inertia of corporate economic development continue essentially unimpeded.

Supporting Community Initiative

A range of ecoforestry tenures could be developed at the community level with tangible local benefits for employment, economic diversification, environmental sustainability, democratic development, and social quality of life. As we have seen, the problem is not the absence of workable possibilities, but the presence of obstacles. Underlying these obstacles runs both a scepticism of alternative economic development strategies (a far larger issue than concerns about the use of economic instruments), and an aversion to alternative regulatory models. Economically, the province's reliance on high levels of output from industrial forestry implicitly justifies the dismissal of alternatives, the escalating encroachment of the "falldown effect" notwithstanding.[74] Politically, government agencies share a concern that community empowerment in the managerial realm will undermine the existing bureaucratic monopoly on decision-making. While it is beyond the scope of this chapter to analyze these economic and management concerns in detail, let us consider them briefly insofar as we are able to discern new economic and regulatory models that facilitate the move to new ecosystem-based forms of forest tenure.

Marketing Economic Transition

As the movement for community-based, sustainable forestry grows, it is accompanied by other changes in the economic landscape. As noted, any transition to ecoforestry would necessarily be gradual, involving some regions and industries willing to embark on new tenure and production arrangements. An abrupt transition strategy for the province as a whole is neither possible nor desirable, in comparison with a strategy that fosters the emergence of a new level of economic activity at a rate, and in places, where the change is manageable. In addition to the author's earlier work,[75] one other study has addressed such a strategy. This study was undertaken by the New York-based consulting group, Environmental Advantage, which specializes in ecologically responsible business development. The study details a seven-step process to develop a sustainable forest industry in British Columbia, the second step (after the formation of a Business/Stakeholder Working Group) being to "expand the land area under sustainable forest industry."[76] While there are many aspects to developing a full-blown strategy, two components of such a strategy are of particular interest because they are already in progress. One innovation is the competitive log market that might maximize the returns to communities for local resources, and make more wood available to entrepreneurial manufacturers. Another is the worldwide movement for the "eco-certification" of products that are produced sustainably.[77] Successful examples of both these economic instruments exist already, but they too have been relegated to the margins and are, in fact, largely in competition with official policy. Bringing them into the mainstream is a critical challenge for reinventing the state in a sustainable form.

Demands for a log market in British Columbia have been long-standing for a number of reasons: to break the stranglehold of corporate tenure holders over the disposition of the wood supply; to price wood more accurately; to remove the corporate subsidies inherent in their monopolistic supply; and to increase access to wood by smaller processors and value-added manufacturers who do not have their own tenure holdings.[78] Under the auspices of the Vernon District Office of the Ministry of Forests, a successful experiment has been taking place with such a market. Utilizing wood available through the Small Business Programme, the Ministry has overseen its own logging operations which have been carried out employing the principles of ecoforestry. The output of these operations has been sold through a local log market run by the Ministry. A Ministry study concluded that the experiment was a resounding success in all its objectives: the application of alternative logging systems; an increase of more than double the level of stumpage paid to the Crown (an average stumpage increase of $39.92 per cubic metre); the creation of new jobs in the woods and in the log market; and, although this point was not reviewed, the increase in wood available to secondary manufacturers.[79] Evidently, these positive findings were unacceptable to the Ministry, which contracted a second review from a less sympathetic source. This second review also found that the experiment had been successful on all counts, although it found a lower increase in the log market's contribution to Crown revenues.[80] Other experiments with small business bidding are occurring in the United States with similar results.[81]

The harshest critics of industrial forestry – members of the environmental community – were also sympathetic to the Vernon experiment. Indeed, two of the strongest antagonists of the industry, Greenpeace and the Silva Foundation, were so impressed that they chose some of the wood coming from the yard as the first timber produced in BC to be independently accredited as "ecologically sustainable."[82] This form of accreditation by non-governmental actors according to substantive ecoforestry criteria is, however, far from accepted by either the forest industry or the provincial government. In fact, despite the potential benefits that could accrue to the jurisdiction that can satisfy such criteria early in the development of this market, the mainstream policy community prefers to support an alternative, more process-oriented, system of accreditation that is controlled by the industry through the Canadian Standards Association, and is being developed in a manner compatible with the widespread continuance of most industrial forestry practices.

Understanding these economic instruments – the log market and eco-certification – is important in relation to the primary focus of this book. The log market is a clear example of a *market*-based instrument, yet it remains suspect to many existing tenure holders because it undermines the justification for the exclusive property rights in the public forest that have been granted to large private corporations. Instead, existing tenures actively avoid reliance on markets in the supply of their timber-processing facilities; in so doing, they create potentially large, hidden subsidies for these companies. Meanwhile, an NGO-based certification regime for forest products is opposed for similar reasons, because it would pressure forest companies to internalize environmental values at the behest of the marketplace. Advocates of industrial forest, instead, press for deregulation (through even stronger corporate tenure rights), and more self-regulation (through an industry-controlled certification process), because these initiatives merely shift the regulatory locus from the pyramid of *public* power to the pyramid of *private* power. In this light, as discussed, couching the debate in the usual dichotomies of economic versus regulatory instruments, and public versus private power, misses the point. These proposals are problematic precisely because they are processed through the very hierarchical structures that are the sources of the ecological problem in the first place.

Alternative Management

Just as a move towards alternative tenures needs a new approach to economic instruments, so too does it demand an informative new approach to the future character of regulatory instruments. As we have seen, the centralist command and control model *is* problematic, whether you are a conservative free marketeer or a communitarian ecologist, a tree farm licensee, or an overburdened woodlot owner. Thus, for example, do the hugely bureaucratic requirements of Ministry oversight pose a significant obstacle to an expanded woodlot program resulting from the inherent "unmanageability" of numerous such small-scale operations. Yet even more regulatory centralism is exactly the direction the government has been following with, for example, the technocratic Forest Practices Code.

Indeed, as the recent decision in the Slocan Valley makes clear, forest policy has, under the NDP, coalesced around the combination of land use zoning/forest practices regulation as a complete management package. Under this paradigm, there is no need for tenure reform, because the environmental and social objectives of alternative tenures can be met by carefully regulated forest practices on so-called low intensity areas or special management zones (none of which has yet been created). The fact that such a package is not workable even in theory (corporate forestry resists uneconomic constraints while complex bureaucratic regulation is too unwieldy) and has not been implemented in practice is not addressed.[83] If the inherent limitations of this package are addressed, however, then a new paradigm becomes mandatory. Ecosystem-based tenure reform combined with a new approach to community-based management is the paradigmatic alternative. And neither is this alternative merely a political choice; it is also mandated by the science of complex systems.[84]

Support for community-based management alternatives does not imply that one is blind to the possibility of abuses at that level where vested economic and political interests can also dominate. The issue is simply the proper democratic design of alternative processes to counter-balance hierarchical power wherever it occurs, centrally or locally. Here then is the political ecological challenge (discussed in my earlier chapter in this book) for "reinventing" the state in the transformative, but protective, mode. While shedding many of its detailed managerial functions, the state maintains a pivotal role in setting minimum common standards, ensuring democratic local processes to facilitate participation and conflict resolution, providing for rights of citizen appeal, and maintaining a continuous oversight role. In short, both local devolution and centralist oversight constitute the essential ingredients for an ecologically based management model for the state. Two new regulatory models emerge from the discussion in this chapter, one based in private self-regulation, and the other based in a new locus of public control.

In the current regulatory regime, some areas of the industry are already self-regulating. As we have seen, a high degree of discretionary deference marks the treatment of corporate decision-making by Ministry officials. One reason for this deference is found in the manner in which control over industrial forestry is exercised through the setting of standards of conduct for all "professional foresters" (whether public or private) through their professional association. This self-regulation is based in a non-governmental association, the Association of BC Professional Foresters. Nevertheless, the model of control is still largely hierarchical, because the practitioners work within hierarchical corporate and bureaucratic institutions, and they are regulated through a centralist professional body. This affiliation with the status quo is common to a number of professions, from foresters to doctors and lawyers. (This structure tends to make all such professions accommodating to corporatist incentives – and very resistant to change. Indeed, the professional monopolization of competence and allegiance to hierarchical interests has been a pervasive factor in the unsustainability of a range of sectors, from energy production to fisheries to architecture.) Instead, community-based forestry points to a new, *network* model of horizontal association that can be applied to any of the alternative tenures considered earlier: woodlots, community forests, First Nations' forests, even independent contractors operating under a competitive bidding system.

The basic orientation of such a system would be to allow each tenure type to be partially self-regulating through an association of tenure holders. This aim could be achieved through a series of clusters of associations organized on the basis of a hierarchy of *geographical networks* rather than of professional, bureaucratic, or corporate organizations. Thus, individual woodlot owners would cluster into a local association such as exists, for example, with the North Island Woodlot Association or Cariboo Horse Loggers Association. Local associations would also be grouped into a regional level association, and these associations would then group together at the provincial level.

Once one has redefined the locus of management, the character of the regulatory instruments used would be redefined as well. An association-based model could rely on a combination of broad forestry *goals* (such as maintaining local biodiversity, or retaining forest structure and function) which each member had to achieve in his or her own way. The association would then license its members to ensure the attainment of these goals in practice. Unlike the complex processes under the Forest Practices Code, this licensing could easily be done on a very adaptive, site-specific basis. The operator's performance would then be reviewed by the association and on the basis of the achievement of goals, rather than compliance with rigid *rules*. As part of this shift would be a shift away from *time-limited* tenures to *performance-limited* tenures, providing the tenure holder with long-term security of tenure conditional upon fulfilment of goal-oriented criteria.

For this new regulatory model, the level of oversight could be reduced dramatically. For example, after demonstrated success, an established operator would move to a higher-level licence that required a reduced level of formal management and oversight. The effect of this shift on the regulatory model would be profound. Overall, the instruments would shift from those derived from the rigid model of an external penal regulator to the more flexible tools that use the *incentives* and *cultural pressures* of membership in a peer group and local community. In addition, the association could also be a vehicle for supervising apprenticeship programs, providing expert advice, coordinating training and extension, sharing technologies, developing management plans, and facilitating cooperative marketing. With such tasks, and as a geographically rooted institution, an association would tend to *embed* the regulatory function in a larger process of *economic/cultural* development around ecosystem-based values and institutions, especially if there are parallel associations for related operations in processing and manufacturing.[85] With the central government overseeing the association as a whole, one would create a healthy balance of horizontal association, and hierarchical oversight.

The ultimate goal of this alternative model of economic instruments and legal regulation bears repeating: *the creation of a territorially based culture of sustainable self-management that depends on neither state nor market discipline to do what is right*. In this way, the regulatory prescription is directed at the real root of the environmental problem in forestry: the inherent unsustainability of the productivist structure of industrial forestry that increasingly dominates the world's forests. This emerging dominance is the true "tragedy of the commons," the erosion at the hands of centralizing institutions of the social capital that allows for collective territorial self-management. Only by rebuilding both the economic and the political foundations for such living communities can this erosion be halted. In the final analysis, nothing else will work.

The level of state involvement can be reduced further by shifting many aspects of regulation to another public locus as well. Parallel to the model of associational (private) self-regulation would thus be a structure of *community-based* (public) management. Tenures, it should be noted, are associated with "private" resource utilization, even if held by a public body. They are not a vehicle for "public" resource management. In any resource development/management regime, both elements must be present, and kept distinct. In fact, in this regard, the Mission and Revelstoke Tree Farm Licences reflect a poor tenure model, because both licences are held by the municipality directly. In such a case, the counter-balancing public management authority is necessarily the provincial government. If one wants both local tenures and local management, the tenure must be held by a separate body. Thus, the Malcolm Island proposal contemplates a specially created, and institutionally separate, cooperative to hold the community tenure. Where this separation of functions is implemented, it is still possible for the broader community to operate as a public manager of the whole range of tenures in the region (from woodlots to community forests to tree farm licences) without engendering a conflict of interest. And there is still room for a reduced level of provincial involvement.

How an authoritative "community forest board" might operate does not require much elucidation, as a range of variations has been well described by many commentators over the years.[86] The Central Region Board in Clayoquot Sound is as close to an operating community resource board as the province has yet come to creating, but this board is far from being embraced as either a model or a precedent, or from being put onto a long-term funding base. A number of criteria for such decentralized management are critical to success, however. Such boards should have a clear mandate to implement ecosystem-based approaches, possess real standard-setting and enforcement authority, embody a structure that is both representative of community interests and is democratically accountable (at least for some proportion of their composition),[87] and have an entitlement to a funding base that is independent of the provincial government.[88] It is through such a comprehensive board that the range of interests in a forest (from mushroom harvesting to logging to recreation) should be regulated in an integrated fashion, not through a fragmented and artificial "share cropping" arrangement as has been suggested in Chapter 6.[89] Again, while the central government would turn over its direct managerial role to these boards, it could still approve the allocation of tenure licences (perhaps on reference from the community board), set minimum standards, retain oversight and appeal powers, and provide technical support through the existing ministries of forests and environment. This balance of central/local authority, and the new dynamic that it would introduce, is the key to sustainability (if not to a sylvan utopia).

Conclusion

After more than two decades, the long-sought dream of community-based forestry in British Columbia remains just that: a dream. If the issue were just the technical design of alternative tenures and management models, the matter would have been resolved long ago. Unfortunately, the issue is far larger – the inability even to think of, let alone attempt, something so structurally different from the status quo. In achieving the change that is needed, the state maintains an important role, if it can be opened up so as to consider integrated solutions. How, for example, can the state act to tie ecoforestry-based tenure reform to new log markets and eco-certification? Or new woodlots to an associational form of self-regulation? In these and other examples, the state is involved with both regulatory and economic instruments, but in a transformative way.[90]

With the authority over the land base vested in the Crown, this necessity points to a fundamental transition, one where the state becomes not a defender of vested institutions, but a facilitator of *cultural* transformation. For this to happen, the state must be transformed itself. The possibility of this transformation is a huge opportunity, and vehicle, presented by First Nations with their legitimacy rooted in a territorial alternative to the Crown. Our failure in British Columbia to embrace this alternative is perhaps the ultimate irony, and final contradiction, which explains why such change does not occur, indeed, why under a social democratic government the grip of bureaucratic and corporate centralism has become even stronger. We need not a timid state that is content to prod and push outmoded institutions of the past, nor a bold one that will "privatize" its functions. Instead, beyond Left and Right, we need a imaginative state that can, in response to social interests, provide an alternative to its own bureaucracies and corporations, and "mandate" community. And that then stays around to make it happen.

In this light, the popular debate that lies at the centre of this book – public versus private rights, regulatory versus economic instruments – is not the central issue. Instead, the relevant spectrum that demands attention is that which spans the abyss between the corporate and bureaucratic forms of contemporary social ordering, and the ecosystem-based transition that looms before us. This is no small task – which explains why there has been so little movement. Despite the lull in recent years, the battle for BC's forests is by no means over. A solution awaits.

Acknowledgments

This chapter is part of a larger research effort on sustainable forestry conducted with the support of the Eco-Research Secretariat, Ottawa. The author would like to thank the following students involved in the project whose work contributed to this chapter: Marne Andrulionis, Vickey Brown, Cheri Burda, Tracey Cook, Deborah Curran, Kristen Lindell, and Catherine Slater. The author would like to note the special contribution of Catherine Slater and Gil Yaron for their research assistance.

Notes

1 Perhaps the leading work on this is E. Ostrom's *Governing the Commons: The Evolution of Institutions for Collective Action* (Cambridge: Cambridge University Press 1990). See also D. Western and R.M. Wright, *Natural Connections: Perspectives in Community-Based Conservation* (Covelo, CA: Island Press 1994); and J. Fox, ed., *Legal Framework for Forest Management in Asia: Case Studies in Community/State Relations,* Occasional Paper 16 (Honolulu: East-West Center 1993).

2 R.M. M'Gonigle and B. Parfitt, *Forestopia: A Practical Guide to the New Forest Economy* (Madeira Park: Harbour Publishing 1994), 54, 55.
3 For a brief review of the current debate, see J.A. Mintz, "Economic reform of environmental law: A brief comment on a recent debate," *Harvard Environmental Law Review* 15 (1991): 149.
4 For a detailed analysis of the existing tenure system, see C. Burda, D. Curran, F. Gale, and M. M'Gonigle, *Forests in Trust: Reforming Forest Tenures for Ecosystem and Community Health*, Discussion Paper 97-2 (Victoria: Eco-Research Chair of Environmental Law and Policy, University of Victoria 1997).
5 One example of this reconfiguration is the common concern to shift from increasing supply (in energy, water, forest products) to "managing demand" (that is, reducing the externalities associated with economic activity itself by reducing overall material flows). This demand management perspective informs BC Hydro's successful PowerSmart program, but, despite the parallels across every resource sector, it has yet to make a dent in forest policy thinking in either academic or policy circles. For one study that does emanate from this perspective, see the author's *Forestopia*, supra, note 2.
6 The British Columbia *Company Act*, R.S.B.C. 1979, c. 59, s. 142, and the *Canada Business Corporations Act*, R.S.C. 1985, c. C-44, s. 122, require a corporate board of directors to act in the best interests of the company. This obligation means responsibly serving the shareholders, which the courts have interpreted to mean managing for profits or business growth so as to increase share value. This interpretation is a serious constraint on the degree to which non-economic objectives can be incorporated in corporate objectives. *Dodge* v. *Ford Motor Co.*, 170 N.W. 668 (1919) (Michigan Supreme Court) still represents the prevailing doctrine in Canada. As Ostrander J. stated in that case: "A business corporation is organized and carried on primarily for the profit of the stockholders. The powers of the directors are to be employed for that end. The discretion of directors is to be exercised in the choice of means to attain that end, and does not extend to a change in the end itself, to the reduction of profits, or to the nondistribution of profits among stockholders in order to devote them to other purposes." Cited in F.H. Buckley, M. Gillen, and R. Yalden, *Corporations Principles and Policies*, 3rd ed. (Toronto: Montgomery Publications Ltd. 1995), 546. There is some limited judicial support for a broader approach to the fiduciary duties of directors in Canada. However, this approach is not yet widespread in Canadian law or policy. Socially responsible activities of corporations which could be argued to be conducive to profit maximization in the long run may be permissible. Berger J. took the broader view in *Teck Corp* v. *Millar* (1972), 33 D.L.R. (3d) 288 (B.C.C.A.), in which he suggested, "If the directors were to consider the consequences to the community of any policy that the company intended to pursue, and were deflected in their commitment to that policy as a result, it could not be said that they had not considered bona fide the interests of the shareholders." Cited in Buckley, Gillen, and Yalden at 560. Berger's is still very much a minority view.
7 Discounting militates against maintaining capital in slow-growing natural assets and discourages long-term silvicultural investments: "In short, it may make 'economic sense' to cut the forest and grow the money because money grows faster in the bank than the value of trees growing in the forest" (B. Evans, *Sustainable Development and Sustainable Forestry: A Review of Concepts, Definitions and Criteria*, prepared for the BC Ministry of Forests, Research Branch 1994, 11). Notes another commentator, "The replanting and management of a northern forest, especially a northern mixed-softwood forest is a costly undertaking. It ties up capital for a very long time ... a condition not likely to attract enthusiastic investors if they have alternatives" [P. Marchak, *Logging the Globe* (Montreal: McGill-Queen's University Press 1995), 34].
8 A 1995 report to the Policy and International Affairs Directorate Canadian Forest Service warns that rapid-growing plantations in countries such as Brazil, Chile, Indonesia, and New Zealand are emerging as competitors to Canada's relatively slow-growing forests. As the report notes, BC second-growth forests are costly to grow and maintain, and more remote access has made them comparatively costly to harvest. Second growth will not grow large or fast enough to provide the quantity of timber the industry depends on. Finally, plantations in BC will not produce the large, tight-grained product that has been the

unique attraction of BC old-growth products in the global marketplace. For a detailed discussion of these issues, see C. Burda, D. Curran, F. Gale, and M. M'Gonigle, supra, note 4. One of the first indications that disinvestment was beginning to occur was found in a leaked 1990 memo of New Zealand-based Fletcher Challenge, which noted that its new "strategy will be designed to reduce the company's exposure to the unique risks facing the forest industry in British Columbia and position it to take advantage of growth opportunities elsewhere in North America" [quoted in "Fletcher Challenge intends to curb investment in BC, confidential memo reveals," *Victoria Times-Colonist* (4 December 1991): D14]. Recent indications of this move to re-invest outside the province are now commonplace, reinvestments that are often blamed on the province's now tighter regulatory requirements (such as the Forest Practices Code), as well as on the reduction in timber potential. See G. Hamilton, "Forest firms seek fortunes outside BC," *Vancouver Sun* (30 March 1996): B1.

9   As one recent press report noted, "shocking logging practices," including extensive clearcutting, caused more than 428 landslides (with extensive damage to public salmon streams) since the 1950s on private forest land owned by MacMillan Bloedel, Pacific Forest Products, and TimberWest. See M. Curtis and L. Leyne, "Shocking report: A river ravaged," *Victoria Times-Colonist* (24 June 1995): A1 and A6; P. Moss and P. Minvielle, "Editorials: Make firms pay for forest mess," *Victoria Times-Colonist* (27 June 1995): A4.

10 For a discussion of the approach of this company, see *The Business of Good Forestry,* Report 95-1, ed. M. M'Gonigle, K. Stratford, and F. Gale (Victoria: Eco-Research Chair of Environmental Law and Policy, University of Victoria 1996).

11 See the discussion by Lignum's chief forester Bill Bourgeois, in *The Business of Good Forestry,* supra, note 10 at 23ff.

12 The proposed amendment was undertaken by the author and David Cohen of the Faculty of Law, University of Victoria. In 1995, the Ministry of Forests replaced 19 of the 34 Tree Farm Licences in BC that came due for replacement (56 percent). The contracts that were renegotiated and signed without fundamental change were Evans Forest Products Ltd., Weldwood of Canada Ltd., Western Forest Products Ltd. (Quatsino), Pope and Talbot Ltd. (Boundary and Arrow Lakes TFLs), International Forest Products Ltd. (Toba), Crestbrook Forest Industries Ltd., Weyerhaeuser Canada Ltd. (Inkaneep and Jamieson Creek TFLs), Northwood Pulp and Timber Ltd., Federated Co-operatives Ltd., Canadian Forest Products Ltd. (Nimpkish), MacMillan Bloedel Ltd. (Alberni and Haida TFLs), West Fraser Mills Ltd., Scott Paper Ltd., TimberWest Forest Ltd. (Duncan Bay) Revelstoke Community Forest Corporation, and Riverside Forest Products Ltd.

13 Tree Farm Licence No. 44, Instrument 19 (Amendment), s. 7.06(e)(i).

14 See Instrument 19 of MacMillan Bloedel's Tree Farm Licence No. 44. The Clayoquot Sound "experiment" has led to a drastic reduction in annual cut (about 75 percent), is heavily subsidized by provincial grants from Forest Renewal BC to laid-off workers, and is supported by the licensee because of the very high-profile character of the issue.

15 Approximately 2.5 million hectares of BC's forest lands are privately owned forest land. About one-third (36 percent, 920,000 hectares) of private forest lands are "managed forest land," and two-thirds are "unmanaged forest land." Managed forests produce about 10 percent of annual timber harvest, 10,000 direct jobs, and contribute over $400 million annually to the public treasury. Most of this output comes from a handful of large corporate lands on Vancouver Island. This section of the chapter will not address these owners, however, as the analysis in the preceding section also applies to this more secure form of tenure, except perhaps in the willingness to make greater silvicultural investments. See D. Zhang and P. Pearse, Forest Economics and Policy Analysis Research Unit, "Effect of Forest Tenure System on Silvicultural Investment in British Columbia," Working Paper, n. 206 (1994).

16 Currently, only Schedule A TFL private lands and private lands in Woodlot Licences are subject to the *Forest Practices Code of British Columbia Act* (s. 1, definition of "forest practice"). The Act allows, per s. 216, the provincial cabinet to make regulations that would subject managed forest lands to the Code. However, to date cabinet has been reluctant to do so. See M. Haley and P. Pearse, *"Regulating private forest: not as easy as it looks," Victoria Times-Colonist* (28 February 1996): A4.

17 The *Assessment Act,* R.S.B.C. 1979, c. 21, s. 29, and Regulation 349/87 (amended by Regulation 341/90), ss. 6 and 10.

18 The private forest owner who applies for this beneficial tax status is required first to have the land reclassified to the Forest Land Reserve. Then, the owner must make a "written commitment to follow and adhere to good forest and resource management practices"; submit a formal Forest Management Plan that includes mapping, forest protection, and reforestation to the free-growing stage (Regulation 349/87, s. 6); and file an annual management report (s. 10) [*B.C. Assessment Fact Sheet 1995. Series 1: Corporate Information No. 1017. Assessing Private land with Trees in British Columbia* (Victoria: Communications Division 1995)].

"Managed forest land" is forest land that is managed in accordance with a forest management plan under the *Assessment Act*, s. 29. Essentially, a beneficial tax classification under the BC Assessment Authority is awarded in exchange for a written commitment to "adhere to good forest and resource management practices of private land used for commercial timber production." The owner of unmanaged forest land must apply for the designation. To be accepted, the forest land must be classified as forest reserve land, be subject to a forest management plan submitted with the application that is in accordance with regulations under the Act and is approved by the assessor, and be managed by that plan (s. 29(4)). Regulation 349/87, s. 6, requires that the management plan contain: a map and site description; objectives for the first five years and a complete growing cycle of the commercial species; and planned undertakings (including methods and practices) to "reforest ... to a density of commercial tree species that anticipates maximum economic harvest ... within five (and 10) years," to maintain and harvest tree crops, to tend the land and seedlings to a free-growing state, to ensure forest protection, and to conduct five-year commercial tree density reviews. Section 10 requires an annual return showing volume scaled, any event that has reduced volume by more than 30 percent, a map of harvesting activity, and progress towards implementation of the filed management plan. Under s. 11, the owner is subject to a demand for particulars of reforestation activities at any time.

Unlike public woodlots, which have vigorous process regulation, oversight by the Assessment Authority is less rigorous and goal-oriented, relying on spot checks to see that management goals are being met. Of course, no AAC requirements apply to managed forest lands. The *Assessment Act*, s. 15, requires the Commissioner not to disclose any information to any person except the landowner. Thus, management plans are confidential information; the *Freedom of Information and Protection of Privacy Act* will not allow access by the public.

19 This laissez-faire approach to private land is in sharp contrast to the Scandinavian countries which have extensive controls on the uses and management of private forest lands. For a detailed discussion of these regimes, see A.J. Grayson, *Private Forestry Policy in Western Europe* (London: CAB International 1993).

20 For a comprehensive set of reforms to the private woodlot system, see the unpublished proposal by Fred Marshall, the president of the BC Woodlot Association, entitled *A Policy for Private Forest Lands in British Columbia* (7 April 1995). Interestingly, many woodlot owners who want to maintain the freedom to practise ecoforestry propose that they be exempted from the Forest Practices Code, and be subject to a new set of regulations that would be more conducive to their special needs. This concern has implications for the appropriate re-design of the regulatory process (discussed later in this chapter). Policy changes suggested by Marshall include exempting managed forest land from the Forest Practices Code, reducing taxes to the equivalent of those of farm lands, eliminating the harvest tax, providing extension services, and allowing any owner of private forest land to choose managed forest land status. The government would then require a comprehensive "resource management plan" to ensure sound stewardship practices. The land could not be changed in status for a minimum of fifteen years. Unmanaged forest landownership would be discouraged by policy that requires Forest Practices Code application, and retains both the current higher tax rates and the current harvesting tax.

21 As noted by Gillespie in his review of the Woodlot Licence Program, two of its major objectives are to "improve the productivity of small Crown and private forest land parcels" and to "increase the amount of private forest land under sustained yield management." See D. Gillespie, *Program Review: Woodlot Program*, Report to the Ministry of Forests 1991.

22 There are numerous criteria by which one could assess an alternative tenure. One recent

study addresses the following issues: the appropriate holder of entitlements; comprehensiveness and exclusiveness; duration and security; transferability; and distribution of economic benefits. See J. Clogg, *Forest Land Use and Decision-making in British Columbia: Toward Greater Sustainability,* MA thesis, York University 1997. See, more generally, E. Schlager and E. Ostrom, "Property rights regimes and natural resources: A conceptual analysis," *Land Economics* 62 (1992): 256.

23 P.N. Duinker et al., "Community forests in Canada: An overview," *Forestry Chronicle* 70(6) (1994): 713.

24 "Bureaucrats have managed to effectively smother BC's woodlot licence program," *Business in Vancouver* (7 September 1996).

25 O. Lynch and K. Talbot, *Balancing Acts: Community-Based Forest Management and National Law in Asia and the Pacific* (Washington, DC: World Resources Institute 1995), 24.

26 Supra, note 22 at 712.

27 *Partnerships for Community Involvement in Forestry: A Comparative Analysis of Community Involvement in Natural Resource Management,* prepared by the Community Forest Project (Ontario: Ministry of Natural Resources 1994), 9.

28 In fact, a 1994 report by the Commission on Resources and the Environment recommended that a substantial portion of this TFL be designated a "special management area," a designation which would reduce the cut level but could also provide the basis for new forms of non-industrial timber management. Instead, given the community's dependence on the cut, the community "loudly rejected the report and ... developed an alternative plan which restricted special management status to a smaller portion of the land base" [D. Weir and C. Pearce, "Revelstoke Community Forest Corporation: A community venture repatriates benefits from local public forests," *Making Waves* 6(4) (1995): 11.] For a more detailed critique, see the article by the author (with C. Burda), "Tree farm ... or community forest? Reflections on the Revelstoke experience," *Making Waves* 7(4) (1996): 16.

29 The North Cowichan Municipal Forest Reserve was established in 1946 by an act of the Municipal Council. In 1981, a Forest Advisory Committee, including six local foresters and engineers, dismantled a system of woodlot agreements with local operators that had practised "diameter limit cutting" (cutting only trees greater than a set diameter) since 1964. The committee believed that this forest practice was downgrading the worth, volume, species, and genetic qualities of the reserve.

30 Non-monetary recreational values such as hiking, horseback riding, motorcycling, hunting, wildlife observation, and scenery are supported. The three areas that have been set aside as ecological reserves are Mt. Prevost Cairn and Wildflower Reserve, Mt. Tzouhalem Ecological Wildflower Reserve, and Maple Mountain Preservation Management Zone. There are 12 kilometres of hiking trails. North Cowichan hosts educational tours for school students and the public and maintains trails at Maple Mountain. Discussed in a memo to J.S. Dias, Administrator, from Darrell J. Frank, RPF, Municipal Forester, "Re. the 1995 Annual Report [of the Forest Reserve]" (29 January 1996); and *Our Communities, Our Forests, Our Future,* pamphlet (Forestry Department: Municipality of North Cowichan 1988).

31 Memo to J.S. Dias from Darrell J. Frank, (29 January 1996).

32 *Malcolm Island Community Forest Feasibility Study* (Vancouver: Robin Clark Inc. 1996), vii. This study provides a good review of many of the components of a community tenure.

33 Ibid., xiii-xiv. For another study that reviews the components of a community tenure, see *Feasibility Study: Prince George Community Forest* (Victoria: Cortex Consultants 1996).

34 Ministry of Forests, Press Release (3 December 1997). The item appeared in the major newspapers on 4 December.

35 Stephen Harvey, The Community Forestry Group, *Ontario Community Forest Pilot Project, Lessons Learned, 1991-1994* (Toronto: Queen's Printer 1995): 29; see also *Partnerships for Community Involvement in Forestry,* supra, note 27.

36 These terms of reference, and the Temagami management experience, are discussed in J. Benidickson, "Co-Management Issues in the Forest Wilderness: Stewardship Council for the Temagami," in *Growing Demands on a Shrinking Heritage: Managing Resource Use Conflicts,* ed. M. Ross and O. Saunders, 256-75 (Calgary: Canadian Institute for Resources Law 1992).

37 J. Dunster, "Managing forests for forest communities: A new way to do forestry," *International Journal of Ecoforestry* 10(1) (1994): 43.

38 R.S.O. 1990, c. 27, amended 1994, c. 27, s. 124, and 1996, c. 1, Schedule M, ss. 40-47. For a detailed assessment of these authorities, see B. Mitchell and D. Shrubsole, *Ontario Conservation Authorities: Myth and Reality* (Waterloo: University of Waterloo Press 1992). See also K.L. Yeager, *Conservation Authorities in Ontario,* Current Issue Paper 115 (Government of Ontario: Legislative Research Service 1991).

39 According to the Act, a conservation authority's objects are to establish and undertake programs "designed to further conservation, restoration, development and management of natural resources other than gas, oil, coal and minerals" (s. 20). In furtherance of these objects, the authority has power "to study and investigate the watershed and to determine a program whereby the natural resources of the watershed may be conserved, restored, developed and managed" (s. 21(a)); "to control the flow of surface waters in order to prevent floods or pollution or to reduce the adverse effects thereof" (s. 21(j)); to alter the course of bodies of water (s. 21(k)); to "plant and produce trees on Crown lands with the consent of the Minister, and on other lands with the consent of the owner, for any purpose" (s. 21(o)); and "generally to do all such acts as are necessary for the due carrying out of any project" (s. 21(q)). Before proceeding with a project, the authority must file plans and a description with the appropriate ministry and obtain the approval of the Minister of Natural Resources (and the Ontario Municipal Board when some of the cost will be raised in subsequent years) (s. 24), particularly for a project on Crown land (s. 32), or when the project may interfere with a public work, including hydro (s. 32(2)) or a highway (s. 32(3)). The authority determines the money required for the capital expenditure for a project and apportions the cost to each participating municipality in accordance with benefit derived from the project (s. 26). Maintenance and administration costs of the conservation authority are apportioned to municipalities (s. 27). The authority may make regulations limiting or regulating the use of or impacts on water bodies (with some restrictions of authority), and prohibiting or regulating dumping of fill. (Appeals of these decisions to the Minister of Natural Resources are allowed.) The authority may also make regulations applicable to its lands regarding use by the public of its property, protecting its property from damage, prescribing fees and permits, regulating vehicular and pedestrian traffic, prohibiting and regulating signage, and regulating access by horses, dogs, and other animals to its lands. Offences are penalized (ss. 28 and 29). The authority may expropriate land (s. 31).

Land vested in an authority is taxable for municipal purposes by levy (s. 33). Under a related Act (the *Conservation Land Act,* R.S.O. 1990, c. 28), the Minister of Natural Resources may make grants to support programs to "recognize, encourage and support the stewardship of conservation land" (s. 2) This section includes conservation authorities' land by definition under the Act.

40 Interview with Stuart Ryan, former Port Hope mayor and co-founder and early member of the Ganaraska River Conservation Authority (personal communication, February 1996).

41 See M. Mittelstaedt, "Bill could land conservation areas on endangered list," *Globe and Mail* (13 January 1996): A10. The Omnibus Bill cut the funding for these areas from $34 million to $10 million, reduced their function to merely flood control, and allowed municipalities to dissolve their authorities and sell off the land for development.

42 Discussed in Duinker et al., supra, note 23 at 715-16.

43 For a recent discussion of these proposed municipal forests, see L. Bouthillier and H. Dionne, *La Forêt à habiter: La Notion de "forêt habiter" et ses critères de mise en oeuvre,* rapport final au Service Canadien des Forêts, Quebec 1995.

44 Bouthillier and Dionne write that, in the inhabited forest, "the forest is no longer viewed solely as an industrial activity sector. Forest developers must respect sustainable development principles and a long-term approach must be adopted for the management of a territory. Thus for management, both territory tenancy and sustainable production objectives have become a top priority guideline. Local populations must derive maximum long-term benefits from their forest resources subject to the various constraints associated with the territory they occupy." To this end, it would be necessary to create a "territorial contract" that balanced functional (production) and territorial (sustainability and community)

objectives. See "A Forest to Inhabit: The Concept of 'Inhabited Forest' and the Criteria for Its Implementation," Executive Summary (English language version), Canadian Forest Service, Quebec, 1996, 2.

45 Duinker et al., supra, note 23 at 716.

46 H. Carey, *Forest Trust Biennial Report 1990-1991* (Santa Fe, NM: Forest Trust 1991); S. Ratner, "Adding value to wood products through flexible manufacturing networks," *Practitioner, Newsletter of the National Network of Forest Practitioners* June (1995): 3.

47 T. Banighen, "An ecoforestry stewardship land trust model," *International Journal of Ecoforestry* 10(4) (1994): 200.

48 P. Marchak, "Global markets in forest products: Sociological impacts on Kyoto Prefecture and British Columbia forest regions," *Journal of Business Administration* 20(1/2) (1991/2): 345.

49 See A. Mallik and H. Rahman, "Community forestry in developed and developing countries: A comparative study," *Forestry Chronicle* 70(6) (1994): 731. The authors conclude that in developing countries, community forestry is "generally small, labour intensive and geared to meeting the basic needs of the community people." For a recent discussion of these issues, see the CPR Forum debate in the *Common Property Resource Digest* (March 1996), School of Forestry and Environmental Studies, Yale University. One author comments:

> If we assume power over and control of a resource to stem solely or even primarily from a concept of property rights, then it is reasonable that JFM [Joint Forest Management] is based on a re-ordering of that concept of property – and that it is "burdened ... by the property focus" ... Yet many forest department officials who ought to be the major proponents of this point of view see JFM primarily as a cooperative form of forest protection (not even cooperative management), and as such, do not see a significant shift in their ownership and control over the resource [J.Y. Campbell, "The power to control versus the need to use: A pragmatic view of joint forest management," *CPR Digest* March (1996): 9].

50 *Case Studies of Community-Based Forestry Enterprises in the Americas, presented at the symposium on forestry in the Americas: Community-Based Management and Sustainability, University of Wisconsin-Madison, February 3-4, 1995* (Madison: Institute for Environmental Studies and the Land Tenure Center 1995).

51 L. Meremo, "Mexico: Community-Based Forests Management Goes Commercial," in *Common Problems, Uncommon Solutions, Proceedings from the NGO Policy Workshop on Strategies for Effectively Promoting Community-Based Management of Tropical Forest Resources: Lessons from Asia and Other Regions,* ed. M.S. Berdan and J.P.A. Pasimio (Washington, DC: World Resources Institute 1994), 25.

52 Lynch and Talbot, supra, note 25 at xv.

53 Ibid., 135. Although these statements apply particularly to traditional peoples, the authors consider non-traditional rural communities within these lessons as well.

54 Duinker, supra, note 23 at 716.

55 Ibid., 718.

56 See Burda, Curran, Gale, and M'Gonigle, supra, note 4.

57 *Silva Forest Foundation, An Ecosystem-based Landscape Plan for the Slocan River Watershed: Part I – Report of Findings,* June 1996. The Silva Forest Foundation, under the guidance of ecoforester, Herb Hammond, is an internationally renowned leader in ecosystem-based forestry. Experience in the valley with watershed-based forest planning goes back to the early 1970s, including the 1974 publication of the landmark study, *The Slocan Valley Community Forest Management Project.*

58 See the press release from the Land Use Co-ordination Office of the provincial government, "Kootenay Land-Use Plan Implementation Unveiled" (21 October 1996).

59 For more detail, see D. Curran and M. M'Gonigle, *Aboriginal Forestry: Community-Based Management as Opportunity and Imperative,* Report 97-7 (Victoria: Eco-Research Chair of Environmental Law and Policy, University of Victoria 1997).

60 In 1997 the Court of Appeal decided that unextinguished title claims could constitute a legal "encumbrance" under the Forest Act so that they might pre-empt forest tenures, at least where this title was specifically established. *Haida Nation* v. *B.C. (Ministry of Forests)* [1988] SCAA No. 1. In addition, the Supreme Court of Canada handed down its landmark

decision in the *Delgamuukw* case ((1997), 153 D.L.R. (4th) 193), in which the Court affirmed unextinguished aboriginal title to much of the BC land base. Although the existence of such title would have to be demonstrated in any particular case, the Court's decision significantly strengthens the position of First Nations in future negotiations over forest rights.

61 National Aboriginal Forestry Association, *A Proposal to First Nations* (Ottawa: NAFA 1994). The text of the draft Act is not publicly available.

62 P. Smith, G. Scott, G. Merkel, *National Aboriginal Forestry Association, Aboriginal Land Management Guidelines: A Community Approach* (Ottawa: NAFA 1995), II-1.

63 *Partners in Forestry: First Nations Forestry Council Strategic Plan* (Victoria: Queen's Printer 1995).

64 25 U.S.C. ss. 3101-3120 (Supp. 1993).

65 P. Huff and M. Pecore, "Menominee Tribal Enterprises: A Case Study," in *Case Studies of Community-Based Forestry Enterprises*, supra, note 50. See also the presentation by Lawrence Wakau, President of Menominee Tribal Enterprises, in *The Business of Good Forestry*, supra, note 10.

66 M. Pecore, "Menominee sustained-yield management: A successful land ethic in practice," *International Journal of Forestry* 90(7) (1992): 12.

67 For a detailed review of this situation, see T. Beckley and D. Korber, *Clearcuts, Conflict and Co-Management: Experiments in Consensus Forest Management in Northwest Saskatchewan* (Edmonton: Northern Forestry Centre 1996). In this unpublished paper, the authors incorrectly refer to public participation in the private planning of the licence holder as "co-management," neglecting an important component of co-management which is the sharing of state (i.e., public) power with public groups.

68 Supra, note 25 at 120 and 24-25.

69 This situation was confirmed by Richard Overstall, Forestry Advisor to the Gitskan (personal communication, 17 July 1996), and John Cowel, a provincial treaty negotiator (personal communication, 16 August 1996). I am indebted to Gil Yaron for much of the research related to this analysis.

70 Ministry of Aboriginal Affairs, *British Columbia's Approach to Treaty Settlements: Lands and Resources* (Victoria: Queen's Printer 1996), 2.

71 See *Nisga'a Treaty Negotiations: Agreement-in-Principle*, issued jointly by the Government of Canada, the Province of British Columbia, and the Nisga'a Tribal Council, 15 February 1996, ss. 76-98, 19-22.

72 See R. Howard, "Native treaty not damaging, BC report says," *Globe and Mail* (17 December 1996): A6.

73 But see Curran and M'Gonigle, supra, note 59.

74 The preferred policy direction is more intensive industrial forestry, however constrained the potential may be in relation to competing jurisdictions where the growing conditions are more favourable. In addition, what is proposed in this and other studies is not the complete dismantling of all major tenures, merely a greater diversification that includes a range of "alternative" non-industrial tenure types where, at present, none exists. Thus, the issue is not the workability of a whole new forest economy for BC, but the ability to embark at all on a gradual transition to a more diversified system of production.

75 See *Forestopia*, supra, note 2, for a proposal to create a "new industrial strategy."

76 See *Seven Recommendations to Develop a Sustainable Forest Industry in British Columbia* (New York: Environmental Advantage 1996). The other five steps are to link value-added manufacturers to certified-sustainable wood supply; to create a task force to develop the sustainable forest industry financial infrastructure; to engage major forest companies in sustainable forest industry demonstration projects; to create an industry development forum for non-timber forest products; and to provide an in-depth financial description of the potential for the sustainable forest industry that reframes the definition of the BC forest economy.

77 For detailed discussion of these new developments, see *The Business of Good Forestry*, supra, note 10.

78 For a detailed discussion, see *Lumber Remanufacturing in British Columbia: First Report* (Victoria: Select Standing Committee on Forests, Energy, Mines and Petroleum Resources, British Columbia 1993). This issue is also discussed at length in *Forestopia*, supra, note 2 at Ch.8.

79 *Report on Special Project: Harvesting Timber and Marketing Logs* (Victoria: Ministry of Forests 1994).

80 Price Waterhouse, *Special Project on Harvesting Timber and Selling Logs* (Vancouver: Price Waterhouse 1995). The variation in the conclusions on stumpage reflects the difference in the reference points chosen for assessment. In the former study, the comparison was made with average stumpage, while Price Waterhouse used the higher stumpage paid under the Small Business Programme as the basis of comparison. As the wood in the log market did come from that program, the latter reference point is technically correct, but it fails to explain the benefits that could accrue should regular tenure holders also be required to market their products through a log market. The costs of silviculture borne by major licensees are not borne by small business operators; they are embedded in the stumpage paid, reducing the surplus to the province. Even when this fact is taken into account, however, the returns to the province are significantly higher. Because of the limited geographical scope of the Vernon experiment, it is not possible to extrapolate from it the exact prices and benefits that would result if a log market were introduced on a wider basis. It would, however, certainly be a boon to untenured manufacturers which presently suffer from both a constrained wood supply and underfinancing linked to an uncertain and limited supply.

81 For a review of these, see J. Fyfe, "Alternative bidding systems in Cascadia," *International Journal of Ecoforestry* 12(2) (1996): 238.

82 *Clearcut-free? Just did it, A Greenpeace Report* (Vancouver: Greenpeace Canada Forests Campaign 1995); "Silva Forest Foundation conducts Canada's first wood certification" (1995) *Silva Forest Foundation Newsletter* 1.

83 For a review of the government's failure here, drawing on a range of studies of the Code's lack of impact on forest practices and cut levels, stream and fish protection, landslide reduction, protective zonations, and so on, see the author's "Behind the green curtain," *Alternatives* 23(4) (1997): 16.

84 For a good review, see J.A. Wilson, J.M. Acheson, M. Metcalfe, and P. Kleban, "Chaos, complexity and community management," *Marine Policy* 18(4) (1994): 291-305. For an excellent prescription for models of resource management based on such insights, see E. Ostrom, "Designing Complexity to Govern Complexity," in *Property Rights and the Environment*, ed. S. Hanna and M. Munasinghe (Washington, DC: Beijer International Institute for Ecological Economics and The World Bank 1995), 33.

85 One model which might be looked at here is SODRA, a cooperative forest service organized by Sweden's 25,000 private foresters. The service is based in sixty-three forest districts, under seven regions. The members elect a board of trustees, and participate in the planning of all activities, in their forest district. SODRA is involved in virtually every aspect of forest management, from silviculture to marketing. Although having very limited authority, another model with some applicability is Quebec's "zones d'exploitation controlée" (ZECs). These zones are reviewed in P. Pearse and J. Wilson, *Local Co-Management of Fish and Wildlife: The Quebec Experience* (Vancouver: Faculty of Forestry, UBC 1996).

86 See, for example, the *Slocan Valley Community Forest Management Project, Final Report,* Winlaw BC, 1975; the *Village of Hazelton's Framework for watershed management*, Hazelton BC, 1991; and the Tin-Wis Coalition's Forestry Working Group, Draft Model Legislation *The Tin Wis Forest Stewardship Act,* Vancouver, 1991. On the latter, see E. Pinkerton, "Co-management efforts as social movements," *Alternatives* 19(3) (1993): 33. The province's existing "community resource boards" do not provide a model as they are purely advisory in function.

87 Bouthillier and Dionne wisely note that "two elements must be integrated into the management framework: debate and decision-making. These can take various forms: the debate aspect comes, first and foremost, under the aegis of municipal assemblies which are open to all citizens" (supra, note 44 at 12).

88 For the CRB's own approach to this, see *From Compromise to Consensus: A Blueprint for Building Sustainable Communities and Protecting Ecosystems in Clayoquot Sound* (Tofino: Central Region Board 1996).

89 While Haley and Luckert's proposal in Chapter 6 is interesting, it is flawed in a number of respects. First, it is highly abstract insofar as it is predicated upon placing monetary values

on a range of intangibles. Second, it fragments responsibility for the diversity of interests in a single forest in an unwieldy fashion, especially considering that the central management authority/authorities are remote from the forest. Third, it entails huge transaction costs. Finally, by asking the Crown to pay for imposing environmental constraints on logging, it effectively provides a subsidy to the status quo of industrial logging. That is, the proposal incorrectly takes industrial logging (with all its ecosystem externalites) as the assumed starting point, rather than mandating logging to fit within ecosystem constraints. We are, after all, dealing with public land.

90 Indeed, the integration of these tools goes even further, requiring a comprehensive reassessment of areas as diverse as taxation policy (taxing resource consumption, not labour), economic development (shifting from trade towards more multiplier-based strategies), international trade (acting to remove the hidden environmental and social subsidies that underpin almost all international trade), and so on.

8

Timber Pricing Policies and Sustainable Forestry

Brian L. Scarfe

Introduction

This chapter considers the interrelationships between timber pricing policies and sustainable forestry. Timber pricing policy refers to the institutional arrangements, and the basic terms and conditions, under which the right to harvest a particular stand of timber is conveyed by the owner of the resource to a forest industry participant. Timber pricing policy is, therefore, integrally related to the forest tenure system through which harvesting rights are conveyed, and associated obligations are acquired.

Timber pricing policies are most often designed with the goal of rent capture in mind. However, insofar as these policies are able to influence the pace at which forest resources are harvested and replenished over time, they clearly can also affect the sustainability of timber resources. Accordingly, this chapter begins with a discussion of timber pricing policy, largely viewed as an instrument for rent capture. It then describes the timber pricing or stumpage appraisal system that currently exists in British Columbia. The issue of whether or not this system leads to the under-collection of forest resource rents lies at the heart of the Canada/United States dispute over softwood lumber trade; therefore, a section on this dispute follows. The chapter then turns to the question of how society should assess forest resource values, i.e., is the sustainability of long-lived renewable resources affected by an inter-temporal market failure that results from market interest rates that significantly exceed the social discount rate?[1] Can alternative timber pricing policies, involving incentive systems that could minimize this sustainability problem, be designed? What implications might these alternative timber pricing policies have for tenure reform and for the allocation of allowable annual cuts?

Timber Pricing and Rent Capture Systems

In British Columbia, as in other Canadian provinces, the vast majority of forest land is owned by the Provincial Crown. Timber is priced, when harvested, through a stumpage appraisal system so as to capture for the Crown a share of the value of the resource. This share has the nature of economic rent, which is the return to the owner of a resource whose value arises either from its basic scarcity or from its differential quality or locational advantage in comparison with other similar resources.

Because both scarcity rents and differential rents may be available to the resource owner, a rent capture system based on two distinct instruments is desirable. For example, mature timber stands could be auctioned off via a cash bonus bidding system to potential harvesters, who would also, when harvesting occurs, pay a market-sensitive royalty based on the prices of semi-processed products that could be manufactured from the timber. In such a system, the differential rent, based on resource quality as perceived ex ante, would be captured by the bonus bidding system, whereas the scarcity rent, based on market performance as measured ex post, would be captured by the royalty system. Although a timber pricing system with these characteristics underlies British Columbia's Small Business Forest Enterprise Program (SBFEP), the province's stumpage appraisal system has generally not evolved along these lines.

In British Columbia, the forest tenure system provides for a variety of licence forms through which access to Crown timber can be established. The volumes of timber harvested in 1993-4 from the most important of these licence forms are outlined in Table 8.1, along with some basic characteristics of each of form.

The forest tenure system in British Columbia has evolved over a century. For example, timber licences are the current manifestation of the old temporary tenures established in British Columbia before 24 December 1907. It is especially important to note that many of the current tenures have evolved in connection with timber-processing investments made by industry participants in appurtenant mills. Sustaining these investments, and the stability of employment they provide to forest-based communities, requires secure long-term supplies of wood fibre. The need for dedicated fibre supply has led to forest tenures with long durations. The imposition of additional obligations for silviculture and reforestation investments on tenure holders has enhanced the need for security of long-term tenure. Such a system cannot easily be replaced by a "market-based" system, which would allocate harvesting rights on shorter-term contracts based on the bids made in some form of "competitive auction," without creating the potential for significant disruptions to rural communities.

Crown timber in British Columbia is priced, when harvested, in accordance with a complicated stumpage appraisal system, known as the Comparative Value Pricing System, which operates separately on the Coast and in the Interior. Under this system, the relative value of each stand of timber being sold depends on estimates of both the total selling price of products which could be generated from the stand and the cost of producing them. Comparative value stumpage prices are then established so that the

Table 8.1

Volumes billed by licence type, 1993-4 (million cubic metres and percentages)

	mm^3	percent	base	duration[b]	obligations	pricing
Forest licences[a]	39.2	50.2	volume	15 years	major	stumpage appraisals
Tree farm licences[a] (Schedule B lands)	12.9	16.6	area	25 years	major	stumpage appraisals
Timber sale licences (SBFEP)	10.2	13.1	area	2 years	minor	stumpage plus bonus bids
Timber licences (royalty bearing)	4.5	5.8	area	until timber removed	major	appraised royalties[c]
All other licences	3.0	3.8	various	various	various	stumpage appraisals
Total for Crown land	69.8	89.5				
Private and other land	8.2	10.5				
Overall total	78.0	100.0				

a Total volume harvested from Tree Farm Licences (TFLs) exceeds the number recorded in this table, because these licences also include additional lands that are either under Timber Licences (TLs) or held privately.

b Most licences are renewable. The duration given refers to the normal initial duration of a licence.

c Royalties are currently (1995-6) set at 60% of appraised stumpage, and will increase in six annual increments to reach 100% of appraised stumpage on 1 April 2001.

average rate charged will approximate, as appropriate, a predetermined Coastal or Interior target rate per cubic metre harvested. The target rates are adjusted quarterly in response to market forces, as measured by two Statistics Canada price indices, one for the Coastal lumber market and the other for the Interior lumber market.

New relationships were established, effective 1 May 1994, between the target rates and the relevant StatsCan index numbers to provide approximately \$400 million annually in additional revenues to the provincial government to finance Forest Renewal BC. These new relationships were based on the optimistic assumption that lumber market prices had moved to a permanently higher plateau, and they increased the sensitivity of the stumpage pricing formulas to lumber price movements. The current formulas that determine the Coastal and the Interior target rates (TRs) are the following rather awkward non-linear functions:

Coast: $$TR = \frac{INDEX}{138.5} \times \$10.59 \text{ if INDEX} \leq 160$$

$$TR = \$12.23 + \frac{INDEX - 160}{25} \times \$9.73 \text{ for } 160 < INDEX \leq 185$$

and

$$TR = \$21.96 + \frac{INDEX - 185}{49} \times \$7.45 \text{ if } 185 < INDEX,$$

where INDEX is the Statistics Canada Softwood Lumber Price Index Number for the BC Coast (Series D613601), and

Interior: $$TR = \frac{INDEX}{139.0} \times \$8.59 \text{ if INDEX} \leq 161$$

$$TR = \$9.95 + \frac{INDEX - 161}{25} \times \$9.35 \text{ for } 161 < INDEX \leq 186$$

and

$$TR = \$19.30 + \frac{INDEX - 186}{49} \times \$6.74 \text{ if } 186 < INDEX,$$

where INDEX is the Statistics Canada Softwood Lumber Price Index Number for the BC Interior (Series D613600).[2]

In the application of the stumpage pricing system, BC also imposes a minimum stumpage rate of \$0.25 per cubic metre on certain low-quality grades and stands of timber. Minimum stumpage rates are often applied to the "guts and feathers" left after the primary harvest of mature timber has occurred. They also apply to timber whose health is endangered by insects and other pest infestations. Accordingly, minimum stumpage rates provide an incentive for companies to harvest marginal but salvageable timber stands so that reforestation can proceed in a timely manner. Although the provision of such an incentive may well lead to an increase in fibre supply to milling operations, a significant proportion of this increased fibre supply is designated for the production of pulp, woodchips, hog fuel, and other products such as shakes and shingles, rather than for softwood lumber production.

In principle, minimum stumpage rates apply to all timber supplies that are uneconomic to produce at higher appraised stumpage rates. An increase in minimum stumpage rates, by itself, would reduce the total timber supply available, but it could nevertheless increase the proportion of the total harvest sold at minimum stumpage rates, depending on the shape of the distribution of appraised stumpage values. The reason for this outcome is that some timber supplies that were previously intra-marginal become marginal when minimum stumpage rates are increased.

Coastal and Interior target rates depend only on softwood lumber market prices. Other end-product prices do not appear in the target rate formulas.[3] The argument for including other end-product prices relates to the notion that when pulp and/or woodchip prices are high relative to softwood lumber prices, domestic firms have an incentive to produce "excessive" amounts of softwood lumber as a byproduct of their woodchip and/or pulp production, which then floods onto the North American softwood lumber market and depresses lumber prices. Weighting woodchip and/or pulp prices into the stumpage formulas would create a disincentive for such "excessive" production to occur at such points during the course of a business cycle that affects pulp and paper markets differentially from softwood lumber markets.

The problem with this proposal is that it shifts the relative burden of stumpage payments over the course of the business cycle between small non-integrated lumber producers, which produce woodchips as a byproduct, and large integrated companies which produce pulp and paper as well as softwood lumber. Pulp prices and woodchip prices are closely correlated from a cyclical perspective, but are often out of phase with softwood lumber prices. Although some form of weighted-average formula might stabilize the burden of stumpage payments in relation to overall profitability for large integrated companies over the course of the business cycle, it could destabilize the situation for small non-integrated companies. At points in the business cycle where woodchip and/or pulp prices are rising relative to lumber prices, these non-integrated companies could be placed at a disadvantage, and thereby rendered vulnerable to takeover bids from large integrated companies. Changing the distribution of the burden of stumpage payments among industry participants over the course of the business cycle could have undesirable side effects on the degree of industrial concentration and the overall competitiveness of the domestic industry.

An alternative approach to increasing the "market-orientation" of the BC stumpage pricing system would be to include in the target rate formulas (using an appropriate weighted-average approach) the most-recently-available average price for stumpage billed (which includes recent bonus bids) from the Small Business Forest Enterprise Program (SBFEP). If this were done, the new stumpage rate formulas for both the Coast and the Interior would take the form:

new target rate = existing target rate + w (SBFEP price − existing target rate),

where SBFEP price is the most-recently-available average price for stumpage billed, which includes recent bonus bids, from the SBFEP, and w is the weight attached to the SBFEP price in the new target rate formulas. A reasonable choice for w would be approximately 0.13, if the weight is intended to measure the proportional contribution of the SBFEP to the total BC harvest volume.

In recent periods of relative lumber market stability, the average SBFEP price for stumpage billed has exceeded the Coastal target rate by about $21.50 per cubic metre and the Interior target rate by about $26.00 per cubic metre. Presumably, these gaps measure "obligation differences" between major tenure holders and participants in the SBFEP. In fact, these two numbers approximate, respectively, the average differences for January 1994, April 1994, October 1994, and April 1995. The first two of these data points precede the introduction of "super-stumpage," associated with the financing of Forest Renewal BC, in spring 1994, and the second two postdate this introduction. If these differences were maintained in the future, then incorporating the SBFEP price into the target rate formulas would increase the Coastal target rate by $2.80 and the Interior target rate by $3.38, from the levels generated by the StatsCan index numbers.

The implementation of this proposal would likely require the BC Ministry of Forests to ensure that the SBFEP portion of the allowable annual cut is maintained and in all probability enhanced, by allocating additional timber stands to the SBFEP whenever it is possible to do so through the application of existing take-back rules when other forms of licence come up for renewal. (Reductions to allowable annual cuts in various timber supply areas restrict the pace at which BC Forests can move in this direction.) The reason for this reallocation would be to increase the probability that SBFEP prices can remain at arm's length from any potentially coercive activity involving the larger industry players, the incentive for which is enhanced by inclusion of the SBFEP price in the target rate formulas.

Whether or not the Comparative Value Pricing System in modified or unmodified form actually captures forestry resource rents in a full and effective manner is not easy to determine. Clearly, however, it provides a reasonably stable fiscal regime for the forest industry in British Columbia. It captures for the Crown a significant proportion of the revenues generated by higher prices for forest products in the marketplace. It allows for species differentials in establishing stand-specific selling prices, and for differential timber quality in the appraisal of harvesting costs relative to the particular stand of timber. It operates differentially between Coastal and Interior forest districts, but in each case in a market-sensitive manner. Nevertheless, the effectiveness of public sector rent capture continues to be an issue that lies at the heart of the softwood lumber dispute.

Timber Pricing and Softwood Lumber Exports

For the past dozen years, the Canadian softwood lumber industry has been an ongoing target for countervailing duty actions by the United States government. In fact, the softwood lumber dispute has recently run through its fourth phase. Canada won the first battle within the US international trade tribunals in 1982-3, bought its way out of the second in 1986-7 by imposing a 15 percent export tax on softwood lumber shipments to the United States (which was later rolled into stumpage increases by provincial governments), and won the third under the dispute settlement mechanisms of the Free Trade Agreement in 1992-4. A fourth countervailing duty action was recently threatened by the US Coalition for Fair Lumber Imports, despite (at least in British Columbia) significant recent increases in stumpage rates and royalty charges.

The Coalition for Fair Lumber Imports alleges that timber policies in various Canadian provinces provide a specific upstream subsidy to softwood lumber producers which allows their exports to penetrate "excessively" the American market, thereby putting American producers under "unfair" or "injurious" competitive pressure, especially at times of falling demand for softwood lumber. However, the transmission processes through which the timber policies of Canadian provinces are alleged to cause "injury" to US softwood lumber producers have never been fully articulated and are tenuous at best.

Nevertheless, a necessary condition for provincial timber policies to cause "injury" to US softwood lumber producers would be for these timber policies to have created price-reducing distortions in softwood lumber markets across North America. Whether or not a stumpage appraisal system fails fully to capture economic rents for the Crown from the harvesting of timber in provincial forests may have little bearing on the price-distortion issue, either because allowable cuts and related timber harvesting restrictions constrain quantitative adjustments in the Canadian supply of softwood lumber, or because forest resource rents that are "left on the table" may simply be dissipated in higher private sector costs, especially wage costs.

Four tests may be used by the US authorities to determine whether or not a specific upstream subsidy exists in the case of a resource input, such as Crown timber, into a manufacturing process, such as softwood lumber production, when the resource input is sold at an administered price by its public sector owner to the private sector manufacturer. The first test is to compare the administered price charged for the resource input with the market prices at which equivalent arm's-length sales are occurring between private sector firms in the marketplace within the same jurisdictional area. The second test makes a similar comparison of the administered price with the market prices of other sales that may be occurring between the Crown and the private sector through a process of competitive bidding, again within that same jurisdictional area. The third test involves a comparison of the administered price with the costs per unit of the resource harvested that are borne by the public sector resource owner in its resource management role. The fourth test involves a cross-border comparison of stumpage prices between the exporting jurisdiction and comparable areas within the United States. Incomparability problems are clearly pervasive in all four kinds of test. Nevertheless, if it had been faced with a fourth countervailing duty action, Canada would have needed to mount an appropriate defence against each of these comparative tests.

The problem with the application of the first test is that private markets for cutting rights are simply undeveloped in Canada because most of the marketable timber is owned by the Crown, in the right of individual Canadian provinces. The second test, at least in the case of British Columbia, would imply a comparison of the administered stumpage prices applied to timber harvested by major tenure holders with the stumpage prices determined by the "bonus bidding" system of the Small Business Forest Enterprise Program. Any such comparison, however, would at least need to recognize that major tenure holders incur substantially larger development costs and silviculture obligations than the firms involved in the SBFEP. Moreover, stumpage prices determined in the SBFEP program are unlikely to be representative of the "competitive stumpages" that might apply should major tenures be awarded through a bidding process.

The third test involves an estimate of the value of the services provided by the relevant forest management agency per unit of harvested timber, and uses this estimate as an offset against the stumpage prices charged when public timber is harvested. In BC's case, it would be important to demonstrate that the new higher stumpage revenues that flow directly to Forest Renewal BC are not fed back to the forest products industry in a way that lowers existing industry costs or obligations for reforestation and silviculture investments. In addition, it might be valuable to point out that in many timber supply areas the US Forest Service does not recover the full costs of the services it provides through the sale of timber to the private sector.

The fourth test, like all cross-border comparisons, is fraught with difficulties. Cross-border comparisons of stumpage prices would at least need to take into account differences in the species composition of harvested timber, differences in netbacked lumber prices resulting from transportation and other costs, differences in development costs and silviculture obligations, differences in the capital costs of timber harvesting and lumber manufacturing, and differences in the timing of stumpage payments. More generally, the application of such tests in the determination of the level of countervailing duties would tend to negate comparative advantages based on natural resources, and thereby undermine the reciprocal gains from trade.

The fourth countervailing duty action recently threatened by the US Coalition for Fair Lumber Imports led officials of the United States government to propose a number of changes in timber policies that Canadian provinces might pursue to reduce the probability of such an action against Canadian softwood lumber exports into the American market. These changes included adjustments to the institutional arrangements under which Crown timber is sold to forest products companies, and adjustments to provincial pricing policies for access to Crown forest resources. In the case of British Columbia, it was suggested by the US trade representative that in the longer term the forest tenure system should be reformed to become more "market-based" by requiring a larger proportion of licensed tenures to be awarded by a process of "competitive bidding." In the interim, the method of stumpage determination should be altered so as to increase significantly the target rates for both the Coast and the Interior, and internal log-trading opportunities should be considerably expanded. Implementing these changes to forest tenure systems and timber pricing policies would clearly have been damaging to provincial autonomy with respect to resource management practices.

In exchange for commitments along these lines, and similar commitments from other provinces, the US Trade Representative would find ways and means to avoid future trade litigation that might otherwise be initiated by the Coalition for Fair Lumber Imports, and thereby provide greater certainty with respect to market access for Canadian softwood lumber. In the end, however, Canada has agreed to implement a set of quantitative border controls to avoid a fourth US countervailing duty action.

Changes in American trade law related to the application of the four upstream subsidy tests have rendered it more difficult than ever for Canada to succeed in a "subsidy determination" case before the International Trade Administration branch of the US Department of Commerce. Moreover, statistics on the share of Canadian lumber in the US softwood lumber market showed this share to be 37.2 percent in the second quarter of 1995. This percentage was the highest quarterly share number on record. At the same time, American lumber market prices fell significantly during 1995 because housing starts decreased in response to the higher real interest rates established in 1994. In consequence, it would have been difficult to defend Canadian lumber producers against circumstantial evidence if the US International Trade Commission had been asked to make an "injury determination" in response to a countervailing duty action initiated by the Coalition for Fair Lumber Imports.

Had a fourth countervailing duty action been initiated, it is highly probable that Canada would have wound up, once again, relying upon the dispute settlement mechanisms of the Free Trade Agreement. A loss on this occasion could well have resulted in a permanent barrier, in the form of a significant ongoing countervailing duty, being erected against Canadian softwood lumber exports. This damaging outcome needed to be avoided, either by negotiating a suitable resolution to the softwood lumber dispute, or by imposing some sort of border measure at the eleventh hour before countervailing duties were imposed.

The real issue in the softwood lumber dispute is not the manner in which forest products companies acquire access to Crown timber resources. Rather, the real issue is the volume of Canadian lumber being sold into the American market as a percentage of the total market size. One of the most important determinants of the share of the American market supplied by imports from Canada is the real exchange rate, which is currently at a level at which Canadian lumber is very competitive in the American market.[4]

If volume is the issue, it is important to address it directly. Changing Canadian stumpage systems to increase their degree of "market sensitivity" is unlikely to be an efficient response to the volume issue. Moreover, British Columbia in particular had, in spring 1994, imposed a major increase in stumpage prices on the forest products industry in connection with the financing of Forest Renewal BC. The province could not have contemplated a further major adjustment in the stumpage system following so closely on the heels of the 1994 adjustments. At most, only minor adjustments to this system would have been feasible. Furthermore, any increase in stumpage rates thereby imposed would apply to all timber harvests within a given provincial jurisdiction, regardless of how the timber was used or where the final products were sold. What is of concern, however, is only the timber harvested to produce softwood lumber that is exported to the United States.

Accordingly, a more appropriate response is to institute some form of Canadian-controlled border measure, which would not directly burden either export shipments of softwood lumber to third countries or production of end-products other than softwood lumber. The absolutely worst form of border measure, from Canada's perspective, is the imposition by the United States of a countervailing duty. Such a measure would lead to deadweight losses in the form of direct transfers from the Canadian industry to the American Treasury. Far better is Canada's own imposition of a quantitative border measure, or export quota system, from which any revenues raised remain within Canada and its provinces. This approach is the solution that has now been implemented.

The implementation of an appropriate border measure was, of course, an option of last resort when US imposition of countervailing duties appeared to be imminent. What Canada obtained in exchange for imposing such a measure is continuous access for all lumber export volumes that remain unfettered by any form of US-based restraints, including countervailing duties, and a written understanding that no further trade harassment will occur through the actions of either the members of the Coalition or the American authorities for the next five years while the border measure remains in place. The global export quota for softwood lumber shipments from Canada's four main producing provinces (British Columbia, Alberta, Ontario, and Quebec) is 14.7 billion board feet per year, and this quota has now been allocated among industry participants. After setting aside a 2 percent reserve for new lumber mills, the federal government has allocated two-thirds of the quota to Western Canada (BC 59.0%, Alberta 7.7%) and one-third to Eastern Canada (Ontario 10.3%, Quebec 23.0%). The quota allocation to BC softwood lumber producers and remanufacturers is, therefore, 8.5 billion board feet per annum. Wholesalers will not be directly allocated export quotas, but they may arrange to acquire export permits charged to the quota allocations held by producers and remanufacturers at the time softwood lumber purchases are made. For reasons of flexibility, quota allocations are also likely to be "rentable" among softwood lumber producers and remanufacturers.

Several strategic points arise from the discussion of border measures. First, unlike stumpage price increases, Canadian-controlled border measures do not burden the Canadian softwood lumber manufacturer on domestic shipments or export shipments to third-party countries. Nor do they burden timber harvested for non-lumber purposes. Second, unlike a countervailing import duty, the proceeds from the export fees that are assessed when quotas are exceeded stay within Canada, and indirectly accrue to the provincial governments from whose timber lands the exported lumber was produced. Third, there may be questions about the legality of the export quotas implemented by Canada as a means of ending the softwood lumber dispute. Indeed, to protect the interests of third-party countries from the trade-diverting effects of export restraint programs, both the General Agreement on Tariffs and Trade and the new World Trade Organization agreement seem to prohibit such restraints. However, it is a moot point whether Canada would be challenged under GATT/WTO rules for imposing export restraints simply to avoid the impact of a (presumably legal) US countervailing duty action which would have the same effects. Fourth, there are serious concerns about the extent to which an export restraint program applied to softwood lumber shipments to the United States creates unfavourable precedents for other sectors of the economy, which would induce the American authorities to challenge further Canada's use of the dispute settlement mechanisms of the North American Free Trade Agreement.

In retrospect, BC should continue to take a position with the US Trade Representative that it will approach long-term tenure reform in its own way, and in its own time, and that it does not "subsidize" its forest industry, but rather recovers through stumpage charges far more than its costs of forest management. BC should be cautious about moving its timber pricing policies closer to those of the US Forest Service, whose expensive timber-auction system neither creates appropriate incentives for private sector management and replenishment of scarce forest resources nor generates sufficient revenues to finance the full costs of public sector management. Rather it should continue to do better than that. Major changes in the tenure system in directions that respond to short-term American demands for a more "market-based" approach must not be inconsistent with BC's fundamental long-term goal to husband and replenish its valuable forest resources. However, the further encouragement of an active log market in the BC Interior does seem to be warranted.

The Vancouver log market has been in operation for many years, and provides an important institutional mechanism for logs of all species and grades to be exchanged among Coastal forest industry participants, including softwood lumber producers. Until the mid-1980s, transaction prices from this market were used to determine stumpage charges for Crown timber harvested in the BC Coastal region. Concerns whether or not these prices actually represented competitive arm's-length transactions led to the replacement of Vancouver log market prices by the relevant Statistics Canada softwood lumber price index in the Coastal stumpage formula, consistent with earlier practice for the BC Interior. Federal and provincial restrictions on log exports tend to enhance these concerns, because foreign buyers are in large part excluded from the marketplace.

Similar concerns could arise with respect to the interpretation of the price quotations that would emanate from an active log market in the BC Interior, so that it is unlikely that such price quotations would be used directly in a stumpage-determination formula. Nevertheless, this fact does not mean that the development and facilitation of such a log market (expanding significantly on the experimental log market that operates in Lumby, BC) would be undesirable. Indeed, the encouragement of expanded log markets should be considered to be a means through which the tightness of the current linkages between the allocation of fibre supplies and the existence of appurtenant mills could be relaxed over time. Such relaxation appears to be a necessary step in the process of tenure reform in British Columbia. Perhaps the BC Ministry of Forests should develop a regulation that requires a certain minimum proportion of the Crown timber harvested in each Timber Supply Area or Tree Farm Licence to be sold through an active regional log market.

Timber Pricing and Sustainability

Two major sources of market failure are likely to be observed in the management of forest lands. Both of these sources of market failure have adverse consequences for sustainability, and therefore for the preservation of the long-term value of these resources. The first source of market failure is that both over-harvesting and under-replenishment will tend to occur if there is a failure by market participants to recognize the non-timber (or multiple use) values of our forest lands in their profit and loss calculations. Thus, the private cost of timber harvesting may be smaller than the social cost, resulting from the negative externality of reduced non-timber values that timber harvesting may cause, whereas the private benefit of timber replenishment may be smaller than the social benefit, resulting from the positive externality of enhanced non-timber values that may not be appropriated by market participants who are contemplating reforestation investments.

Timber pricing policies can go some way towards overcoming these externalities. A significant increase in stumpage charges may help to internalize the cost of reduced non-timber values to those undertaking timber harvests, whereas a program that supports investments in forest renewal may help to offset the disincentives associated with replenishment. Indeed, the increased stumpage charges can be used to create a dedicated fund for reforestation investments and watershed/habitat restoration projects. At one basic level, this aim is what Forest Renewal BC is all about. Essentially, it is a scheme designed both to offset the negative externalities associated with timber harvesting and to encourage the positive externalities associated with timber replenishment and the restoration of forest lands and adjacent watersheds/habitats. Of course, Forest Renewal BC also has other objectives, largely associated with the need to maintain employment in the forest sector as it undergoes adjustment to the reduced fibre supplies that may result from the Protected Areas Strategy and the CORE process (which take additional lands out of active forestry), the reassessment of allowable annual cuts in various timber supply areas, and the application of the Forest Practices Code. Nevertheless, it is interesting to note that the creation of a fund through which revenues raised from timber harvests would be dedicated to timber replenishment was originally proposed by Fred Fulton in his 1910 Royal Commission report.

The second source of market failure involves inter-temporal considerations rather than multiple use externalities. Fundamentally, for long-lived renewable natural resources like BC's forest resources, inter-temporal market failure is virtually inevitable. Given the length of rotation periods, of 100 years and more in British Columbia, there are no effective futures markets whose prices could provide adequate incentives for the private sector to undertake sufficient investments in second- and third-growth timber stands to replace BC's old-growth forests, even if forest tenures were reformed to provide additional security to longer-term area-based tenures. Moreover, even if such futures markets existed, current market levels for real interest rates would deter ecologically sound investments in timber replenishment. Given the long rotation periods of BC forests and the current level of interest rates, public sector involvement that both regulates and subsidizes reforestation and silviculture investments, and the restoration of adjacent watersheds/habitats, becomes the natural accompaniment of inter-temporal market failure.

Another way of expressing this point is to argue that, in real terms, market interest rates overstate the social rate of discount that should be applied in the management of long-lived renewable forest resources. The appropriate real rate of interest for discounting the future net benefit stream that may be associated with forest restoration and replenishment investments combines additively a basic time-preference element with a specific risk-premium. However, a case can be made for setting at least one of these two elements equal to zero. In a general sense, the value to society of a sound ecosystem ought to increase gradually over time in a way that offsets the effect of discounting on the present value of the future stream of net benefits that may accrue from investments in ecosystem restoration that are made today. In particular, there may be a general gain in the form of a "public good externality" that accrues from these investments through their potential to preserve and/or enhance the biodiversity of the environment. Put differently, there is a general ecosystem risk to society if forest replenishment and restoration investments are not made and the environment is allowed to deteriorate. This general risk may be accommodated in present value calculations by dropping either the time-preference element or the specific risk-premium.

Nevertheless, investments in forest replenishment made today have outcomes that not only accrue in the future but also are to some degree uncertain. Accordingly, a real discount rate that reflects their future orientation and their basic uncertainty should be applied to these investments; one which takes into account the offsetting general riskiness of not undertaking investments in ecosystem health, or, alternatively, the increasing value of ecosystem health to society. Thus, it would be appropriate to apply a real discount rate of perhaps 3 percent per annum to the net present value calculations that necessarily underlie the application of a cost-benefit criterion to the evaluation of forest restoration projects. The application of significantly higher real discount rates (e.g., 6 percent to 8 percent) would not be appropriate in this context.

The economic and environmental benefits that enter the determination of net present value using the suggested 3 percent real discount rate are likely to be multi-faceted in the case of forest restoration projects. Forest values may be enhanced, if appropriate replenishment and silviculture investments enable harvestable timber to regenerate more quickly. Recreational values may be enhanced by an increased abundance and variety of the flora and fauna of the forest lands that are restored. There may also be a potential gain from the renewed availability of clean water, from both instream and consumptive use perspectives, as well as a potential gain from the associated recovery of the fish stocks made available through the rehabilitation of adjacent watersheds. These non-timber values should be recognized together with the timber values that accrue to successful forest replenishment investments.

In Chapter 6 of this book, David Haley and Martin Luckert advocate a "sharecropping" approach to timber replenishment investments that recognizes the non-timber values associated with healthy, growing forests. This approach would lower the capital costs of timber replenishment to private sector forest industry participants because part of the costs would be incurred by the public sector. In many ways, the suggestion that a low social discount rate of, say, 3 percent in real terms should be applied to forest restoration investments and watershed/habitat restoration projects is complementary to the "sharecropping" proposal. Of course, by no means all of the investments made by Forest Renewal BC will fall into the forest restoration category.

Conclusions

The overarching objective of BC forests policy seems to be sustainability of forest resources, non-timber resource values, forest-based communities, and timber harvest levels. The potential for conflict among these four sustainabilities is self-evident. Regulating the allowable annual cut in each timber supply area is central to these issues of sustainability.

Timber pricing policies and tenure systems are also fundamental to the issues of sustainability that are so paramount in the management and the protection of our forest-based resources, including those which represent non-timber values. Tenure systems should continue to evolve towards longer-term, area-based tenures that are renewable only when clearly specified obligations for forward-looking replenishment investments, made in accordance with "best practice" reforestation and silviculture techniques, have been fully met by the licensee. However, even with long-term renewable area-based tenures, such as the Tree Farm Licences that provide exclusive harvesting rights to timber stands on Crown lands, property rights to future timber stands are not sufficiently vested in private corporations for these corporations to have the incentive to replenish fully the timber they harvest. But there is certainly no public sentiment towards further alienation of Crown timber resources to forest product companies. Accordingly, forest resource depletion remains a major issue in British Columbia, and an issue which will continue to require regulatory solutions and public sector involvement because of the inter-temporal market failure problem.

Nevertheless, sensitive timber pricing policies can influence the pace at which our forest resources are harvested and replenished over time, and they must therefore be included in the set of economic instruments required for improved stewardship over our natural resource endowment. In particular, stumpage charges should remain high to reflect the economic value of timber resources to society, but incentives to replenish our forest inventories after harvesting has occurred should also be maintained at a generous level. If it is well managed, Forest Renewal BC should be a step in the right direction. However, as Fred Fulton long ago feared, a raid on this dedicated fund is already under way, so that there is now considerable danger the endowment it represented will be dissipated in favour of other, "more pressing" government priorities.

Notes

1 Social discount rate refers to the opportunity cost of capital to society: "Discounting is the principal analytical tool economists use to compare economic effects that occur at different points in time ... The higher the discount rate, the less future benefits and the more costs matter in the analysis. Selection of a social discount rate is also a question of values since it inherently relates the costs of present measures, to possible damages suffered by future generations if no action is taken" (from *Climate Change: Evidence and Implications, Report of the Intergovernmental Panel on Climate Change (IPCC) Summaries for Policymakers. Working Group III: Technical Assessments of the Socioeconomics of Impacts, Adaptation, and Mitigation of Climate Change Over Both the Short and Long Term and at the Regional and Global Levels,* published in 1997 in *Foreign Affairs* and available on the Web at www.foreignaffairs.org/envoy/documents/v7n2_clim.html).

2 Effective 1 June 1998, the formulas were adjusted so as to reduce the Coastal and Interior target rates by $8.10 per cubic metre and $3.50 per cubic metre respectively. As a result, the Coastal target rate fell to $24.97 per cubic metre and the Interior to $21.40. These reductions go some way to alleviate the "cost-price squeeze" on the industry that has resulted from the implementation of the Forest Practices Code and the collapse of Pacific Rim markets.

3 In the 1 June 1998 adjustments, woodchip prices were linked into the softwood lumber price index numbers to provide a somewhat broader base for target rate determination. However, the weight given to woodchip prices remains small relative to the weight given to softwood lumber prices in the target rate formulas.

4 Over the twenty-one years before 1995, the Canadian share of the US softwood lumber market varied from a low of 18.0% in 1975 to a high of 33.4% in 1994. Much of the growth in this share can be attributed to timber harvests in provinces east of the Rocky Mountains, because over the same period the BC share has varied from a low of 14.3% in 1975 to a high of 19.8% in 1978. The average Canadian share for 1974 to 1994 inclusive is 27.6%, and the average BC share is 17.9%. However, as noted earlier, in the second quarter of 1995, the Canadian share reached a record high level of 37.2%; in this same quarter, the corresponding BC share was 20.3%.

In fact, changes in the real exchange rate explain over three-quarters (79%) of the variance in the Canadian share of the US lumber market over the past twenty-one years. Moreover, each 1% increase in the real exchange rate (expressed in Canadian dollars per American dollar) increases the Canadian market share by about 0.4 percentage points. See

D.M. Adams, B.A. McCarl, and L. Homayounfarrokh (1986). The Adams article used annual data from 1950 to 1983. Using more recent annual data from 1974 to 1994, the simple linear regression equation that relates the Canadian percentage share of the US softwood lumber market (S) to the real exchange rate (E) expressed in Canadian dollars per US dollar is

$$S = 27.6 + \frac{7.7}{0.234} (E - 1.167)$$

For this regression equation, $R^2 = 0.789$, and $\bar{R}^2$ (which corrects for degrees of freedom) is 0.714. At the mean value for the real exchange rate, each 1% increase in the real exchange rate increases the market share by 0.384 percentage points (consistent with the number 0.4 calculated by Adams, McCarl, and Homayounfarrokh in their 1986 article). The regression equation reported above, however, shows a high degree of first-order autocorrelation in the residuals, because the Durbin-Watson statistic is 0.354. This finding may suggest the omission from the regression equation of other explanatory variables, such as US housing starts, which would explicitly capture cyclical movements in US softwood lumber demand. Further analytical work is required to improve the specification of the regression equation. But if the real problem as perceived by the Coalition for Fair Lumber Imports is the share of the US market supplied by imports from Canada, tinkering with stumpage formulas to somehow increase the degree of "market sensitivity" of Canada's provincial stumpage systems is unlikely to provide an adequate response. Instead, it might be preferable to impose a sliding-scale export tax on softwood lumber shipments to the United States, which would offset the effects of movements in the real exchange rate on the Canadian market share.

References

Adams, D.M., B.A. McCarl, and L. Homayounfarrokh. 1986. "The role of exchange rates in Canadian lumber trade." *Forest Science* 32(4): 973-88
BC Forest Resources Commission. 1991. *The Future of Our Forests*. Victoria: BC Forest Resources Commission
Boardman, A.E., D.M. Greenberg, A.R. Vining, and D.L. Weimer. 1996. *Cost-Benefit Analysis: Concepts and Practice*. Upper Saddle River, NJ: Prentice-Hall
BriMar Consultants Ltd. 1995. *An Independent Review of Timber Royalty Rates in British Columbia*. Victoria: Ministry of Forests
–. 1996a. *Project Screening Criteria for the Watershed Restoration Program: Fish Habitat Rehabilitation*. Victoria: Ministry of Environment, Lands and Parks
–. 1996b. *The Allocation of Softwood Lumber Quotas: A Softwood Lumber Quota Allocation System*. Victoria: Ministry of Forests
–. 1997. *The B.C. Forests Products Industry: Current Status and Future Prospects*. Victoria: Ministry of Forests
–. 1998. *Forest Resource Rents and the B.C. Timber Pricing System*. Victoria: Ministry of Forests
Field, B.C., and N. Olewiler. 1995. *Environmental Economics*. 1st ed. Toronto: McGraw-Hill Ryerson
Fulton, F.J. 1910. *Final Report of the Royal Commission of Inquiry on Timber and Forestry, 1909-1910*. Victoria: King's Printer
Neher, P.H. 1990. *Natural Resource Economics: Conservation and Exploitation*. Cambridge: Cambridge University Press
Ostrom, E. 1990. *Governing the Commons, The Evolution of Institutions for Collective Action*. Cambridge: Cambridge University Press
Pearce, D.W., and R.K. Turner. 1990. *Economics of Natural Resources and the Environment*. Baltimore: Johns Hopkins University Press
Pearse, P.H. 1976. *Timber Rights and Forest Policy in British Columbia*. Report of the Royal Commission on Forest Resources. Vols. 1 and 2. Victoria: Queen's Printer
–. 1990. *Introduction to Forestry Economics*. Vancouver: UBC Press
Percy, M.B., and C. Yoder. 1987. *The Softwood Lumber Dispute and Canada-US Trade in*

Natural Resources. Halifax: Institute for Research on Public Policy

Price Waterhouse. 1995. *Analysis of Recent British Columbia Government Forest Policy and Land Use Initiatives*. Vancouver: Forest Alliance of British Columbia

Scarfe, B.L. "Financing First Nations Treaty Settlements." In *Prospering Together: The Economic Impact of Aboriginal Title Settlements in BC*, ed. R. Kunin, 275-304. Vancouver: The Laurier Institute

Schwindt, R., and T. Heaps. 1996. *Chopping Up the Money Tree: Distributing the Wealth from British Columbia's Forests*. Vancouver: David Suzuki Foundation

9
Sustainable Practices? An Analysis of BC's Forest Practices Code
Tracey Cook

Introduction

On 15 June 1995, the Forest Practices Code of British Columbia Act (the Code) came into effect, premised on the declaration that "British Columbians desire sustainable use of the forests they hold in trust for future generations."[1] This declaration is not contentious – organizations on both sides of the forestry debate would agree on it as a simple truth. There is much disagreement, however, over what means should be used to best meet the desired end of sustainable forestry in BC.

Before the enactment of the Code, the provincial government promoted it as "legislation that will change the way we manage our forests in BC."[2] Indeed, this aim has proven true – by setting new and comparatively strict standards for forest practices, the Code represents a step (how big is debatable) towards improved forest management. But rules are only as good as the methods used to enforce them, and while it contains new initiatives in compliance and enforcement, the Code has held fast to the traditional policy tools of strict command and control. In light of the dissatisfaction from all sides of the forestry debate over the management of provincial forests, it is time to look to alternative regulatory mechanisms for their potential to ensure sustainable forest practices in British Columbia. Bill 47,[3] the government's first attempt at amending the Code, appears to be an initial effort to do just that, and the movement away from interventionist regulation of forest practices may be gaining momentum.

This chapter considers the broad principles of the command and control approach as compared with regulation based on incentives, and presents a variety of perspectives on the current debate over which better advances environmental goals. The chapter evaluates the regulatory provisions of the Forest Practices Code in the context of this debate, focusing on the issue of whether and how economic incentives and other market mechanisms should be used to promote compliance with the priorities for forest practices as laid out in the Code[4] (whether those priorities are appropriate is certainly arguable, but that debate is beyond the scope of this chapter). The chapter concludes with some recommendations on which combination of economic incentives and commands and controls might be used to attain those priorities most efficiently, effectively, and equitably – the buzz-words of the current "market versus regulation" debate.

Regulation by Command and Control

The command and control approach to regulation is a familiar one: "the affected segment of the public is told that it must behave in a specified manner (the command), failing which a penalty will be exacted (the control)."[5] In the environmental context, regulation by command and control aims to direct the environmental performance by requiring or proscribing specific conduct, and using such techniques as licensing, standard-setting, and zoning to limit the discretion that may be exercised by the regulated. There is no choice left to the regulated party, who must comply with the minimum acceptable standards, or face penalties. Clearly, the command and control approach to regulation will fail when the "controls" are not sufficiently stringent. That is, when the cost of non-compliance is less than the cost of compliance, marginal penalties or fines may be accepted as a cost of doing business.

This interventionist approach to environmental regulation is justified by the well-worn observation that unregulated private actors are motivated chiefly by a desire for short-term profit, and tend to discount or ignore the effects their actions may have on third parties such as the environment and future generations. The market cannot account for those costs, because those who bear the "externalities" are not party to the transaction. Government intervention is necessary to limit these externalities by regulating the behaviour of private actors in the interest of society as a whole.

Regulation by Incentive

The orthodox command and control approach to regulation in the environmental field has been challenged in recent years. Among the critics are not only those whose activities have been regulated, but also environmentalists who are dissatisfied with what they generally perceive to be a generally low degree of efficacy in environmental regulation. The former group points out that the interventionist approach is always expensive and bureaucracy-intensive, often inefficient, and frequently inequitable. Free market environmentalists argue that environmental regulations are not often written strongly enough to protect the environment, and that the regulators are commonly lax in enforcing those regulations that are on the books. Both groups have advocated deregulation, promoting instead a "hands-off" approach which would assign a large part of the responsibility for environmental regulation to the ingenuity of entrepreneurs and the forces of the private market system. Free market environmentalists hope to reconcile economic growth with environmental protection by promoting reliance on the same self-interested behaviour that leads to development for the purpose of slowing the process of environmental degradation.

Even where the free market is relied on as a principal regulator, government officials must intervene with its workings in the initial stages of standard-setting. Thereafter, the market may be left to guide the governed in choosing the manner in which they will reach those goals. Whereas an interventionist approach to regulation is process-oriented (for example, specifying just how cutting should occur near riparian areas in order to prevent siltation and destruction of fish habitat), regulation by incentive is concerned with results, and spurs businesses to search for their own low-cost methods of achieving the same environmental standards.

An incentive is simply "that which influences or encourages action; a motive; spur; stimulus."[6] A wide array of economic incentives may encourage corporations and individuals to avoid abuse of publicly owned resources and to provide goods and services that would not otherwise offer a private return.[7] These tools share the common feature of promoting gains in environmental quality at the lowest possible cost. Economic incentives may be characterized as either positive (a benefit offered to modify environmentally harmful behaviour), or negative (the threat of punishment if regulated standards are not achieved).[8]

Free market environmentalists claim that a natural resource management system that makes extensive use of incentives has several advantages over more intrusive regulatory intervention. Perhaps the advantage most commonly noted is that it can offset the high price of government monitoring. In theory, an adaptive market can also regulate more efficiently than can government officials, because prices provide immediate information and incentive for action as soon as changes are seen, without first having to transmit all relevant information to a centralized decision maker who must convince voters of the wisdom of a particular course of action, and communicate the resulting decision to those affected by it.[9] Further, the market is praised for fostering creative and diverse solutions to environmental problems, because "there is no single, centralized decision maker, but many asset owners and entrepreneurs each of whom can exercise his [or her] own vision"[10] in identifying the most cost-efficient method of meeting environmental standards, and in developing new technologies which may actually exceed legislated standards. Finally, free market environmentalists see the market as a more democratic method of regulation, allowing for greater individual freedom than does the paternalistic command and control system – those who choose to use a resource pay for it by sacrificing some of their own wealth on the basis of willing consent.[11]

In the conceptual debate over interventionist versus market approaches to regulation, there is ostensibly a clear dichotomy between the two categories of policy instrument. In fact, regulations and incentives are not so obviously distinct. Economic incentives require government intervention for their creation, and regulators must set standards for the behaviour that those incentives are designed to encourage. From this perspective incentives are seen as "a less direct form of regulation [that] allow more scope for initiative on the part of the persons regulated, and provide financial rewards for those who exceed the minimum standards. Economic incentives are intended to reduce the overall costs of any given level of environmental protection when compared with other methods [of regulation]."[12] Rather than embracing one approach to the exclusion of the other, hybrid regulatory systems that involve a combination of regulatory and market instruments are in fact the norm rather than the exception.

In this sense (if no other) the Forest Practices Code of British Columbia Act, as enacted in summer 1995, must be considered exceptional. The Code is the quintessence of command and control, designed to regulate virtually all aspects of forest management activities in the province, and makes only the slightest use of incentives. It is this use of "tough enforcement mechanisms" that is generally identified in government communications as the Code's principal contribution to forest management.[13]

Regulation of Forest Practices in BC Before the Code

Before the coming into force of the Forest Practices Code of British Columbia Act, forest practices in BC were governed by six federal and twenty provincial statutes, and upwards of 700 federal and provincial regulations which were not bound by clear policy and presented overlapping and sometimes contradictory guidelines.[14] The provincial government has refined and consolidated these guidelines in the Code.

While the approach to regulation and enforcement before the Code has been described as "hands-on," the rules were insufficient and unclear, and the government was not adequately empowered to effectively control forest practices. In some instances, the government did not have the ability to penalize a licensee for non-compliance where violations of management requirements were found. And where penalties were assessed, they were not always adequate to deter future contraventions.[15] Moreover, pre-Code legislation presented an inadequate array of responses to non-compliance, leaving enforcement officials few alternatives with which to address the various contraventions that might occur in provincial forests.

The forest tenure system, developed over the years by government and industry to allocate and manage provincial forests, places a great amount of responsibility with forest companies to plan their specific operations and monitor and report on their own performance in meeting forest practices requirements. While the tenure system does alleviate the need for close government supervision at all times in all aspects of industrial operations, it has been criticized for "placing industry in the driver's seat and leaving regulators essentially a rubber stamp role."[16] The Code has introduced new strategies for supervision and monitoring which the Ministry claims will "ensure that problems with forest practices are discovered quickly, corrected where possible, penalized where needed, and avoided in the future,"[17] but the tenure system itself remains essentially unchanged.[18]

The Forest Practices Code of British Columbia[19]

The Forest Práctices Code is not a single document. It is described as a "cascading set of laws and rules" which begins with the Forest Practices Code of British Columbia Act. The Act is the legal umbrella by which the component parts of the Code falling beneath it are made law. It establishes the overarching principles of the Code, outlines how the legislation will be administered, and authorizes the specific rules and penalties associated with contraventions of the Code.

Beneath the Act are the Regulations that establish fundamental province-wide standards for the whole range of forest practices from timber harvesting to road use and construction. The Regulations also outline the contents of the various plans and prescriptions that must be in place before harvest may proceed.

Below the Regulations, there is a place for the implementation of what are termed "Chief Forester's Standards." It is envisioned that these will be legally enforceable, area-specific standards set by the Chief Forester to expand on and provide some degree of site-specific variability of the forest practices laid out in the Act and Regulations. Although no Standards have actually been set at the time of writing this chapter, they could provide the level of detail and the precise numbers and measurements required to regulate very specific topics such as maximum clearcut size, green-up specifications, and minimum culvert size in particular biogeoclimatic zones.

Finally, falling beneath the Standards are guidebooks which provide site-specific interpretation to assist in the application of the Regulations and Standards. The guidebooks are effectively "how-to" manuals, proffering what the government considers to be the management practices by which operators can best meet Code requirements. Although the guidebooks "will be indispensable to the Code's success,"[20] they have no standing in law, and only become legally enforceable as a matter of contract when incorporated into licence documents.

A Vancouver lawyer has observed, "If this was all there was to the Code, one might conclude that it is an unremarkable piece of legislative and regulatory housekeeping. But what makes it eventful is its central theme: the command-control regulation of our relationship with the forest environment."[21] In the Code's four components, the provincial government has drawn a detailed picture of forest practices in BC. It has chosen to rely on the power of its own administration to enforce these new standards for forest practices, virtually ignoring the applicability of market forces and economic incentives as tools to encourage compliance.

Regulatory Mechanisms in the Code

Changing the way we manage our forests. Tough enforcement.[22]

Commands and Controls

Planning
Under the Forest Practices Code, all licensees are obliged to prepare and submit to the Ministry their Operational Plans which detail the way in which the licensee proposes to manage each stage of planning and operation on its respective tenure holding, consistent with the provisions of the Code and what the Code refers to as "higher level plans."[23] These Operational Plans attempt to "incorporate environmental values into up-front planning before operations begin on the ground"[24] in a proactive approach to prevent degradation – which makes more long-term economic and ecological sense than does rehabilitation after the fact. This planning technique is designed to share the public cost of forest administration with the tenure licensees. Rather than having centralized Ministry staff develop a unique Operational Plan for each tenure agreement, the licensees themselves formulate their own plans for all aspects of harvesting and silvicultural operations. At the time this chapter was written, depending on the conditions of the specific tenure agreement, Ministry approval of six different operational plans and prescriptions may be necessary before the required permits to proceed with the harvest will be granted.

Operational planning under the Forest Act was more primitive than under the Code, and the terms of approved plans were enforceable simply as a matter of private contract between the government and the licensee, not as a matter of law. The difference between enforcement by law or contract is crucial: "members of the public have legal remedies available if an official does not follow a law, [but] a contractual agreement made pursuant to legislation is only enforceable by the parties to the contract."[25] Under the Forest Act, then, the public as third party could not directly seek to enforce the terms of an Operational Plan if the licensee breached its terms, but instead would have had to compel the Ministry of Forests to take action. The Code allows the public to lodge such complaints directly with the Forest Practices Board, which can then request a review or appeal on their behalf.

Monitoring
There has been an undertaking of "joint enforcement" of the Code in an effort to achieve better monitoring of forest practices in provincial forests.

To supplement the general ongoing efforts of the Ministry of Forests, staff from the Ministry of Environment, Lands, and Parks (MELP), and the Ministry of Employment and Investment, Energy and Minerals Division (MEI, now Ministry of Energy and Mines), are now authorized to monitor and enforce those provisions of the Code that may have some bearing on their respective portfolios.[26] At first glance, the joint enforcement concept appears to be a major departure from previous legislation (whereby other ministries referred all instances of suspected non-compliance with the Forest Act or Range Act to Ministry of Forest officials), but there has been little effective change in monitoring and enforcing forest practices in the time that the three ministries have been "working together" under the Code. The MEI seems to have been severely limited in the resources it could channel to Code enforcement, and MELP's focus has been on the resources management planning and operational provisions of the Code. This has left the Ministry of Forests to carry out the great bulk of monitoring and enforcement activities.[27]

The Forest Practices Board is also involved in monitoring the implementation of the Code. Promoted as "the public's watchdog on effective forest management,"[28] the Board is charged with overseeing the activities of both industry and the government in relation to the Code by performing periodic independent audits, and reporting the results to the Minister and the general public. The Board is also empowered to investigate public complaints regarding the application of the Code, and publicize its results. Where it finds, through its own independent audits or as a result of the investigation of public complaints, that the intent of the Code has not been achieved, the Board may initiate an administrative review or appeal, conduct a special investigation, make recommendations, or publish special reports about all matters of application and implementation of the Code that are within its jurisdiction.[29]

Enforcement

The Code provides a broad range of administrative remedies: monetary penalties, stop-work orders, remediation orders, suspension and cancellation of burning permits, penalties for unauthorized timber harvesting, and seizure and sale (forfeiture) of timber.[30] These act as a system of progressive penalties that becomes more harsh in accordance with the severity and frequency of the contravention, culminating in a maximum administrative monetary penalty of $50,000 per day.[31] Under the Forest Act, only contraventions that could be categorized as trespass were subject to administrative penalties, while the Code allows for staff to levy these administrative penalties for a greater range of infractions. Nearly all administrative penalties under the Code are subject to review and appeal.[32]

In addition to administrative penalties, the Code provides for quasi-criminal penalties to be levied against certain infractions.[33] Quasi-criminal penalties are most often court-ordered, but they are sometimes issued by enforcement staff in the form of tickets (which bypass cumbersome criminal procedure unless a defence is raised). While administrative penalties pose an alternative to the criminal system, ticketing is merely a procedural reform to the criminal system, intended to facilitate – and thereby increase – the laying of charges.[34] The Code increases the number of ticketable offences from 43 under the Forest Act and its Regulations to 107.

Upon detecting a ticketable offence, enforcement staff may issue one of two types of ticket: a violation ticket or a warning ticket. The former is an expedient remedy for contraventions that are not considered by the enforcement official to be very serious. For every section of the Code or its Regulations that is ticketable in this manner, however, it is possible to issue a warning ticket instead. Warning tickets merely acknowledge that a violation has occurred, and do not carry a fine. Normally, warning tickets are used for first infractions that are seen by the enforcement officer as relatively insignificant.

Court-ordered penalties exist under the Code to apply to more serious instances of non-compliance. When enforcement staff uncover an offence of sufficient severity, they may refer the matter to Crown counsel, who then decide whether or not to proceed with prosecution.

Convicted quasi-criminal offenders will be subject to stiffer sentences than existed pre-Code. The Forest Practices Code has increased the maximum amount of fines for offences from $2,000 to $1 million per day for each day the contravention continues. And because these fines may be imposed in addition to any costs that the court may order be paid for rehabilitation, the resultant financial penalty may actually be higher than the legislated limit. Also notable is the "Court Orders" provision in the Code, s. 155, which gives the court the broad ability to make orders to prevent repetition of the offence, and to direct the violator to publish the facts relating to the conviction.

Review and Appeal

This greater range of penalties and fines in the Code is accompanied by an enhanced statutory administrative review and appeal process. Nearly all determinations[35] made under the Code are subject to the review and appeal process laid out in the Act and Regulations.[36] The Code allows for the subject of a determination or the Forest Practices Board (of its own accord or on behalf of a member of the general public) to request an administrative review or appeal. The Board may also request a review of a situation in which no determination has been made. The Ministry, however, may not request a review of an original determination or a review decision, presumably because it must not be perceived to be challenging its own determinations.

The administrative review is the first avenue for challenging a determination. Reviews are held before a panel of one or more independent government employees appointed for the task by a Ministry of Forests review official. The reviewer(s) consider written and/or oral submissions to decide whether the official who issued the impugned determination properly considered the law and all relevant information in making that determination. Reviewer(s) are authorized to confirm, rescind, or vary the original determination; make a new determination; or refer the case back to the original decision maker for reconsideration, with or without instructions.

After completion of the administrative review, either the party who requested that review or the Forest Practices Board may submit a request for appeal to the Forest Appeals Commission (FAC), an independent tribunal designated to hear appeals under the Code. As the Commission may require witnesses to attend appeals and give evidence under oath, and can also choose to invite the participation of intervenors, hearings before the FAC may be slightly more "judicial" in nature than are administrative reviews. Like review panels, the Commission has the power to confirm, rescind, or vary the original determination, to make an entirely new determination, or to refer the case back to the determining official for reconsideration.

The Minister of Forests or any party to an appeal may apply to have a decision of the Commission appealed to the BC Supreme Court on questions of law or jurisdiction. That court's decision may, with leave of the court, be appealed to the Court of Appeal for British Columbia.

Prosecutions under the Code, including tickets, may be challenged in the usual manner in the provincial courts. Where there are no statutory avenues for review and appeal, and the appellant is granted standing, he or she may still apply to the BC Supreme Court for a judicial review under the Judicial Review Procedure Act.

Incentives in the Code

The Ministry of Forests has enumerated eight objectives for the new Forest Practices Code, one of which is "to provide an administrative system that is effective through a comprehensive system of incentives and penalties."[37] The penalty provisions, as discussed here, are comprehensive indeed. The Code's system of incentives, however, is decidedly less so.

Performance-Based Harvesting

There is a hierarchy of enforcement options under the Code. The most serious sanction is the Performance Based Harvesting Regulation.[38] The only mechanism which may be properly categorized as an "incentive" in the Code,[39] this regulation expands on s. 63.1 of the *Forest Act* concerning eligibility for a cutting permit or forest tenure agreement application, and makes the approval of future logging activities contingent upon the licensee's level of performance during current operations. Under this regime, companies that have successfully complied with the requirements of their individual operational plans and the general requirements of the Act, Regulations, and Standards, can expect to have new harvesting approvals issued with the usual degree of review and monitoring. Companies that seriously violate the Code may have further harvesting approvals withheld, and for extreme or repeated violations, may have their cutting privileges revoked.

Other Incentives?

Related to the Performance Based Harvesting Regulation is s. 117(4) of the Code, which allows decision makers to consider previous contraventions by a licensee in coming to subsequent penalty determinations. This provision is not an unfamiliar incentive against contravening a regulation. Under the Code, however, the incentive is weakened because determinations are stayed until all avenues for review and appeal have been exhausted. Until the time limit for challenging prior determinations has expired, or until those determinations have been considered by the Forest Appeals Commission, the Ministry's position is that they should not be thought of as "prior contraventions" for the purposes of making future determinations.[40] This policy provides a great incentive for licensees to appeal all determinations made against them.

The provincial government predicts that "the very existence of a legally enforceable Code with strong penalties and independent auditing will encourage most operators to perform to higher standards."[41] Much the same claim can be made about any legislation with offence or penalty provisions, however, and advocates of market regulation would suggest that more effective incentives might be used in the administration of the Code.

Critique of the Regulatory Mechanisms in the Code

The Code is costly. It is burdensome and bureaucratic. The pages of rules and regulations to which loggers must now adhere add up to a meter-high stack of paper.[42]

Perhaps the most frequently cited criticisms of command and control regulation are that it is both expensive and inefficient to administer as compared to the market approach, and that the outcome is frequently inequitable. There is some merit in each of these criticisms when tested against the regulatory approach adopted in the Forest Practices Code, which demands from the government functions that are arguably too large, too complex, and too widely dispersed to be properly carried out by it alone. In the short time the Code has been in force, it has become apparent that with this legislation, government regulation of the forest industry "like the mastodon, [has] become handicapped by its own dimensions."[43]

High Cost of Administration

The cost of government administration will clearly increase under the Code, as more public servants are needed to supervise the detailed planning process and to enforce more regulations. Indeed, when the Code was introduced, the provincial government promised to dramatically expand its enforcement presence in the field. Unfortunately (but predictably), it has become apparent that the budget will not permit the creation of the number of new enforcement positions required. Instead, the Ministry of Forests has attempted to organize a more efficient bureaucracy, shifting the structure of the Ministry by transferring 200 central positions from Victoria and regional offices to the districts where the groundwork is done. Field staff are now being provided with wireless communication devices and mobile computers to make them more efficient on the job. In addition, the Ministry is reportedly developing a new inspections policy which may provide shortcuts to time-pressed staff. Because limited human resources dictate that there cannot be frequent and comprehensive inspection of all sites, the policy would see that high-risk areas are designated on forest plans so that inspections could be concentrated in those areas.[44] The question remains as to whether the government's efforts should be focused on cell phones and modest policy changes when the larger forces of the market are waiting to be harnessed.

All British Columbians ultimately pay the costs of administering the Code. University of British Columbia forest economist David Haley has estimated the annual cost of the Code to the BC forest economy will be approximately $2.12 billion – an amount equal to 40 percent of the province's entire education budget.[45] The provincial government has not denied that there will be significant costs to both taxpayers and industry in fully implementing the Code, but in an independent study commissioned by the provincial government in January 1994, the estimated cost of the proposed Code was $304 to $486 million annually – notably less than Haley's more recent estimates. The government has rationalized the hefty expenditure by admonishing that the alternative – "continuing with inadequate forest practices – is unacceptable to British Columbians."[46] This statement is arguably somewhat misleading. While it is true that continuing with inadequate forest practices in this province is unacceptable, that is not the only alternative to the costly regulatory scheme in the Code. Incentives for compliance might be implemented at less expense.

Having said this, the only true incentives in the Code, found in the Performance Based Harvesting Regulation, have resulted in significant expense to the Ministry by clogging the arteries for review and appeal. This has resulted from the potential economic significance to licensees of being found guilty of violating the Code. Illustrative is an appeal heard by the Forest Appeals Commission in January 1997.[47] In this case, the Commission was asked by a logging company to consider a review decision that affirmed a determination by which it had been penalized for slightly damaging the bark on thirty-two trees just outside the boundary of its cutblock. The likely loss of volume to the injured trees was reported by the Ministry to be in the amount of three cubic metres, and the company was penalized $304.14. For its part, the company brought the case to review, and then to appeal, clearly not because of the dollar amount of the penalty, but rather to keep its record clean in light of the Performance Based Harvesting Regulation.

Each review and appeal is costly in terms of bureaucratic time and public money. There is the predictable expense associated with organizing and administering these hearings, but more important is the potentially higher price of removing enforcement staff from the field so that they may attend hearings as witnesses or review panel members. Each day enforcement officials spend in a hearing room is a day some licensees will go without inspection.

Beyond the implications of the Code on the public purse, the new legislation has had a tremendous impact on the cost of logging to industry. A report in the *Vancouver Sun* captures the extent of economic repercussions of the Code to forest companies: "BC logging costs are now among the highest in the world. BC pulp is now the world's most costly to manufacture ... In 1995, 67 million cubic metres were harvested from Crown land. It cost the companies $600 million to comply with the Code, half as much as their earnings of 1.28 billion."[48] All British Columbians have cause to be uneasy about these figures. Multinational forest companies are the engine of the forest industry in British Columbia, and in their drive for profit, they may abandon this province in pursuit of cheaper timber and less regulation in other jurisdictions. Although these multinationals themselves are undeniably at the root of many of the problems in the provincial forest industry, the immediate result of any significant withdrawal of corporate investment in this sector would clearly be devastating to forest communities and the provincial economy.

Inefficient Administration

As the market fails, so may government administration. Several instances of what we might call "government failure" can be found in the implementation of the Code. As discussed, for example, the review and appeal procedure for administrative penalties has proven to be an administrative quagmire, because licensees have discovered that by challenging almost any determination made against them, they can tie up field staff with paperwork in the office, leaving fewer in the field to detect contraventions. One must hope that FAC decisions will provide some guidance for the exercise of Ministerial discretion, and thus spare some of the costs of this cumbersome procedure when licensees see the futility of some potential appeals.

At present, field officials have every incentive to avoid the backlogged review and appeal processes. The process can be avoided in part by making effective determinations in the first instance, but officials recognize that industry is challenging even ostensibly good determinations. The best remedial option for any given contravention, then, might appear to be one that carries no right to review. Warning tickets and verbal instructions, which do not demand the involvement of district managers and do not spur the review process – but which have no standing in law and are unenforceable – are commonly chosen options for enforcement. Statistics from the Ministry show that between 15 June 1995 and 31 March 1996, 49 percent of all enforcement action taken was by instruction[49] (analogous to a peace officer admonishing the driver of a speeding car to "slow down" rather than issuing a ticket). The Ministry of Forests Compliance and Enforcement Branch is attempting to educate field staff that administrative efficiency is not an acceptable reason for choosing one enforcement option over another.

Inequitable Administration
One goal of any enforcement system must be to ensure that similar contraventions receive similar punishment; government failure can be said to occur when this goal is not met.

Given the broad array of enforcement options under the Code, and the discretion forest officials have to choose the "appropriate" option in any given circumstance,[50] the equitable application of remedial options has become a problem under the Code. For many contraventions, penalties may be issued under the Administrative Remedies Regulation or pursuant to the "Offences and Court Orders" provisions of the Act. While allowing officials more flexibility in responding to the different situations they face, the range of options also leaves an expansive grey area where any number of actions might be taken for a single offence.[51] A trend analysis undertaken by the Compliance and Enforcement Branch confirms the problem. Ministry of Forests statistics show that in the Boundary Forest District, between 15 June 1995 and 31 March 1996, 75 percent of all enforcement actions taken were informal instructions, while violation tickets made up 10 percent of the total. In the Fort St. John District for that same period, almost the reverse was true: 65 percent of all enforcement action taken were violation tickets, while only 13 percent were instructions.[52]

The Forest Practices Code is also inequitable as between different forest tenure holders. Although opportunities for small forestry business are increasing as more of the timber resource is gradually becoming accessible to them, big business remains the norm in the BC forest industry, and the Code is predominantly geared towards businesses of large scale. "Since it is largely process oriented, [the Code] forces everyone to achieve the same goal, and also requires the same methods to arrive there."[53] The local horse logger may be asked to prepare the same extensive operational plans as a large multinational firm. This criticism is not to suggest that the standards that apply to local operations should be lower than those that apply to large corporations; they should, however, be realistic given the resources available to small businesses. In short, all tenure holders are not equal, but because the Code applies equally to each, inequity may result.

In Support of the Code's Regulatory Approach

Response to Market Failure

> Forestry is an activity which contains a large portion of public goods and services. Further, the long rotation period gives plenty of room for incomplete information on prices and biological effects. Accordingly, a free market system in forestry is likely to produce market failures of significant magnitudes. These are the reasons for the existence of and need for forest policy.[54]

Externalities, the "fatal flaw of the free market,"[55] occur when the effects of production and consumption, which are not priced in the effort of one group of producers, affect the activities of another group.[56] Spurred by the immediate and personal costs and benefits of production alternatives, and ignoring the wider environmental and social consequences of its decision, the "rational" logger will continue business as usual, sloughing off the broader concerns, because it is in its own best interests, economically speaking, to do so.[57] Regulations restricting the size, shape, timing, and location of clearcuts, or logging in riparian areas, are intended to influence the decision-making of the logger and polluter by pricing the externality and shifting the cost back on them. This manner of government intervention is a generally accepted response to market failure, because it "singles out those segments of society deemed to be responsible for our problems and provides direct limits on their offending behaviour."[58]

Government intervention is also warranted when polluters choose to disregard incentives that market theory predicts they will act upon. "Unlike the strict regulatory approach, incentives do not require all (or perhaps any) polluters to undertake the maximum abatement that is technically possible. Some polluters may choose to pay the charge or forego the subsidy rather than to engage in abatement"[59] – another instance of market failure. The government, as regulator, is empowered to coerce – more than simply encourage – desired behaviour with legal responses that cannot be simply ignored.

The free market functions optimally when consumers have perfect information. The market fails, however, and government intervention is warranted, where insufficient information obscures consumer choice. For example, information necessary for environmentally responsible consumptive decisions is particularly scarce,[60] so that whatever their preferences, typical consumers simply do not know what the environmental consequences of their discrete market decisions might be:

> Direct environmental consumption is often frustrated because people cannot convert general aspirations for environmental consumption into informed commodity choices. Environmentalists want to enjoy aesthetic and recreational experiences, to protect endangered species, to live among natural surroundings, and to avoid exposure to pollution. But people cannot purchase preservation in a market; they must buy particular resources in particular locations to protect particular species ... and this requires specific and detailed knowledge that people do not possess in the great majority of environmental contexts.[61]

Ill-informed consumers may be saved from their own poor decisions if their (presumably) better informed political representatives are compelled to make better choices.

This assumption has been criticized for implying that "people who make unwise decisions in the market place make wise decisions at the ballot box."[62] Because our role as political citizens does not always coincide with our role as consumers, however, it may often be true that centralized decision-making based on politically expressed preferences may be more sound from an ecological perspective than the aggregate of decentralized decisions in the market. Many of us support organic farming in principle, for example, yet when in the supermarket we opt for "cheaper" non-organic produce, even if we recognize its high ecological price. By the same reasoning, if preservationists were asked to use their own resources to purchase pristine wilderness on the market to protect it (and indeed, to pay more than what a multinational corporation might offer for permission to log that land), the sheer expense of the proposition would make it impossible for most environmental organizations to do so. But when society as a whole is asked to pay the price of preservation through tax dollars, the notion of "protected areas" suddenly has a broader appeal. This mentality can be conceived of as an inversion of Hardin's tragedy of the commons. While decentralized decision makers might prefer to see the remaining old growth left untouched, and would support preservation if it was financed by society as a whole, they would be unprepared personally to foot the bill to the benefit of the rest of the free-riding public. Government intervention is thus warranted to motivate the preservationist inclination in the marketplace where too much information is required for decentralized consumers to make appropriate decisions, and also in instances where centralized decision-making can unite and satisfy the discrete preferences of individuals.

The Language of Economics: "Productivity" and "Efficiency"

According to conventional economic theory, free competition among producers will lead to "efficient" methods of production and resource use. Economic efficiency is achieved when the maximum level of production is obtained while maintaining the input of resources at a minimum. Unregulated, the market will translate this simple formula for efficiency into an ever-increasing volume of timber recovered from the forest, coupled with a constant tapering off of the amount of labour necessary to extract it – a trend which has been destabilizing forest communities in BC for the past two decades or more.[63] But the amount of resources that can be removed from forests in the quest for "productivity" before ecosystems begin to break down is limited. Moreover, although economic efficiencies can arguably be achieved by replacing labour with capital, this efficiency comes at the peril of forest workers, their families, and the communities in which they live. The further this phenomenon of exchanging labour for capital is allowed to progress, the more difficult it will be to reverse. That is, as industry becomes more heavily invested in machinery specialized for industrial forestry, the prospect of disinvestment appears increasingly less feasible. Governments can intervene in the workings of the forest industry to reverse this "old-styled productivity trap"[64] by redefining the classical economic concepts of "productivity" and "efficiency" to reflect current social values, such as the inherent value of protecting natural spaces.

Promoting Social Goods

Classical economic theory holds that economic viability alone drives a market left to its own devices. In the forests (as elsewhere), this means that social concerns get short shrift. If we are to achieve sustainable forestry practices in BC, economic viability must be considered in the context of public priorities such as ecosystem preservation and job security. "Private operation [in the market] reduces responsiveness to politically expressed desires,"[65] because the market will not produce social goods unless they also happen to stimulate the economy. Thus, "regulation may be considered superior to the use of economic incentives when the regulatory goal is non-monetary in nature."[66]

Valuing Forests

The free market recognizes only goods and services to which a monetary value has been assigned; others (women's traditional work in the home, for instance, or a deep breath of clean air) simply do not register. To create a "level playing field" between dimension lumber and standing old-growth, then, means assigning a price to the latter which can be weighed against that of the former.

But surrogate valuation of environmental goods is no easy task. "A deficiency of the price mechanism involves its inability to detect long-term problems. The price system operates on signals generated from a market and the time horizon for a market may be too short to take account of changes in resource availability or of environmental adaptability that will not occur for 10 to 25 years in the future."[67] As a result, natural resources are underpriced and are being exploited at unsustainable rates. Government intervention is necessary to plan for the future, and the onus is on the public to demand that political representatives begin thinking beyond the next election.

An Emotional Appeal

Beyond the sheer difficulty of putting a price on some environmental amenities, the fact that one would even attempt this exercise offends many who oppose free market environmentalism. These critics contend that because only a fraction of forest biodiversity has been investigated and recorded, the true value of forest ecosystems cannot be known, and any assigned dollar figure is meaningless. Moreover, environmental values are commodified when they are assigned a price, and removed from the realm of specially valued things to become just another item for sale in the marketplace. It is further argued that Nature is so sacred that its elements should not, indeed, cannot, be priced, and in any event, incentives with appeal to self-interest are not the appropriate tool with which to address environmental and other social issues.[68]

Such arguments certainly do have an emotional appeal, but can in fact be counter-productive if they prevent an appreciation of the merits of regulation by incentive. "Neither regulation nor deregulation ought to be undertaken purely on ideological grounds. The selection of the appropriate mechanism for social choice should turn on a realistic evaluation of the advantages and constraints associated with alternative decision making systems."[69] And although economists are commonly criticized for attaching dollar figures to those things that really cannot be priced, it should be recognized that this is not something done only by economists:

> Whether the policy-maker likes it or not, [she or] he is putting price tags on non-priced goods (and services) when [she or] he is formulating [her or] his policy solutions ... Accordingly, the difference between the policy-maker and the economist is that the former sets the price by making policy and the latter tries to reveal the (optimal) price first, and then makes policy recommendations.[70]

Conclusions

Recognizing that sustainable forest practices are imperative in light of the dwindling timber supply, and recognizing as well the many and varied non-timber values of provincial forests, the government has raised considerably the standards for forest practices in enacting the Forest Practices Code in 1995. But since that time, the simple failure to adequately enforce these standards has hampered the intended move towards sustainable forestry.

Simply put, it is unlikely that the provincial government can afford to properly enforce the standards in the Code without looking to incentive mechanisms to supplement its limited regulatory power. Enforcement under the Code, then, should make use of a variety of mechanisms (both "carrots" and "sticks"), tailored to better meet the requirements of different provisions in the legislation and related documents. The following section of this chapter points to some areas in which regulation and incentives might be used in concert to achieve a more efficient, effective, and equitable scheme for regulating forest practices in BC, lowering the overall cost of monitoring and enforcement without lowering the standards set in the Code.

Recommendations: Making the Most of the Market[71]

Make Better Use of Negative and Positive Incentives
Governments in some nations have used incentive programs to subsidize agricultural settlement of forested areas. The Brazilian government, for instance, has become notorious for having implemented agricultural subsidies and tax credits in a massive effort to colonize its tropical forests with cattle ranchers who feed international beef markets. The power of these incentives is illustrated in the resulting deforestation of enormous areas in Brazil; similar techniques could assist in the opposite move towards sustainable forestry in British Columbia.

Negative Incentives
The negative incentives in the Performance Based Harvesting Regulation should be reworked so that they actually discourage poor forest practices as opposed to simply encouraging frivolous requests for review and appeal. The Performance Based Harvesting Regulation should provide a "gradated approach" for imposing performance-based penalties.[72] That is, penalties under the Regulation should begin with increased monitoring, and if difficulties with the licensee continue, monitoring and reporting requirements should become progressively more stringent, culminating in the withholding of approvals and suspensions of permits. The disincentive to poor forest practices would remain with gradated penalties, but because licensees would be allowed slightly more slack before large penalties are imposed, they would not be as likely to challenge each determination made against them.

The Performance Based Harvesting Regulation should also give some guidance to district managers on the weight they should place on contraventions of varying degrees of severity. Currently, this decision is entirely discretionary. While it may seem overly simplistic to legislate that less severe contraventions should be given less weight in applying the Performance Based Harvesting Regulation (and this practice may be what is happening anyway), having the policy in writing might discourage companies from challenging $314 penalties in an effort to keep a spotless record.

Finally, the Performance Based Harvesting Regulation should be area-specific in penalizing the operator for prior contraventions, because errors in forest practices may not be occurring in all locations where the forest operator is active. The BC Council of Forest Industries (COFI) suggests that the Regulation should incorporate a "targeting provision" which would single out and penalize poor operations while allowing the forest operator's good operations within the same tenure to continue.[73] While focusing as narrowly as COFI has suggested may weaken the Regulation too dramatically, licensees may be less likely to challenge determinations made against them if they knew the contravention would show only against their operations in the particular region or district in which it occurred.

Positive Incentives

There are no positive incentives written into the Forest Practices Code's compliance and enforcement regime. While it is clearly necessary to penalize poor performance, negative incentives will not promote the goal of performance above and beyond regulated standards. Exemplary standards will only be achieved if they are rewarded. That is, positive incentives which make compliance less costly, or even financially rewarding (such as some combination of grants, soft loans, tax exemptions, and rebates), can be offered to corporations that exceed or are in consistent compliance with the Code.

Prospects of a streamlined administrative process may also act as a positive incentive for forestry companies, especially in light of the increased administrative workload for licensees under the Code.[74] Deadlines for government approval of applications could be shortened, the intensity of monitoring and auditing could be lessened for companies in consistent compliance, and the amount of reporting generally required could be decreased in instances of continued good performance. Like penalties, rewards should be gradated, increasing in amount the longer exemplary performance is maintained. Such rewards obviously must be administered with caution, but they should be achievable within a realistic time period if they are to provide a true incentive to corporations.

Public recognition of good forest practices can also act as a positive incentive by bolstering a corporation's client base and boosting its market share.[75] By retaining a third party to publish an annual report on corporate compliance, the government could ensure that the media and the public have a resource that identifies which corporations are the most diligent in their compliance efforts. The public could then make purchasing and investment decisions accordingly.

"Eco-certification" is another means by which discriminating consumers may be provided with the information they need to purchase only forest products that have been produced in an environmentally responsible manner. Generally speaking, "environmental labels are intended to provide consumers with easily recognizable, qualified assessments of the environmental worthiness of products and promote the production and consumption of more desirable goods."[76] To be credible, certification programs must be administered by independent certifiers capable of adequately monitoring forest practices, and of tracking the chain of custody of timber from the certified stand through transport, processing, manufacturing, and retail distribution. Naturally, the wider the objectives are, the more complex and costly the certification system will become, but this expense need not directly fall to the public. Independent third party agencies should be encouraged to implement certification programs.[77]

Whatever type of positive incentives may be implemented in the Code, if they are to have a real chance of enhancing performance, the rewards must be clearly legislated and automatically forthcoming when reasonably attainable performance standards are met.[78]

Use Economic Theory as a Tool to Forge the Best Forest Policy

> Increasingly, the state of the environment will determine future economic growth, and environmental costs will be an essential element of economic calculations. Social, economic, and environmental needs will require careful balancing and frequent adjustments.[79]

The optimal use of regulation and economic incentives will depend on a systematic, open, and transparent cost-benefit analysis of each of the alternatives for allocating scarce forest resources. While environmental economists often criticize cost-benefit analysis and the contingent valuation methods it entails,[80] the fact remains that money talks (while appeals to intergenerational equity and the existence value of plants and non-human animals go largely unheard). In a market economy, putting a dollar value on nature may help to protect it by enabling the forces of the market to assist in our preservation efforts.

In undertaking an economic analysis, then, the full costs and benefits to ecosystems and future generations must be accounted for to the greatest degree possible and recognized as relevant to the evaluation. If the analysis points to the use of free market mechanisms in forest policy, those mechanisms should be applied. If the analysis indicates a need for greater government intervention, the exercise of applying economic theory to forest practices will have nonetheless illuminated the logical reasons behind more intrusive regulation, providing solid ground for that approach. "There is no contradiction between economic theory and forest policy. Economic theory is simply a tool that can be used to find the best policy ... When used with care, economic theory is a good servant to the makers of forest policy."[81]

Give Local Communities the Incentive to Assist in Enforcement

It has become clear in the Code's first few years of operation that the provincial government may have overestimated its own enforcement capability. Rather than regulating the use of forest resources across the province from central offices in Victoria, the provincial government should seek to enlist the aid of community-based institutions or individual citizens in this regard. The common argument is that those who are closer to the forests might be more sensitive to the need to protect them, or at least in a better position to do so.

The government of British Columbia might take a first concrete step towards this goal by offering financial incentives to individuals across the province to keep a vigilant eye on local forest practices. This proposition need not be expensive. "Community watchdogs" could be compensated with a proportion of any fines or penalties collected from the operations that they had discovered to be in contravention. Because members of forest communities are often employed by the corporation that holds the local tenure agreement, such a provision would complement s. 173 of the Code, which currently offers protection to employees who "blow the whistle" on their employer.

Regulate the "Ends" Rather Than the "Means"

A results-based – rather than a systems-based – approach to regulation should be considered. This approach is rumoured to be behind the developing regulation or legislation that will apply to forest practices on private land.[82] To respect as much as possible the private property rights of landowners, any new rules to address this issue will differ from the Forest Practices Code by setting out the desired ends, but leaving the landowners to their own devices to attain those ends. The same results-oriented approach might be adopted in parts of the Code to save on the costs of monitoring and administration and to allow for the development of innovative forest practices.

Maintain a Regulatory Approach Where the Market Will Not Promote Social Goods

Although less interventionist governance can be effective and less costly in some instances, the command and control approach should continue to be used to promote social goods that are not recognized in the market.

Because some social and environmental goods (such as stability in forest communities or healthy ecosystems in which the full range of biodiversity thrives) generally have no value in the free market, promoting or protecting them has no clear and immediate economic return. "Rational" market actors will not promote these goods as ends in and of themselves, unless they are rewarded to do so. Hands-on regulation must continue to address social and ecological problems, and would be best used in conjunction with market instruments to promote sustainable forest practices in British Columbia.

Postscript: New Initiatives in Streamlining the Code

This chapter was written before the coming into force of Bill 47, the Forests Statutes Amendment Act, 1997, and more recent regulatory changes to the Code that have clearly been fashioned with attention to one of the buzzwords used throughout this chapter – efficiency. It is interesting to note now how some of the recommendations for reform outlined above can be recognized in these initiatives to streamline the Code.

While industry has generally responded favourably to these initiatives, the environmental community has been dismayed by what they perceive as a step backwards on the path to sustainable forestry. This latter reaction is not attributable to the trend towards deregulation itself, but rather to the fact that Bill 47 and related streamlining initiatives have relaxed the environmental standards laid out in the Code and closed some windows for public participation in the planning process.

It has been argued in this chapter that a less interventionist approach to regulation does not necessarily spell disaster, and conversely, that more regulation will not always bring improved environmental protection. The foregoing discussion of the Code has illustrated how implementation of such a highly regulatory statute in times of fiscal restraint brings with it enforcement difficulties, which in turn can lead to inefficiencies and inequity at the expense of the environment. Further reforms to the Code seem inevitable. Looking ahead one must hope that these reforms do not compromise environmental standards through mere regulation but instead employ innovative compliance and enforcement strategies which will make more realistically achievable the sustainable forestry philosophy that the Code was originally promoted as embracing.

Notes

1 See the Preamble to the Code, R.S.B.C. 1996, c. 159. There is some debate over what constitutes "sustainable forestry." For the purposes of this chapter, the meaning of "sustainable use of forests" corresponds to the definition given in the Preamble to the Code:

Whereas sustainable use includes
(a) managing forests to meet present needs without compromising the need of future generations,
(b) providing stewardship of forests based on an ethic of respect for the land,

(c) balancing productive, spiritual, ecological and recreational values of forests to meet the economic and cultural needs of peoples and communities, including First Nations,

(d) conserving biological diversity, soil, water, fish, wildlife, scenic diversity and other forest resources, and

(e) restoring damaged ecologies.

2   BC Ministry of Forests Ferret Infobase, Legislative Module Resource Book 2 (as updated to February 1996).

3   Bill 47, *Forest Statutes Amendment Act, 1997*, S.B.C. 1997, c. 48.

4   Readers should bear in mind that whatever value the market may have as a tool to promote sustainable forestry, orthodox economic theory can only offer what Garrett Hardin refers to as a "technical solution" to environmental problems. That is, free market environmentalism is a solution that requires a change only in the techniques of the approach to regulation, demanding little or nothing in the way of change in human values.

Classical market theory is rooted in anthropocentrism. The anthropocentric bias is most apparent in the classical liberal economic assumption that the primary aim of human behaviour should be efficiency, because efficient arrangements increase the total welfare to human beings. The main defence of the market thus rests on the utilitarian assumption that human affairs should be arranged so as to maximize human utility. At this juncture, it should be noted that if the anthropocentric assumption embodied in orthodox economic theory is indefensible, then the theory itself – which regards only benefit and harm to human beings as having moral significance – is unacceptable. Many environmental ethicists have made strong cases against the anthropocentric perspective: see the writings of Peter Singer, Tom Regan, Paul W. Taylor, Aldo Leopold, and Arne Naess in this regard. But rightly or wrongly, the anthropocentric bias still holds the monopoly, and classical liberal economic theory still appeals to the general sensibilities of the North American population. Notwithstanding that it might seem counter-intuitive to take a new approach to environmental regulation based on the instrumental view of nature (which is commonly recognized as one reason for its destruction), and notwithstanding that to adopt the "nature as capital" perspective inherent in economic theory will further entrench the normative anthropocentrism that currently underlies destructive forest practices, this chapter takes the position that it is not morally wrong to use the market to protect the environment.

The recommendations offered here do not reach what can be seen to be the "meta-root" of all symptoms of environmental degradation – misplaced human values. But because change in a democratic society comes only in increments as society is willing to embrace it (and it is fair comment to say that society is not generally ready to recognize that trees have any value beyond their instrumental value to humans), regulators should consider taking what opportunities currently exist to promote sustainable forest practices, and harness the waiting forces of the free market.

5   P. Cassidy, "BC's forestry code adopting suspect regulatory approach," *Environmental Policy and Law* (December 1994): 135.

6   J.A. Cassils, *Exploring Incentives: An Introduction to Incentives and Economic Instruments for Sustainable Development* (Victoria: Socio-Economic Impact Committee of the National Round Table on the Environment and the Economy 1991), 4.

7   Ibid., 9-12, for an overview of the categories of economic instruments and incentives that have been used for environmental regulation in various contexts.

8   D.G. McFetridge, "The Economic Approach to Environmental Issues," in *The Environmental Imperative: Market Approaches to the Greening of Canada*, ed. B. Doern (Toronto: C.D. Howe Institute 1990), 93, notes the difference between negative and positive incentives in comparing emissions charges to subsidies for abatement. He states that the choice between the two is equivalent to deciding who owns the common pool resource. If polluters have the right to discharge waste into the water or air, a positive incentive such as an abatement subsidy paid to the polluter would be employed. On the other hand, if other users had the right to be free of pollution, an emissions charge, tax, or the like to be paid by the polluter would be used as a disincentive from polluting. Thus, deciding which form of incentive is best suited to any given circumstance involves economic, legal, political, and ethical considerations.

9 R. Stroup and J. Baden, "Externality, property rights, and the management of our national forests," *Journal of Law and Economics* 16 (1993): 306.

10 Ibid. For further discussion on this point, see J.F. Chant, D.G. McFetridge, and D.A. Smith, "The Economics of a Conserver Society," in *Economics and the Environment: A Reconciliation*, ed. W.E. Block (Vancouver: Fraser Institute 1990), 8.

11 Stroup and Baden, supra, note 9 at 307. Note that this notion of equity is daunting to some conservationists who worry that fewer parks would exist if the relatively few wilderness "users" had to pay for the resources that are withheld from other users in order to preserve a pristine wild space.

12 Cassils, supra, note 6 at 4.

13 For example, see British Columbia Ministry of Forests, *The British Columbia Forest Practices Code Discussion Paper* (Victoria: Ministry of Forests 1993), 1.

14 Ibid., 7.

15 BC Ministry of Forests, supra, note 13 at 8. The maximum amount violators could be fined for most offences, regardless of severity, was $2,000.

16 M. Haddock, "Intent of Code lost in bureaucratic wrangling," *RPF Forum* 4(3) (1996): 12.

17 BC Ministry of Forests, supra, note 13 at 8.

18 In Ch.5, infra, Michael M'Gonigle observes that to change the rules of forestry in BC without first changing the institutional structure is to "put the cart before the horse."

19 S.B.C. 1994, c. 41, now R.S.B.C. 1996, c. 159. The debate over forestry in BC occurs at two distinct levels. First, there is the question of land designation and allocation, which considers the use to which provincial forest land will be put, such as recreation, timber harvesting, wilderness, and protected areas. Second, there is the issue of actual forest practices: what are the requirements that should govern forest activities on forest lands? The Code addresses the latter issue, but works in conjunction with other government initiatives such as CORE and the Protected Areas Strategy which address the first.

20 BC Ministry of Forests, supra, note 13 at 6.

21 Cassidy, supra, note 5.

22 BC Ministry of Forests, 1993, supra note 13.

23 Higher level plans may include regional CORE recommendations and sub-regional LRMP processes.

24 BC Ministry of Forests, supra, note 13 at 13.

25 J.E. Vance, *Tree Planning: A Guide to Public Involvement in Forest Stewardship* (Vancouver: Public Interest Advocacy Centre 1990), 18.

26 Responsibility for enforcement of the Forest Act and Range Act remains with the Ministry of Forests alone.

27 A Ministry of Forests Compliance and Enforcement official estimates that the Ministry detects 95 percent of the contraventions of the Code, while 4 percent are caught by MELP officials, and 1 percent by MEI (personal communication, 10 May 1996).

28 So described by Andrew Petter, then Minister of Forests, and Moe Sihota, then Minister of Environment, Lands and Parks, as they announced the appointment of the Board on 21 December 1994.

29 The Forest Practices Board Regulation, B.C. Reg. 170/95, describes how the Board will carry out its mandate, details how auditors are selected and appointed, how audits are managed, and how public complaints will be heard.

30 The Administrative Remedies Regulation, B.C. Reg. 166/95, sets out the details regarding certain administrative enforcement provisions of the Act and Regulations, and a schedule to the Regulation lists the maximum administrative penalties that a senior official may levy for any given contravention.

31 Although administrative monetary penalties (AMPs) are common in non-environmental regulatory statutes, P. Cassidy [in "AMPs 'flavour of the month' for environmental regulators," *Environmental Policy and Law* 6, 12 (1996): 349] points out that administrative monetary penalties are relatively new on the environmental scene. Cassidy suggests that AMPs are beginning to take the place of criminal and quasi-criminal offences in environmental statutes because they allow regulators to circumvent the cumbersome court process. Moreover, Cassidy observes that "perhaps the most important feature of the AMP scheme is its simplification of the evidentiary requirements necessary to non-compliance."

Whereas success in a criminal or quasi-criminal prosecution demands the Crown prove guilt beyond a reasonable doubt (and demands that the regulator undertake significant investigations to collect the evidence necessary to meet this onus), the burden of proof in the AMP scheme is completely reversed, placing the onus on the defendant to demonstrate that the penalty imposed was unfounded (and making the regulator's job easier).

It has been argued that such large administrative monetary penalties as the Code prescribes might provide the basis of a Constitutional challenge. This proposition is contested by C. Rolfe and L. Nowlan, in *Economic Instruments and the Environment: Selected Legal Issues,* ed. A. Hillyer (Vancouver: West Coast Environmental Law Research Foundation 1993), 66-71. The authors suggest that the two largest potential Charter challenges to administrative monetary penalties are, first, that the ability to impose a penalty with no trial might violate the s. 11(d) right to be presumed innocent until proven guilty; and second, that absolute liability for administrative offences may infringe s. 7 of the Charter, which grants the right to life, liberty, and security of person. With regard to s. 11(d), which applies only to public hearings if the processes are considered to be proceedings in relation to a criminal offence, Rolfe and Nowlan review the relevant decisions and conclude that "so long as an administrative penalty process is carried out in a largely private manner as opposed to involving a court hearing where a finding or fine is disputed, it will not be considered criminal in nature," and s. 11(d) of the Charter will not be invoked. Addressing the possibility of a s. 7 challenge, Rolfe and Nowlan conclude that "it will not be viewed as an infringement of the Charter to impose fines or restrict economic liberty on an absolute liability basis" when such fines do not carry the potential for imprisonment.

32 Seizure and sale of timber or livestock under ss. 115 and 116, and suspension or cancellation of a burning permit under s. 124, are the only exceptions. While maximizing fairness to industry and the general public, application of the Code's liberal review and appeal process has been a burden on the Ministry of Forests administration; see the discussion of reviews and appeals in "Regulatory Mechanisms in the Code," a later section in this chapter.

33 Such infractions are not considered to be "true crimes" because they do not involve prosecution under the Criminal Code of Canada. Prosecution may be an enforcement option for certain violations under the Forest Practices Code, however, if the contravention of the latter also amounts to a breach of the Criminal Code.

34 Rolfe and Nowlan, supra, note 31 at 72.

35 "Determination" is defined in s. 1 of the Code as "any act, omission, decision, procedure, levy, order or other determination made under the code, its regulations or standards by an official or senior official of the Ministry of Forests, Ministry of Environment, Lands and Parks, and Ministry of Energy, Mines and Petroleum Resources."

36 Reviews and appeals under the Forest Act and Range Act remain separate and distinct from those under the Code. For a description of this process, see *Reviews and Appeals under the Forest Act and Range Act* (Victoria: Ministry of Forests Compliance and Enforcement Branch 1996).

37 BC Ministry of Forests, supra, note 13 at 4.

38 B.C. Reg. 175/95.

39 As discussed, problems arise in categorizing regulatory initiatives as either interventionist or non-interventionist, because the distinction is not always obvious. Indeed, the *Economic Instruments for Environmental Protection* (Paris: OECD 1989), 13, notes that "certain instruments have grown to be considered as economic or regulatory, whereas their performance might warrant the use of the other label." The Performance Based Harvesting Regulation, which works in conjunction with the commands and controls in the Code, has been appropriately labelled a negative incentive. There are no "commands" in the Regulation itself; it simply poses another disincentive, on top of the controls already in the Code, to deter repeated contraventions of existing regulatory standards.

Since a poor compliance record can lead to a licensee being denied valuable future logging rights, the Performance Based Harvesting Regulation can be characterized as an economic incentive. The financial effects of the Code's non-compliance penalties and fines may also be severe – up to $1 million per day. Theoretically, fines and penalties may also be considered an economic incentive designed to deter behaviour, because these also provide an economic rationale for compliance where non-compliance is considered to be an alternative. The more common conceptualization of monetary penalties and fines is as an intrusive form of command and control, implicated when a regulated standard is not met. This conception holds true especially where the penalty is so strong that it effectively leaves no option to the subject of the regulation.

40 As indicated by a Ministry of Forests official, personal communication, May 1996.

41 BC Ministry of Forests, supra, note 22 at 12.

42 G. Hamilton, "Forest Code: a year's vicissitudes," *Vancouver Sun* (15 June 1996): B1, B7.

43 Aldo Leopold, "The Land Ethic," in *People, Penguins and Plastic Trees: Basic Issues in Environmental Ethics*, 2nd ed., ed. Christine Pierce and Donald VanDeVeer (Wadsworth 1995), 142-50.

44 Such a policy would mean that licensees whose tenure agreements are not in a "high-risk area" have less chance of being inspected than those whose agreements are. Because fewer inspections will be carried out on the former group, their records are more likely to be "clean," giving them a distinct advantage under the performance based harvesting rules. Under the current process, it is not clear whether licensees with clean records have actually passed inspections or simply have not been inspected at all. In an effort to achieve equity among licensees, enforcement staff should now undertake to investigate areas that appear to be in compliance, in addition to those that do not, and tally the results of both to achieve an equitable application of the Performance Based Harvesting Regulation.

45 D. Haley, "Paying the piper: The cost of the British Columbia *Forest Practices Code*," paper presented at the conference Working with the BC Forest Practices Code, Vancouver, 1996. In a breakdown of the $2.12 billion figure, Haley predicts that the largest costs will arise from lost revenues resulting from a forecasted 6 percent reduction in annual harvest (as a consequence of the Code's biodiversity requirements). In Haley's estimate, actual implementation of the Code's regulations will be minor factor in the the overall expense of the Code.

46 BC Ministry of Forests, supra, note 13 at 18.

47 Appeal 96/08.

48 Hamilton, supra, note 42 at B7.

49 Statistics compiled by the Ministry of Forests Compliance and Enforcement Branch, May 1996. Note that the over-use of verbal instructions can be explained by the fact that enforcement officers are easing their way into the application of the new Code, and that this percentage may not be indicative of a future trend in enforcement efforts.

50 In a province as ecologically diverse as British Columbia, a certain degree of discretion in the application of regulations may be warranted, especially when those regulations are not site-specific, but are written to apply consistently across the province. Discretion may avoid what Haddock (supra, note 16 at 12) has referred to as a "rigid and legalistic approach to forestry." The alternative would be to draft a discrete set of regulations for each ecosystem, which would add yet more paper to the existing stack which is the Forest Practices Code. There exists the opportunity for site-specific variability in the regulations in the Chief Forester's Standards (see this chapter, "Introduction to the Forest Practices Code"), which to date have not been set.

   None of the discretionary provisions in the Code provides forest officials with the discretion to apply standards that are more strict than those regulated, unless the licensee agrees to incorporate them into its operational plan. Haddock (supra, note 16 at 12) explains that as a result, a "debate over what prescription is best for a site turns into a debate over what the Code requires ... A standard seems to have developed where the only thing the resource managers have to manage for is consistency within the Code's minimum requirements, and nothing further. Public comments are occasionally dismissed on the grounds that the proposed cutblock is 'up to Code' because it is 39.5 or 59.5 hectares in size, or because there is the 'correct' riparian management area next to streams flowing through the block. If it is not against the law, end of discussion."

51 The same was not a real concern under the Forest Act, because the remedial options consisted almost entirely of fines.

52 Data obtained from the Compliance and Enforcement Branch, May 1996.

53 D.A. Routledge, "Jury is still out on whether Code meets essential seven principles," *RPF Forum* 4(3) (1996): 7.

54 S. Wibe, "Economic Theory and Forest Policy," *Journal of Forest Economics* 1(3) (1995): 271.

55 Cassils, supra, note 6 at 6.

56 The unhappy aesthetic of clearcut logging is an externality – the visible scars on the land-scape being a private cost of harvesting which is shifted from the logger to society. Similarly, downstream fishers or communities that depend on a body of water pay the price of siltation caused by poor logging practices in riparian areas.

57 So goes classical liberal economic theory, premised, as it is, on a distinctly individualistic definition of rationality.

58 Chant, McFetridge, and Smith, supra, note 10 at 64. Presenting a slightly different twist on government intervention is Cassils, supra, note 6 at 6, who suggests that "by increasing the cost of using the environment, for example, by taxing damaging products and services, the government is not interfering with the marketplace as much as applying better information so that prices more accurately reflect social costs."

59 McFetridge, supra, note 8 at 93.

60 H. Latin ["Environmental Deregulation and Consumer Decisionmaking under Uncertainty," *Harvard Environmental Law Review* 1(6) (1982): 187] blames this scarcity of information on such factors as the inherent complexity, differentiation, interdependence, local variability, and non-linearity of effects in the environment.

61 Ibid., 196.

62 Stroup and Baden, supra, note 9 at 308.

63 Acknowledged by the Ministry of Forests in the 1994 *Forest, Range and Recreation Analysis* (Victoria: Ministry of Forests 1994), 273: "The forest sector, driven by market pressure to remain competitive, improved efficiency by mechanizing and implementing new technologies. This, however, resulted in reduced employment. The ratio of employment to volume harvested had been steadily declining since the 1960s and dropped more sharply in the early 1980s ... During this period, the rate of harvest on regulated lands continued to rise despite the threat of falldown."

64 A phrase used by M. M'Gonigle and B. Parfitt in *Forestopia: A Practical Guide to the New Forest Economy* (Madeira Park, BC: Harbour Publishing 1994).

65 Stroup and Baden, supra, note 9 at 310.

66 Cassils, supra, note 6 at 8.

67 Chant, McFetridge, and Smith, supra, note 10 at 5, quoting from Brooks, *Conserver Society Notes*, winter/spring (1977): 29.

68 For further discussion, see R.G. Lipsey, "Greening by Market or Command?" in *The Environmental Imperative: Market Approaches to the Greening of Canada*, ed. B. Doern (Toronto: C.D. Howe Institute 1990), 128-32.

69 Latin, supra, note 60 at 189.

70 Wibe, supra, note 54.

71 The following recommendations address improved enforcement of the Forest Practices Code. While effective implementation of the Code is a necessary part of the move towards sustainable forestry in this province, it must be embraced as a small accompaniment to many more fundamental changes to the structure of the forest industry in British Columbia, many of which are outlined in other chapters in this book, and foremost among which is the reformation of the tenure system and the reduction of the AAC.

72 As recommended by the Council of Forest Industries of BC, in "Responding to the British Columbia *Forest Practices Code*: Promoting good forestry, improving knowledge, building confidence, *Making It Work*" (December 1993): 40-42.

73 Ibid., 41.

74 Ibid., 44.

75 Cassils, supra, note 6 at 12.

76 B. Ghazali and M. Simula, *Certification Schemes for all Timber and Timber Products* (Yokohama, Japan: International Tropical Timber Organization 1994), 7.

77 For a detailed discussion of eco-certification and the prospects for implementing such a system in British Columbia, see F. Gale and C. Burda, infra, Ch.12.

78 COFI, supra, note 72 at 46.

79 Cassils, supra, note 6 at 4.

80 See S. Kelman, "Cost-Benefit Analysis: An Ethical Critique," in *People, Penguins and Plastic*

Trees: Basic Issues in Environmental Ethics, 2nd ed., ed. C. Pierce and D. VanDeVeer (Wadsworth 1995), 384-90.

81 Wibe, supra, note 54 at 272.
82 Currently, the Code does not apply to private land (except that which has been incorporated into a TFL or woodlot licence) even where public resources could be negatively affected by private harvesting. There is much debate between those who believe that the Code should apply consistently to all forest practices, and others who are wary of forcing private landowners to comply with such highly interventionist regulation. The rationale for not doing so is that the standards in the Code are seen to be too restrictive for general application on private land where there must be a balance between the interests of the private landowner and the broader public interest in sustainably managed forests. But the Ministry also recognizes that forest practices on private land should be overseen in some manner.

10
Priority-Use Zoning: Sustainable Solution or Symbolic Politics?
Jeremy Rayner

In a number of jurisdictions, public lands forest policy is currently undergoing an unusual convergence on the idea of zoning land for priority uses. However, as is almost inevitably the case when a solution to a pressing policy problem is adopted by a number of diverging interests, different sectors of the forest policy community have very different expectations of priority-use zoning. In fact, as this chapter argues, there are at least two quite different conceptions at work, based on diametrically opposed perceptions of exactly which forest policy problem zoning is supposed to be addressing. While short-term political capital can certainly be made out of deliberately covering over these differences, any evaluation of the potential effectiveness of zoning as a policy instrument demands clarity about the goals zoning is trying to achieve. Of course, it is perfectly possible for a single policy instrument to address multiple objectives. However, when those objectives come into conflict, as they inevitably will, hard decisions will have to be taken about which one is to be given priority and, at this point, clarity about goals becomes a virtue. This chapter will try to determine what goals priority-use zoning is aimed at, what kinds of conflicts remain unresolved, and what kinds of political and administrative mechanisms are emerging in a belated attempt to address these conflicts.

Like most good ideas, priority-use zoning, although described in a British Columbia report as part of a "new management paradigm" (LIARC 1995), is not particularly new. In brief, it is simply the idea that, though most forest land can support a wide variety of different uses, there are advantages to devoting areas of the forest to single uses especially suited to those areas. The resulting land use decisions can be formalized in a land use plan, the specialized use being referred to as the "priority use" for the land use area in which it is to take precedence over other uses in case of conflict. The only novelty is that such specialization flies in the face of an important forest management orthodoxy that has developed over the past thirty years or more: that public forest land is best managed for the widest possible variety of sustainable uses across the entire landscape, and that failure to manage in this way will discriminate against one or more legitimate public constituencies. The move to priority-use zoning is prompted by widespread dissatisfaction with this "multiple use-sustained yield" (MUSY) paradigm and constitutes, in effect, a rejection of its fundamental assumptions.

Intriguingly, agreement on priority-use zoning is more than purely formal. Whether land use plans have been drawn up by panels of scientists or by extensive interest group consultation and "shared decision-making," the substance of the land use recommendations has been remarkably similar. With some minor variations, such plans tend to revolve around landscape *triage*, in which land protected from consumptive uses is set in a matrix of working forest, itself divided into more and less intensive forestry zones. This pattern can be found in the options developed by the Forest Ecosystem Management Assessment Team (FEMAT) for the "owl forests" of the Pacific Northwest, where the central ecological idea behind priority use zoning – that a system of isolated reserves managed without reference to the rest of the landscape has very little likelihood of achieving conservation goals – is first articulated (FEMAT 1993). However, it can also be found in the land use designation developed through negotiation by the BC Commission on Resources and the Environment (CORE) and subsequently adopted, with modifications, by the BC government (Rayner 1996).

Differing perspectives on the purpose of priority-use zoning begin to emerge when the actual division of the landscape according to the agreed-upon designations has to take place. A cynic might have predicted such disagreement, observing that this point is where real land use decisions with real impacts on current users must be made. The two major alternative perspectives will be considered in more detail below, but their main outlines can be easily sketched.[1] On one view, zoning is an economic device, designed to promote the most efficient use of public lands in the absence of a fully functioning market for all of the goods that such lands can produce. Though not really an "economic instrument" in the technical sense of that phrase, zoning is seen as an essential precondition for other economic instruments to function effectively. In this view, the drawing of the zones becomes an exercise in applying the theory of comparative advantage (Vincent and Binkley 1993). Highly productive land should be zoned for intensive forest management with a minimum of constraints (to respect the concerns of other users). The pay-off is that, because higher volumes of fibre can now be produced on a smaller land base, more land becomes available to devote exclusively to other uses or for carefully managed compatible uses. These three management regimes – dominant fibre production, dominant other use, and multiple compatible uses – provide the basis for the *triage*.

In an alternative view, zoning serves ecological goals. Its function is to address the damage done to native biodiversity by many decades of uncontrolled development. This goal can be achieved by preserving areas of land large enough to permit representative native ecosystems to function as far as possible undisturbed by human intervention. Such large protected areas need to be buffered from activities in the consumptive use zone by an intermediate zone in which low-impact uses will be permitted. Looked at in this light, the convergence on landscape *triage* is more apparent than real. Very different criteria will be employed to determine what land should have which priority-use designation. In fact, each perspective's favoured land use is the residual land use in the other perspective. The surprising fact is that the consensus around zoning has been achieved at all. To understand why this should be and where we go from here requires a closer look at both the problem and its proposed solution.

Multiple Use as Myth and Reality

As foresters have pointed out, the management of forest land is de facto multiple use management (Clawson 1975: 41). It is impossible to imagine a forest, actively managed or otherwise, that produces one and only one good. Nearly all forests are watersheds, even the most intensively managed forest is home to some species other than commercial trees, and so on. However, in the real world of forest policy, not all uses are created equal. The 1960 Multiple Use Sustained Yield Act (MUSYA), the classic American statute of the early phase of multiple use forest management, charges the USDA Forest Service with "the management of all the various renewable resources of the National Forests so that they are utilized in the combination that will best meet the needs of the American people; making the most judicious use of the land for some or all of these resources or related services" (16th US Congress, §531(a)). Three contemporary commentators have observed that "few federal agencies are guided by a fundamental principle which is so ambiguous and devoid of explicit substantive standards" (Alverson, Kuhlmann, and Waller 1994: 136-7).[2] The result in the US was to allow the Forest Service considerable discretion to determine what multiple use means in practice, exercising this discretion in accordance with some fairly rigid organizational and professional norms (Twight and Lyden 1988; Daniels 1987) and interpreting disagreement as an attack on its professional competence. In effect, because any management regime inevitably produces multiple benefits, it became possible to justify any management plan by listing such benefits as conforming to the Forest Service's multiple use mandate.

In Canada, managing public forests primarily for wood fibre did not have to be smuggled in disguised as multiple use. It was a central plank of the policy eventually pursued by all the provinces of converting existing, naturally regenerated forests into faster-growing, actively managed forests with the goal of increasing forest capital. Lacking the resources to reach this goal themselves, governments' chosen policy instrument has been the long-term area-based forest management licence in which forest products companies are given some security of tenure on Crown land in return for a commitment to practise sustained yield timber management (Howlett and Rayner 1995). This policy has run into some well-documented difficulties.

First, for the companies themselves, if not for society as a whole, the economics of the first pass are quite different from those of the second and subsequent ones. In the first pass, the forest comes for the cost of the licence plus the expenses associated with access and removal, together with protection for remoter forests until they can be reached. In the second pass, the costs include the expense of restocking and stand tending, costs that are paid upfront for trees that will not be commercially useful for many decades. If taken on purely financial grounds, the decision on whether to keep forest land in production after the first cut depends heavily on factors such as the productivity of the site and its ease of access to processing facilities. Using these criteria, land that is economic to log once would be abandoned or sold by a private owner after the original forest cover had been removed. This option is not available for public land, so a very simple regulatory framework arose – not without resistance from the licensees – designed to force licensees to restock cutovers as a condition of retaining their licence. To make this economically feasible, the standard of satisfactory restocking has been made an average one ("basic silviculture" in BC) that has the economically perverse result of licensees doing more work on marginal land than they otherwise would and less on highly productive sites. Licensees have responded by including marginal and remote or rugged sites in their inventories to inflate the size of their total allowable cut, while concentrating their cutting activities in the high productivity and accessible locations, a perfectly rational strategy but one that has made timber supply planning extremely difficult (BC Ministry of Forests 1991).

Second, "multiple use" played little part in the original design of this licensing system, which was directed largely at management for fibre production in perpetuity. Although some consideration was given to recreation, fish, and wildlife values, directing licensees to manage for multiple use is, in effect, to reduce the value of the licence by removing some timber from the inventory or placing potentially costly and burdensome constraints on timber management in areas where options for other uses are to be retained. Moreover, third, the long planning horizon of sustained yield timber management makes the whole process extremely vulnerable to unpredictable changes in the basic assumptions of the planning model, chiefly the total operable land base available for timber harvesting. Both the outright withdrawal of land from the working forest and the imposition of regulatory constraints on how much timber can be removed in any one pass can have serious consequences on current harvest levels.[3]

Ironically, the pressure for withdrawals and constraints to protect other values is in part a consequence of the increasing sophistication and public accessibility of management plans which reveal in detail where logging is going to take place for many years to come. While logging went on in remote and relatively inaccessible locations or close to timber-dependent communities, and while it appeared that there was plenty of forest land for everyone, the problem of multiple and incompatible uses was solved in practice by spatial segregation or de facto zoning. By the 1970s, however, a number of related developments were undermining this comfortable modus vivendi. Increasing interest in tourism and outdoor recreation brought larger numbers of people into contact with active logging operations at the same time that these operations were expanding into new areas and licensees were announcing their intention in management plans to expand into still more areas. Even for those who did not actually participate in back country recreation, the growth of environmental awareness put many of the cosy practices of Canadian logging under a glare of unwelcome publicity. Confronted with the evidence of lax regulation, non-regenerating cutovers, washouts of poorly constructed roads, and other examples of unacceptable impacts for non-timber forest users, the forest policy community indulged in an orgy of mutual recrimination: small companies blamed large ones, large ones blamed contractors, and everybody blamed the government.

In Canada, government reacted in predictable ways. The federal government attempted, with some considerable success, to use its spending power to invade yet another area of provincial jurisdiction through Federal Resource Development Agreements (FRDA), turning a technical issue into a constitutional one, while the provincial governments engaged in a hasty round of symbolic politics, reassuring the public that "something was being done" while not actually doing it. That classic instrument of symbolic politics, the Royal Commission, was rolled out in province after province, and laws and regulations multiplied in inverse proportion to the capacity of understaffed provincial forest services to implement or enforce them

The key idea that emerged was integrated resource management (IRM). IRM involved, first, the public acknowledgment of the importance of managing Crown forests for values other than timber production, and, second, the development of planning processes designed to identify and capture those values in response to public demand. The policy regime that resulted was not at all unlike that envisaged a decade or more earlier by the MUSYA, granting extensive discretion to forest managers to determine which uses should be integrated and how to do so. Like the MUSY policy, IRM succeeded admirably in keeping up cut levels, but signally failed to diminish public controversy, in large part because the planning process remained resolutely "top-down" and driven by timber targets such as licensees' allowable annual cut (AAC). On the basis of recommendations from Peter Pearse's Royal Commission, the BC Forest Act of 1978 placed a statutory responsibility on the Chief Forester to determine the AAC by taking into account a very wide range of forest policy goals – a discretionary decision that was in practice much influenced by fairly short-term political and economic considerations (Hoberg 1993; Dellert 1994). This level of cut then had to be "found" in the planning area, greatly limiting the options of the various lower-level planning processes that were supposed to be considering IRM. While it is doubtful that anyone was really taken in by publicity that suggested competing uses could actually be carried on simultaneously across the landscape, there was considerable frustration at the lack of flexibility shown by the Ministry of Forests and its major licensees, frustration that was expressed at a number of public hearings in BC throughout the 1980s and into the 1990s.

In the United States, meanwhile, there was similar dissatisfaction with the implementation of the MUSYA on the part of interests who felt that their legitimate concerns were not being heard, particularly in the controversy over the effects of clearcutting on other forest users. This dissatisfaction resulted in the 1976 passage of the National Forest Management Act (NFMA). The NFMA attempted to "fence in" the broad discretionary powers given to the USDA Forest Service in the MUSYA by setting out statutory standards for forest practices and by beefing up the planning requirements for National Forests (Wilkinson and Anderson 1985). The USDA Forest Service now had to show that it had actively solicited input from users and responded to their concerns.

The impact of the NFMA on Canadian forest policy remains largely unstudied, and there is even a tendency to downplay its impact by focusing on the very different legal and constitutional context for forestry in the two countries. However, there are at least four areas in which the NFMA has been influential. First, notwithstanding the well-merited objections to applying national standards to very different forest ecosystems, these standards, especially the maximum size of a clearcut, became benchmarks against which Canadian practice was measured and, in some eyes, found wanting. Second, and a related point, as US environmental organizations began to use appeals and court challenges to enforce their interpretation of the NFMA, they also began to develop the argument that other countries, especially Canada, should not gain a trade advantage by operating under less stringent rules. In consequence, US-based environmental organizations began to take an increasingly active role in forest policy disputes north of the border. Third, the growing Canadian environmental law community, frustrated by the limited opportunities for legal challenges to Canadian forest management practices, looked enviously at the NFMA and began to raise the issue of discretionary management and to propose statutory forest practices in Canadian jurisdictions. Finally, and perhaps most important of all, the NFMA greatly increased the role and prestige of scientists in forest policy. This development did not happen all at once and its implications remained hidden for a decade or more, even after a blue-ribbon committee of scientists had been called upon to draw up NFMA regulations. It eventually became clear from American experience that the greatest threat to forest managers' discretion is the deference of judges to scientific testimony about the alleged environmental impact of forest management plans.

In response to the perception that IRM was simply business as usual under a new name and that important lessons could be learned from experience south of the border, Canadian environmental interests successfully organized around two new policy goals: the complete withdrawal of more land from timber management to be preserved as park and wilderness and stricter legal regulation of logging practices to minimize the impact on other uses in the working forest and beyond. In BC, the Protected Area Strategy (PAS) and the Forest Practices Code were the result. Under PAS, the province committed itself to raising the proportion of land in protected area status to 12 percent from its then 6.5 percent. The Code, as its name implies, is effectively a codification of existing forest management practices, putting some limits on managers' discretion and giving extra attention to high-profile problems such as soil conservation and the protection of fish-bearing streams. Moreover, the province also attempted to address the deep distrust that had accumulated between opposing elements of the forest policy community by adopting a novel "shared decision-making" land use planning exercise under the aegis of CORE. The CORE process, to some extent at least, circumvented the well-established lines of communication between industry and government, and combined with the other initiatives to create the impression that a much more radical policy shift was envisaged than had been undertaken in the past.

The moral of this brief narrative history of postwar forest policy in North America is that multiple use as actually adopted was a code word for constrained timber production, recognizing the actual primacy of timber production as the historic goal of forest policy. The modifications to the original policy marked by the passage of the NFMA in the United States and the increasing focus on IRM in Canada were at first a merely symbolic commitment to constrain timber production in order to produce more of the other values to which organized interests were laying claim. After a long and at times bitter battle on both sides of the border, these other interests have at length been able to require that this commitment be honoured in more tangible ways (Hays 1988; Lertzman, Rayner, and Wilson 1996). With forests fully committed to existing processing operations, severe dislocation and resulting political conflict can only be avoided if some way is found to break out of the zero-sum game where each new withdrawal or constraint means a reduction in timber supply. No more land is available, so offsets can only come from more intensive production and a "flexible" interpretation of the constraints.

Zoning as an Economic Instrument
It is against this background that the twin claims now being heard that multiple use/IRM has failed and that zoning is superior should be assessed.

It is important to note that both the failure of multiple use and the superiority of zoning are judgments relative to the goals of particular interests. From the perspective of industrial interests, the combination of further withdrawals and increasingly onerous constraints apparently means that continued access to the entire forest landscape is no longer worth fighting for. Because of the reverse allowable cut effect, the combination of more protected areas and more complicated timber management constraints has the potential for immediate and dramatic reductions in timber supply. In BC, estimates in the order of a one-third reduction or more were at one time in circulation. Faced with this unacceptable outcome, industry had to abandon its general strategy of restricting pressure for change to easily digested incremental steps under the broad banner of implementing IRM. All at once, the steps were too large and too fast. Instead, we began to hear from industry that IRM had been a costly and inefficient mistake.

This about-face serves to obscure a crucial ambiguity that was always present in the concept of multiple use: the scale at which multiple use is to be practised. Do we mean the production of the maximum feasible outputs from every part of the land under management (extensive multiple use), or do we mean maximizing production of multiple outputs from the land base taken as a whole (intensive multiple use)? For most private landowners, this issue is a distinction without a difference, but for public lands, where tens of millions of hectares are under management, the distinction is crucial. Maximizing multiple outputs from the whole land base does not exclude the option of devoting considerable areas to specialized production of one output and, indeed, may even require that this be done. Sahajananthan (1994) notes that diverging opinions on this question could be observed as far back as the early 1940s, but he suggests that the concept of extensive multiple use or multiple use of every hectare – a concept championed by S.T. Dana during his editorship of the *Journal of Forestry* during the Second World War – won the day.

While not exactly wrong, since, as we have seen, multiple use did come to symbolize extensive management for multiple values, this way of characterizing the issue contributes to an important myth that needs to be debunked if we are to understand the lessons of the past for the options that we have now. As we have seen, even in the United States, where extensive multiple use was written into the mandate of the USDA Forest Service in the MUSYA, it was never actually practised on the ground. The last fifty years of forest management are best characterized as constrained timber production, which is not quite the same thing. It is not that IRM or extensive multiple use "didn't work" – however one might try to assess such a claim. Rather, the increasing number of constraints and the worrisome tendency of regulatory agencies to try to enforce them, spurred by courts in the US and international public opinion in Canada, means that the policy can no longer achieve its symbolic goal of public reassurance while continuing business as usual. The game is no longer worth the candle and a new game has to be invented.

Moreover, the alternative conception of multiple use – maximizing outputs by specializing – periodically resurfaced even during the heyday of IRM and MUSY. Almost as soon as the United States began to experience pressure from wilderness preservation advocates, the idea of intensive multiple use was put forward as a potential solution. Clawson, for example, produced a very rough and ready calculation in his popular treatment of US forest policy (1975), showing that by concentrating industrial forestry on the better site classes in both public and private hands, timber production would exceed then current levels using only 40 percent of the commercial forest acreage. The rest could be deferred or permanently set aside, giving policymakers enormous flexibility in dealing with recreation and wilderness demands. The calculation, he admitted, was "suggestive rather than definitive," but "with such very great potential for increased timber output from smaller areas by means of intensive forestry, the possibilities of reconciling competing demands for forest land are significantly greater than would be the case if timber output were constrained to present levels" (1975: 107). Despite Clawson's advocacy, the NFMA, then being steered through Congress, opted for new constraints on timber production right across the landscape, with the consequences that we have already observed.

In declaring that IRM across the landscape is a mistake, zoning proposes to concentrate industrial forestry in areas with high timber values; to permit forestry under a variety of carefully specified conditions where other conflicting values exist; and to forbid it in those areas where incompatible values are deemed worthy of complete protection. Used in this sense, priority-use zoning can be treated as a precondition for the effective use of economic policy instruments in two respects. First, it is proposed on "efficiency" grounds, and efficiency in the sense of Pareto optimality or some more relaxed modern variant is put forward as a criterion for determining the zones themselves. The primary allocation rule is that, if the net present value of the fibre that could be produced outweighs the value of potentially conflicting other uses, land should be zoned for intensive forestry (Stanbury, Vertinsky, and Thille 1991).

Second, zoning also creates or restructures the incentives for people to behave in market-oriented ways. Within the areas managed for timber production, management decisions will become market-driven and largely free from burdensome constraints designed to mitigate their impact on other uses. Freedom to manage – the holy grail of the forest manager since the rise of the environmental movement – will at last be achieved. In addition, uncertainty about timber supply is diminished, opportunities for profitable long-term silvicultural investment are identified, and conflicts between resource uses put on the same level playing field where each has to be justified on the grounds of efficiency. Thus, priority-use zoning can be, and in BC almost certainly will be, linked to the perennial issue of security of tenure for licensees and promoted as a potential way of diminishing the very marked difference between the levels of silvicultural investment on private lands and those on Crown lands of the same productivity (BC Forest Resources Commission 1991: 39). There will be pressure for increased security of tenure in intensive management zones to protect silvicultural investments, while the possibility of putting additional uses on a similar footing by granting tenures for uses other than timber in the general management areas will also appear on the policy agenda (Pearse 1993; Haley and Luckert 1998).

However, as the BC experience clearly shows, the path towards economic zoning of the forest is not a straight or an easy one. The difficulties are not technical: site productivity inventories exist in reasonably detailed form and simulations like Sahajananthan's continue to produce results remarkably close to Clawson's rough calculation that present levels of timber production could be maintained on 40 percent to 50 percent of the current land base if concentrated on high productivity sites with IRM constraints removed. The obstacle is a different vision of the objectives of zoning, one which competes with the economic view for determining how zoning is to be carried out on the ground.

Zoning as an Ecological Instrument
This other major interest in zoning comes from a quite different direction, namely, conservation biology. In effect, of course, the traditional designation of parks, with different categories of parks each involving different levels of permitted resource exploration and extraction, and the treatment of the working forest as a residual after deducting these areas constituted a rough and ready land use zoning scheme. However, the intensifying pressure for new areas to be added to the park system that continued throughout the 1980s and beyond also spawned new attempts to explain the rationale for parks based on conservation biology rather than on the needs of the tourism and outdoor recreation industries. It was at this time that the concept of biodiversity in its current, extended sense came to prominence, together with the idea of managing the entire landscape to conserve and restore native biodiversity. As Noss and Cooperrider remark in their popular introduction to the subject, "a little more than a decade ago, biodiversity progressed from a shorthand expression for species diversity into a powerful symbol for the full richness of life on earth" (1994: 3). That is, a concern for biodiversity in this sense symbolizes a commitment to the idea that undisturbed ("natural") ecosystems are in some way qualitatively better than disturbed or managed ones. If forced to explain the quality that makes an undisturbed ecosystem better, accounts go in two different, but not necessarily contradictory, directions. In one direction, identified by

Goodin (1992) as the central "green" value, undisturbed nature simply has value because it is undisturbed. It exists in contrast to our own efforts at modifying and managing and constitutes an essential dimension against which we can continue to measure and understand ourselves. To lose it would be to lose one of the "horizons of meaning" that we use to construct our very sense of self. In another direction, following some famous remarks of Aldo Leopold, more stress is put on the complexity of natural processes, a complexity that currently, and perhaps forever, lies beyond our comprehension (Grumbine 1994; Stanley 1995). In this view, it is simply prudent to retain functioning, undisturbed ecosystems if they produce things that we value and we have, as yet, no very clear sense of the long-term consequences of our ability to reproduce these values through active management. In theory, of course, these two variants would come into conflict if it could be shown that we did completely understand ecosystem structure and function and could reproduce them. Enough uncertainty currently exists, or is believed to exist, that such conflicts rarely arise.[4] The symbolic dimension of biodiversity is crucially important, giving the project of biodiversity conservation extraordinary political resonance with a wide variety of groups.

Unfortunately, this resonance of "biodiversity" in the extended sense used by Noss and Cooperrider underwrites a quite different approach to zoning than the economic approach. It should be noted at once that this difference does not occur because conservation biology is in some way ignorant of or indifferent to the idea of scarcity. Land and money for conservation and restoration is certainly scarce, and a variety of criteria has been developed to produce informed land use choices, for example, the Nature Conservancy's "last of the least and the best of the rest" ranking (Jenkins 1988) and the increasingly popular gap analysis. In the latter case, an ecological classification system is used to identify "gaps," ecosystem types that are currently underrepresented in protected areas, and efforts are focused on filling the gaps. CORE used gap analysis as one criterion for identifying priority protected areas, but it was only partially successful in closing the gaps in its land use options. CORE's difficulties are significant but hardly surprising. After all, the main reason that gaps exist is usually that a particular ecosystem type is especially valued for commodity production. However, the effect of applying these criteria has been to produce a consensus around landscape triage that appears superficially similar to the approach taken by priority-use zoning. Some areas will merit complete protection from human disturbance while others are already so far removed from their natural state that it is not worth bothering with them. Between these two extremes lies a potentially shifting and fluctuating boundary which must be used to buffer the protected areas from the impact of human use but which can also support some innovative and experimental forms of management that do not necessarily exclude resource extraction. More controversially, the buffer zones might also provide "connectivity" between protected areas, preventing them from becoming isolated islands of diminishing genetic diversity even while natural species diversity is maintained.

Finding a politically acceptable balance between reserves and the "matrix" in which they are set has proved difficult. While politicians can no longer dismiss estimates of reserve size based on ecological criteria as completely absurd, ecologically based proposals remain, for the moment at least, apparently beyond the bounds of political possibility. Noss's study of the Oregon Coast Range indicates a need for 23 percent of the region to be placed in reserves where a high priority would be given to conservation and restoration, and a further 26 percent in reserves where some compatible human uses might be permitted. Surveying other US studies, Noss and Cooperrider conclude: "Our estimate of the area needed in reserves is an order of magnitude beyond what is currently protected in most regions"(1994: 68). Hunter and Calhoun (1994) have proposed converting triage to triad: three roughly equal parts of protected areas, lightly managed forest, and intensive timber plantations.

Conservation biology's emphasis on protecting large, undisturbed ecosystems has serious implications for forest management in those jurisdictions, pre-eminently British Columbia, where substantial areas of old-growth forest remain intact. As noted, timber supply calculations have already taken these areas of high-volume, high-quality timber into account in setting past and present AACs and licensees have made production and investment decisions accordingly. As the simulations that are run when AAC is set for Tree Farm Licences show, removing old growth from harvesting schedules will have dramatic consequences that last for decades while second-growth forests, currently in immature stages, reach cutting age. Commenting on the results of simulating the complete setting aside of old-growth timber from the working forest, the authors of the draft Management and Working Plan for Tree Farm Licence No. 44 on the west coast of Vancouver Island commented:

> In this case, the immediate impact is extreme. Harvest for 1991-1995 would be only 15% or 405,000 m^3 [compared with] the 2,700,000 m^3 for the preferred option. This reflects the interruption in the orderly process of conversion from unmanaged to managed forest. Specifically, very little of the new forest is old enough to harvest and a semblance of balanced forest ages is not achieved until about 2040 (MacMillan Bloedel 1991: 28).

For this reason alone, there is not going to be a simple trade-off in which high-quality sites are zoned for intensive timber production and the remaining 60 percent of the forest goes to other users.

Second, conservation biology is also concerned with ecosystem integrity at very large scales, larger in some cases than existing administrative and political jurisdictions. The most familiar aspect of this concern has been the debate about the "fragmentation" of old-growth ecosystems, even where substantial total areas remain, and the consequent need to maintain or reintroduce connectivity across the entire landscape. Although this goal could be achieved in a dynamic way – by deferring cutting in some areas until other cutovers have reached a desired future state where they could assume connectivity functions – the resulting adjacency constraints have further substantial impacts on present cut levels. Indeed, in reintroducing complex constraints, requirements for landscape-level connectivity threaten one of the central selling points of economic zoning: the promise of reducing regulations and freeing up managers to manage. Unsurprisingly, industry has mounted a sustained attack on the scientific credibility of the theory of island biogeography that underwrites connectivity recommendations (Preston Thorgrimson 1991). On Vancouver Island, at least, this attack seems to have carried the day.

Third, concerns about the ecological impacts of silviculture at the stand level are not going to go away. Some of these concerns have the backing of important economic interests with their own bureaucratic allies: for example, riparian zone management and the fishing industry. Others, such as the greater attention being paid to soil conservation, command wide scientific support and are periodically brought to public attention by media-worthy washouts and water-quality problems. Still others, such as stand-level biodiversity constraints, may acquire greater scientific credibility as more research is carried out. In each of these cases, the preferred policy instrument is regulation, because conservation biology presents the issue as preserving a floor beneath which practices cannot be allowed to fall without irreparable damage to the environment. In this respect, there is a distinction to be made between using zoning to avoid a "worst case scenario" of burdensome constraints across the entire landscape and using zoning as an economic policy instrument to produce optimal production and investment decisions.

The Economist as Soi-Disant Arbitrator

The bringing together of these two perspectives on zoning in a unified vision of forest management is a political task. Nothing in this chapter should be read as suggesting that such a task is impossible. Arguably, in British Columbia at least, we have rarely been closer to achieving a solution to outstanding forest policy problems that commands such widespread legitimacy – a political solution in the largest sense. Encouraged by partisan policy analysts, however, neither side has entirely abandoned hope that its favourite analytical tools could, at this eleventh hour, provide the much-sought-after knock-down blow to the other side's pretensions. This is a vain hope, for each side's analysis is inevitably refracted through the distorting lens of its own conception of what zoning is trying to achieve. The ecologists' claims that science can definitively tell us how much land to devote to each use and what practices are appropriate in each zone are as fundamentally misconceived as the economists' belief that zones should be drawn according to calculations of economic efficiency. Both ecology and economics can provide information to guide policy choices. The ultimate triumph of one vision of zoning over the other, however, will depend on its ideological appeal.

To understand the ideological appeal of zoning as an economic policy instrument, it is important to disentangle two related but distinct elements of this appeal. On the one hand, there is the efficiency argument: economic instruments are proposed as more efficient than alternative instruments in two senses. First, by using market mechanisms wherever possible, economic instruments are efficient in the sense that markets are efficient, their outcome is thus "optimal" in the economic sense of that term. Second, economic instruments are efficient in the sense of cost effective. Because, in theory, at least, economic instruments achieve their aims by structuring incentives for individuals to act in the desired way, they can dispense with a costly apparatus of surveillance and enforcement. On the other hand, there is the freedom argument. Economic instruments, in contrast to the so-called command and control instruments used in regulatory approaches, are supposedly less intrusive and preserve a greater amount of individual freedom. Undergraduates are often taught to locate alternative policy instruments on a scale of less to more coercive, with the assumption, usually unexamined, that less coercive instruments are to be preferred to more coercive ones, other things being equal (Trebilcock, Pritchard, Hartle, and Dewees 1982). Where, as in the case of economic zoning on public land, we aim to simulate the result that a market for public land would have produced if it existed, the freedom argument suggests that there is some special quality about the outcome beyond its economic efficiency, because it is the result that rational people would have chosen if they had been given the choice. In assessing the relative merits of economic and other policy instruments, it is important to keep distinct the cost effectiveness or efficiency argument, and the argument that economic instruments promote freedom.

The key political argument that we have heard in favour of using zoning as an economic instrument in BC is a negative variant of the efficiency argument. It is claimed that by failing to treat zoning as an economic instrument, by using ecological or, more darkly, "political" criteria to determine the extent and location of specific uses of public forests, we have produced a sub-optimal outcome. The major proponent of this argument has been van Kooten who, in a series of articles and reports (van Kooten 1994a, 1994b, 1995), attempted to work out the opportunity cost of not logging mature timber in BC. On his calculations, using some fairly heroic assumptions to value preservation and biodiversity benefits, van Kooten argued that we had already preserved too much coastal old-growth by 1991, and that the proposed increase from the then in effect 6 percent to 12 percent under the Protected Areas Strategy would diminish the net present value of Crown land by around $7 billion or nearly $300 million annually when discounted.

As van Kooten notes, the main weakness of his analysis is the possibility that he has failed to account for some of the benefits of preservation. In particular, he recognizes the argument made by some economists that preservation of old-growth forests captures a quasi-option value – the value of not foreclosing future options by setting in motion irreversible changes. In a separate study of this issue, he argues that the value of biodiversity preservation can be handled this way. Because there is no market for quasi-options, the best we can do, he claims, is to assume that these benefits are "substantial" but not infinite. At some point, the burden of forgoing advantages now will outweigh the future benefits of keeping our options open. We cannot even begin to discuss this question sensibly, however, unless we have some sense of the costs that we are imposing on ourselves by our forbearance. We must put the cost of preservation squarely before the members of the public and ask them if they are willing to bear this cost in order to keep these options open for themselves and for future generations. A referendum might be appropriate. For the moment, however, politicians seem unwilling even to discuss the costs of preservation, with the result that the issue becomes "politicized."[5]

Van Kooten, however, notes an alternative approach to valuing biodiversity benefits. Instead of allowing present generations to make the decision on the value of options for future generations, ecological integrity could be put beyond cost-benefit calculations as a floor beneath which trade-offs of commodity values against preservation values are not allowed (Bishop 1993). This Safe Minimum Standard (SMS) approach is not without its own drawbacks. Like the valuation of quasi-options, there has to be a non-market standard applied to determine when the costs of SMS become "intolerable" so that the present generation could legitimately foreclose the options of future generations (for example, to preserve itself). Presented this way, the difference between assuming a "substantial" value for the preservation of quasi-options and using the SMS to hold biodiversity beyond trade-off except in the case of "intolerable" costs is more apparent than real. Some person or institution has to decide what is substantial or intolerable and the process by which this decision is made in liberal democracies is what we call politics. We can only make sense of van Kooten's distaste for the politicization of the issue when we understand that for him, as for many other economists, the freedom argument, though unacknowledged, plays a role that is every bit as important as the efficiency argument.

To see this point clearly, and to see the limits of partisan economic analysis, we need to distinguish between the two kinds of political argument at work here, helpfully labelled by Barry as "want-regarding" and "ideal-regarding" arguments (1965). Want-regarding principles are familiar to economists precisely because they take into account only people's preferences without discrimination. (And, it should be added, economic analysis is congenial to people who are persuaded by want-regarding arguments, because it proceeds in this way.) If there is no market, as in the case of biodiversity conservation, there must be a simple decision principle such as Bentham's dictum that "each is to count for one and no more than one" in a majority vote. The vote would determine the costs that people were prepared to bear without reference to any value other than the value that wants are to be respected; hence, van Kooten's conclusion that additional protected areas should only be added after a referendum (1995: 58).

Barry treats ideal-regarding principles as the exhaustive alternative to want-regarding ones. He notes that the simplest deviation from a want-regarding principle is to allow that policies be judged according to how well they satisfy people's preferences, and to deny that each person's own ranking of the relative importance of different wants be taken as definitive. In fact, both SMS and quasi-option costing can be treated this way. That is, instead of letting the market decide which combination of biodiversity and commodity values is desirable, or taking a vote on the question, someone is empowered to set relative values, by deciding, for example, if the subsequent decline of economic activity is adequately compensated for by the preservation of biodiversity. Of course, such a decision is greatly aided by accurate information about both the economic and the ecological consequences of our actions and will be completely arbitrary in the absence of such information, but, unlike the case of the want-regarding decision, someone has the authority to decide what is an "intolerable" or a "substantial" cost, and he or she must do so by reference to an ideal.

An alternative understanding of SMS that departs further from a want-regarding principle is to assert that some wants have no value at all, so they can be left out of calculation. More radical a departure still, as Barry observes, is to regard the satisfaction of some wants as intrinsically bad and hence their suppression as a benefit rather than a cost. Much of what we have come to understand by "environmentalism" involves valuations of this kind; for example the belief that any extinction is not merely a potential, if uncertain cost, but an evil, a failure to discharge our responsibility of stewardship towards nature, and to be avoided whatever the cost.

We can observe two things about this slide from a want-regarding to an ideal-regarding justification. First, any attempt to argue that we shouldn't take the first step that leads from want-regarding to ideal-regarding justifications is self-defeating, because it must itself be a form of ideal-regarding argument. The ideal at stake is none other than the foundation of the second of the two preferences for economic policy instruments noted earlier, i.e., the freedom argument. Want-regarding arguments are held to be superior because they respect freedom and autonomy by not empowering anyone to "second-guess" individual preferences. There is no doubt that this powerful argument has deep resonance in our society. The values of respect for individual choice and a self-directed life are undeniably important. But recommending them is an ideal-regarding argument nonetheless and cannot itself be decided by appealing to the preferences of the majority, even if we are more confident than John Stuart Mill that a majority, if given the choice, would actually opt for self-direction rather than mindless conformity. Second, many economists engaged in partisan policy analysis, when they bother to make the distinction between want-regarding and ideal-regarding principles at all, are inclined to let their ideology show by denouncing public policy based on ideal-regarding principles as the slippery slope that leads to totalitarianism. It is easy to mock this sort of thing – today the Forest Practices Code, tomorrow the gulag – but there is an important issue at stake. The authoritative imposition of values through law remains a generally legitimate government function,[6] and this fact explains why, other things being equal, a more coercive policy instrument is often in practice preferred to a less coercive one that would do the same job. The mere fact that an economic approach to land use zoning is less coercive than an approach that puts greater stress on regulation will convince someone of the merits of economic zoning only if that person is already convinced of the superiority of want-regarding over ideal-regarding political arguments.

The Politics of Priority-Use Zoning

If forest land use zoning is primarily a political rather than an economic problem, some indication of how the conflict of interpretation may play out in practice and some suggestions for how to resolve this conflict in a productive way are provided by the land use planning exercise on Vancouver Island that began with the CORE process. CORE was given statutory authority to engage in regional-level interest-based consultation aimed at writing land use plans for areas of the province where conflicts had proved particularly intractable. In the event, CORE undertook an ambitious exercise in "shared decision-making" in which regional tables of interest-based representatives negotiated directly on everything from table membership to land use designation. Inevitably, given its ambitions, CORE's record was a mixed one. Of the four regional tables eventually put in place, only one achieved something like consensus on a land use plan; two more, including Vancouver Island, were able to give some guidance to the Commissioner in writing a plan; and one collapsed after failing to agree on some basic planning area boundary decisions. The failure to reach consensus on Vancouver Island is particularly instructive, because the most intractable problem was precisely the meaning of zoning. The table could only agree to put forward two quite different reports, and left the Commissioner, and ultimately the provincial government, to draw the boundaries of the zones (CORE 1994: Vol.III).

The first proposal, supported by major forest companies, some forestry independents, forest industry workers, timber-dependent community rep resentatives, municipalities, mining, and agriculture, saw the regional table's role as setting a system of land use classification with indicative criteria on how to apply the classification. In their view, the actual job of designating particular lands should be handed over to sub-regional processes in which local community interests would play a dominant role. A major part of sub-regional planning would be to split three ways the "multi-resource use land" (itself a residual category consisting of all Crown land except settlement and protected areas), according to the proposed intensity of resource extraction based on existing resource values. The alternative, proposed by the conservation sector, accepted the essentially same land use designation system but saw handing over the actual designation decisions to sub-regional planning processes (likely to be driven by the basis of existing resource values) as a trap in which overall conservation goals would be sacrificed to local commodity interests. Examples of conservation goals cited by the conservation sector, clearly influenced by conservation biology, were the proposed use of low-intensity areas for connectivity, buffers, and restoration, and the designation of general management areas as a "forest ecosystem management zone," in which the maintenance of biodiversity and long-term productivity would be priority goals. For the "high-intensity" zone, the conservation sector explicitly restricted use of the designation to "the most productive growing sites on the Island *that are now immature or second growth forests* [emphasis added] ... where the options of maintaining natural forest ecosystems have already been lost" (CORE 1994: III-6).

Faced with the failure of consensus, and noisy protests from forest industry workers when the Commissioner's land use plan was finally unveiled, the BC government sought a political solution. In public, it made a symbolic endorsement of the plan while accepting that some marginal trade-offs would make the plan more attractive to those interests that could effectively block implementation if they chose. In particular, the cabinet has approved a number of figures, apparently plucked from the air, which set upper limits on the impact of new measures on timber supply. Designed to reduce uncertainty for industry, their effect is to underline in the strongest possible way that resource targets based ultimately on forest industry demand still drive forest management.

To meet these targets, representatives of key interests were brought together in various "teams" and working groups, wider in membership than the old forest policy network but distinctly more focused than the impressionistic categories favoured by CORE. The Protected Areas Boundaries Adjustment Team, an internal government group but with membership from seven ministries including Aboriginal Affairs and Environment, Lands and Parks in addition to Forests, removed some high value forestry land from proposed protected areas status while maintaining the overall target for protection. A Low Intensity Area (LIA) Review Committee, another inter-ministerial working group which took public submissions, made some more extensive changes to the areas proposed by CORE, and rejected the argument that LIAs should be used for landscape level connectivity: "Connectivity is a concept that some interests claim may have merit but at this date is without substantial scientific validation" (LIARC 1995: 16). More significantly still, the Review Committee rejected the idea that extra constraints beyond those found in the Forest Practices Code should by applied to LIAs. Some work was done on high-intensity areas, stressing the importance of abandoning IRM, but the enthusiasm of some zoning proponents for using HIAs to sustain existing timber harvest levels threatened to reopen public conflict (Jeffery 1995). This work seems to have been downplayed in the public process to date. Finally, a Resource Targets Technical Team – with representation from Forests, Environment, Lands and Parks, industry, and consultants acceptable to the environmental movement – has mapped proposed resource management zones for the rest of the island, replacing the low-, medium-, and high-intensity zones proposed by CORE with the more anodyne designators special, general, and enhanced management (VIRTTT 1996). The team has amplified the LIA Review Committee's comments about management constraints by emphasizing that the zoning system will take advantage of the "flexibility" to be found in the Forest Practices Code. Enhanced management zones (formerly high-intensity areas) may dispense entirely with less important constraints (e.g., visual-quality objectives) and be managed for a low biodiversity emphasis, but they will continue to be subject to regulations (e.g., for riparian zone management) designed to maintain a floor of basic ecological protection. This designation is an unfortunate and presumably unconscious echo of the much-maligned "basic silviculture." And it may result in similar distortions where an average level of protection will waste resources on some sites while ignoring possibilities for restoration on others that would help fill conservation gaps or perform landscape functions.

The outcome, then, is that neither vision of priority-use zoning has been adopted in its pure form. Additional areas of old-growth forest have been protected, despite van Kooten's conclusion that protecting any more coastal old-growth is uneconomical (and increasingly uneconomical, as more is protected). On the other hand, the Resource Targets Technical Team has recommended that a number of areas with more than two-thirds mature forest cover be zoned for enhanced management, including a low biodiversity emphasis. Far from implementing a "new resource management vision," the changes, real enough, are going to be incrementally grafted onto the main body of forest policy after all.

Conclusions: The Myth and Reality of Zoning

The current interest in zoning for priority uses is supported by two quite distinct perspectives, only one of which sees priority-use zoning as primarily an economic instrument designed to achieve the most efficient use of public lands for the production of commodities.

This clash of perspectives cannot be resolved by demonstrating that a proposed land use scheme is economically inefficient, because the ecological perspective does not consider consumer preferences to be sovereign. In addition, the perspective makes common cause with arguments that question the justice of the present distribution of resources. To the extent that analyses based on setting the opportunity cost of forgone timber production against the monetized value of the benefits from preservation and less intensive timber management actually *do* succeed in capturing the relevant costs and benefits, they can provide background information useful to decision makers. In the end, however, some political decision-making procedure has to be in place to determine whether these costs are reasonable or not. The experience of implementing the CORE land use plan on Vancouver Island suggests that these decisions will continue to be made by relatively closed technical consultation between government and those interests with veto power over implementation. At the time of writing, a last-minute effort was being mounted by the environmental movement to bring this process out into the open, where the unresolved underlying conflict between different visions of zoning would become apparent again. It is by no means clear that zoning is going to be politically achievable (Riddell, Smith, Street, and Hoffmann 1997).

The potential for priority-use zoning to reduce the need for command and control regulation has also been exaggerated, although here the present position is murkier still. While the government seems to have made a public commitment to something like the Safe Minimum Standard approach, as embodied in the Forest Practices Code, it is unwilling to pay the political and economic costs of setting that standard at the level demanded by most conservation biologists. However, even in enhanced management zones, some practices are not open to flexible interpretation, and these practices will continue to be embodied in command and control instruments such as regulations and prescriptive standards. For example, it is inconceivable that licensees operating in enhanced management zones will be allowed to set targets for fish production and left to meet those targets however they choose. Instead, standards for leave strips varying by stream class will be applied in the traditional way.

Although priority-use zoning achieved by negotiation and consensus has some potential to reduce future land use conflict, this point is true of any policy instrument adopted as the outcome of negotiation and consensus. Priority-use zoning, however, has the additional burden of being presented as the philosopher's stone of forest policy: the miraculous way to get more out of less. It is not clear that any of the paper calculations and computer simulations of the gains that could accrue from zoning for priority use can survive the transition to the real world of forest policy. In this world, the struggle between opposing visions of a sustainable forest policy remains completely unresolved. Priority-use zoning may create a little breathing space, but it is not the ultimate solution.

Realistically, zoning should operate in conjunction with other proposals for greater use of economic incentives discussed by the contributors to this book. Zoning has a modest role to play in making tenure reform more attractive and promoting economically rational investment in forest resources. It can also help in determining priority areas for protection. Proponents of extreme positions will continue to be disappointed, however, by the political compromises that will be made over actual zoning decisions. To ensure that these compromises reflect social preferences as accurately as possible, governments need to continue to experiment with processes that attempt to balance scientific and technical input with group representation. As always, greater transparency will enhance the legitimacy of zoning decisions, especially when the decisions are contentious. The practice of striking relatively small and effective working groups that accept public submissions and publish extensive technical justifications for their decisions – as in the implementation of Vancouver Island land use decisions – is a step in the right direction.

Notes

1 Space precludes discussion of at least two other perspectives on zoning that further complicate the issue. The first, and more important, is a concern for the impacts of zoning on the distribution of income – a view that holds that even if society as a whole would be better off after a particular land use zoning scheme, there might be considerations of distributive justice that would argue against adopting it. The second is a familiar concern with protection of "favourite places," regardless of ecological or economic benefits.

2 The commentators are unaware or perhaps remain diplomatically silent about standard Canadian practice in this regard.

3 These consequences are the often-neglected other side of the so-called allowable cut effect, where silvicultural investment that improves yields in the future has an immediate impact on current cut levels (Bell, Flight, and Randall 1975).

4 Goodin's discussion of the difference, if any, between an original work of art and a perfect reproduction is helpful in capturing the difference between natural and managed landscapes that is at issue here.

5 Happily, this situation was remedied by the Forest Alliance of BC, whose report on the impact of the first NDP government's land use initiatives (Price Waterhouse 1995) included a prominent discussion of van Kooten's analysis.

6 Although discussion of this point would take us too far afield, the major attempt in recent political philosophy to demonstrate otherwise – John Rawls's claim that a liberal government is one that observes strict neutrality towards the values held by its citizens – is generally regarded as a noble failure.

References

Alverson, W.S., W. Kuhlmann, and D.M. Waller. 1994. *Wild Forests: Conservation Biology and Public Policy.* Washington, DC: Island Press

BC Forest Resources Commission. 1991. *The Future of Our Forests.* Victoria: Queen's Printer

BC Ministry of Forests. 1991. *Review of the Timber Supply Analysis Process for BC Timber Supply Areas.* Final Report. Victoria: Ministry of Forests

Barry, B. 1965. *Political Argument.* London: Routledge Kegan Paul

Bell, E., R. Flight, and R. Randall. 1975. "ACE: The two-edged sword." *Journal of Forestry* 73(10): 642-3

Bishop, R.C. 1993. "Economic efficiency, sustainability, and biodiversity." *Ambio* 22: 69-73

Clawson, M. 1975. *Forests for Whom and for What?* Baltimore: Johns Hopkins University Press

Commission on Resources and the Environment (CORE). 1994. *Vancouver Island Land Use Plan.* 3 vols. Victoria: CORE

Daniels, S.E. 1987. "Rethinking dominant-use management in the forest planning era." *Environmental Law* 17: 483-505

Dellert, L. 1994. *Sustained Yield Forestry in British Columbia: The Making and Breaking of a Policy.* MA thesis, York University

Forest Ecosystem Management Assessment Team (FEMAT). 1993. *Forest Ecosystem Management: An Ecological, Economic, and Social Assessment.* Washington, DC: US Government Printing Office

Goodin, R.E. 1992. *Green Political Theory.* Oxford: Polity Press

Grumbine, R.E. 1994. "What is ecosystem management?" *Conservation Biology* 8: 27-38

Haley, D., and M. Luckert. 1998. "Tenures as Economic Instruments for Achieving Objectives of Public Forest Policy." In *The Wealth of Forests: Markets, Regulation, and Sustainable Forestry,* ed. Chris Tollefson. Vancouver: UBC Press

Hays, S.P. 1988. "The new environmental forest." *University of Colorado Law Review* 59: 517-50

Hoberg, G. 1993. *Regulating Forestry: A Comparison of Institutions and Policies in BC and the US Pacific Northwest.* Vancouver: Forest Economics and Policy Analysis Research Unit, UBC

Howlett, M., and J. Rayner. 1995. "The Policy Framework of Forest Management in Canada." In *Forest Management in Canada,* ed. M. Ross, 43-107. Calgary: Canadian Institute of Resources Law

Hunter, M.L., and A. Calhoun. 1994. "A Triad Approach to Land Use Allocation." In *Biodiversity in Managed Landscapes,* ed. R. Szaro, 477-91. Oxford: Oxford University Press

Jeffery, R.M. 1995. "Memorandum of Analysis Submitted to Rudi Mayser." BC Ministry of Forests, Vancouver Region, June 14

Jenkins, R.E. 1988. "Information Management for the Conservation of Biodiversity." In *BioDiversity,* ed. E.O. Wilson, 231-39. Washington DC: National Academy Press

Lertzman, K., J. Rayner, and J. Wilson. 1996. "Learning and change in the British Columbia forest policy sector: A consideration of Sabatier's advocacy coalition framework." *Canadian Journal of Political Science* 29(1): 111-33

Low Intensity Area Review Committee (LIARC). 1995. *Low Intensity Areas for the Vancouver Island Region: Exploring a New Resource Management Vision.* Victoria: LIARC

MacMillan Bloedel. 1991. *Final Draft Management and Working Plan no. 2 for Tree Farm Licence no. 44 and Managed Forest Unit no. 74.* Vancouver: MacMillan Bloedel

Noss, R.F., and A.Y. Cooperrider. 1994. *Saving Nature's Legacy: Protecting and Restoring Biodiversity.* Washington, DC: Island Press

Pearse, P. 1993. "Forest Tenure, Management Incentives and the Search for Sustainable Development Policies." In *Forestry and the Environment: Economic Perspectives,* ed. W.L. Adamowicz, W. White, and W.E. Phillips, 77-96. Wallingford: C.A.B. International

Preston Thorgrimson Shidler Gates and Ellis. 1991. *A Facade of Science: An Analysis of the Jack Ward Thomas Report Based on Sworn Testimony of Members of the Thomas Committee.* A Report for the Association of O & C Counties and the Northwest Forest Resource Council

Price Waterhouse. 1995. *Analysis of Recent British Columbia Government Forest Policy and Land Use Initiatives.* Report for the Forest Alliance of BC. Vancouver: Price Waterhouse

Rayner, J. 1996. "Implementing sustainability in West Coast Forests: CORE and FEMAT as experiments in process." *Journal of Canadian Studies* 31(1): 82-101

Riddell, D., M. Smith, T. Street, and J. Hoffman. 1997. *Beyond Timber Targets: A Balanced Vision for Vancouver Island.* Victoria: Sierra Club of British Columbia

Sahajananthan, S. 1994. *Single and Multiple Use of Forest Lands in British Columbia: The Case of the Revelstoke Forest District.* Nelson, BC: Ministry of Forests, Nelson Region

Stanbury, W.T., I. Vertinsky, and H. Thille. 1991. *The Use of Cost Benefit Analysis to Allocate Forest Lands among Alternative Uses.* Background Papers, Vol. 2. Victoria: BC Forest Resources Commission

Stanley, T.R. 1995. "Ecosystem management and the arrogance of humanism." *Conservation Biology* 9(2): 255-62

Trebilcock, M.J, R.S. Pritchard, D.G. Hartle, and D.N. Dewees. 1982. *The Choice of Governing Instruments.* Ottawa. Ministry of Supply and Services

Twight, B.W., and F.J. Lyden. 1988. "Multiple use vs. organizational commitment." *Forest Science* 14(2): 474-86

van Kooten, G.C. 1994a. *Economics of Biodiversity and Preservation of Forestlands in British Columbia.* FRDA Report. Victoria: Forestry Canada and the BC Ministry of Forests

–. 1994b. *Nonmarket Values of Forestlands: What Are the Potential Indicators?* Working Paper 200. Vancouver: Forest Economics and Policy Analysis Research Group, UBC

–. 1995. "Economics of protecting wilderness areas and old-growth timber in British Columbia." *Forestry Chronicle* 71(1): 52-58

Vancouver Island Resource Targets Technical Team (VIRTTT). 1996. *Resource Management Zones for Vancouver Island.* Interim Technical Report: Discussion Paper. Nanaimo, BC: Ministry of Forests, Vancouver District

Vincent, J.R., and C.S. Binkley. 1993. "Efficient multiple use forestry may require land-use specialization." *Land Economics* 69(4): 370-6

Wilkinson, C.F., and H.M. Anderson. 1985. "Land and resource planning in the National Forests." *Oregon Law Review* 64: 1-375

11
Sustained Yield: Why Has It Failed to Achieve Sustainability?
Lois H. Dellert

Introduction

British Columbia has long recognized the need to sustain its vast forest wealth for both current and future generations. The main mechanism by which sustainability was to be achieved was by regulating the annual rate of harvest to ensure a continuous supply of mature timber on a crop rotation-basis. It was believed that a steady supply of wood was the essential factor necessary for long-term community and economic stability.

Current circumstances raise serious doubts about the effectiveness of harvest regulation in achieving sustainability. The rate of harvest has declined by as much as 25 percent in some regions as the Chief Forester manages the transition to lower-volume second-growth harvesting and takes into account ecological and other resource values; the ratio of employment to cubic metres of wood harvested has dropped dramatically in the last two decades and communities are facing the possibility of mill closures as the industry rationalizes its operations to deal with excess production capacity; and there is concern for environmental degradation with the reduction in biological diversity and the loss of irreplaceable old-growth forests. British Columbia's forests, communities, and economy have not been sustained.

Why has harvest regulation failed to achieve sustainability? Critics point to too much state regulation: weak or unclear private property rights to Crown timber, limited competition for timber, administered pricing, and environmental regulations have interfered with the ability of market-based pricing signals and economic scarcity to regulate supply and demand in a sustainable manner (Vincent and Binkley 1992; Haley and Luckert 1989, 1993).

These conclusions have been drawn from conventional state-market analysis that categorizes harvest regulation as a form of state regulation and critiques it from the perspective of market-based economics. There are two significant limitations to taking this approach.

First, state-market analysis overlooks the important regulatory role accorded to science. The harvest was regulated by the Chief Forester – not by the state or by the market. Forestry expertise played a much more prominent role than either politics or economics. A critique of harvest regulation cannot be complete without considering how science and expertise contributed to its failure to achieve sustainability.

A second limitation of conventional state-market policy analysis is its failure to tackle more substantive questions. It is primarily concerned with a critique of policy instruments and fails to address whether the policy itself was appropriate. Regardless of which mechanism was best, was sustained yield an appropriate policy for achieving sustainability? Was timber an appropriate indicator of sustainability?

In this chapter, I do not debate the relative merits of state-based versus market-based regulation. This has already been done. Instead, I first explore the role of forestry expertise in the failure of harvest regulation to achieve sustainability. Second, I tackle the appropriateness of harvest regulation as a policy to achieve sustainability. My purpose is to offer insights into the failure of sustained yield and to provide guidance in the redesigning of forest policy.

Scientific Regulation: Optimistic Limits

The need for setting limits to protect future generations from short-term economic exploitation was not a new idea to foresters when it was popularized through the work of the 1987 World Commission on Resources and the Environment. Setting limits on timber cutting to prevent over-harvesting was the raison d'être of harvest regulation. Why then, if the purpose of harvest regulation was to guard against short-term depletion, is there a timber supply crisis in British Columbia today? The most obvious explanation would be that the harvest limit was set too high. Foresters find this explanation difficult to accept.

Scientifically based harvest regulation was implemented to protect against the short-term political influence of single interests and the short-term exploitative forces of economic development. It was believed that British Columbia's forests would be managed objectively to maximize long-term benefits to society. The Chief Forester, "a first-class scientific man"[1] (Fulton 1910: 67), was given the legal authority to set an allowable annual cut (AAC) of timber from Crown land.

The correct rate was to be set by the Chief Forester using the best data and technical expertise available. However, obtaining reliable data was problematic: British Columbia's forest was vast and diverse, making it extremely difficult to inventory accurately; the long crop rotation of timber made it very difficult to forecast future yields; and the complexity of forest ecosystems made it difficult to model. The best data were statistical estimates and mathematical projections of a complex biological system. Add to this the unpredictability in social and economic systems and it becomes obvious that the selection of the AAC was a subjective choice within a wide range of uncertainty.

History shows that the Chief Forester's choice was strongly influenced by the economic values embedded in sustained yield. In 1955 and again as recently as 1992, the Chief Forester was thwarted in his attempt to take a more conservationist approach to setting the AAC.

In 1955, the Royal Commission of Inquiry into Forest Resources heard evidence that the Chief Forester was acting too cautiously in setting the AAC. The Commission agreed. Chief Justice Sloan believed that the Chief Forester should not react too conservatively in setting the AAC, even in the face of incomplete data. Too much caution would lead to undercutting and the build-up of excess growing stock. Production would be decreased by slowing down the conversion to a fast-growing young forest and the economic potential of the forest would not be fully used. Further, restricting the cut to a highly competitive industry could have disastrous implications to the economy (Sloan 1956: 240).

A similar argument was made in the 1992 decision regarding an appeal by MacMillan Bloedel of the Chief Forester's decision to reduce the AAC for its Tree Farm Licence No. 24 (Appeal Board 1992: 44). The Appeal Board, established by the Minister of Forests to hear the appeal, reinstated the higher AAC and justified its decision based on the opinion that the harvest rate should be kept as high as possible to maximize the production of wood and its resultant social and economic benefits. It felt that penalties – a reduction in the AAC – should not be imposed in the short-term because of uncertain future outcomes (1992: 47).

These decisions show that economic values embedded in sustained yield mandated taking an optimistic approach to uncertainty. The Chief Forester consistently set the AAC at the high end of the range of uncertainty. In doing so, there was substantial risk of negative long-term impacts. This risk was not ignored. A re-planning strategy was implemented to minimize the long-term risk of optimistic decision-making: the AAC was to be reviewed on a periodic basis and adjusted, up or down, when and if new data and better information became available. It was believed that because the reviews were conducted regularly, any negative impact resulting from an optimistic assumption in the previous period would be minimized and easily absorbed over the much longer rotation period (from 60 to 120 years).

The assumption implicit in using a re-planning strategy to minimize the risk of optimistic decision-making was that, if required, a reduction of the AAC could be implemented without significant conflict or impact. Current circumstances show this has not been the case: reductions in the AAC have left the forest sector with excess mill capacity; the industry is challenging the Chief Forester; mills are scrambling to acquire wood from outside the province; and some companies are trying to rationalize their operations by purchasing additional timber rights and/or closing other mills.

Given the obvious economic and social impacts associated with reducing the AAC, it can only be assumed that foresters had a great deal of confidence the AAC would never have to be reduced. This confidence can be attributed to foresters having a considerable faith in technology to generate an increase in, or, in the worse case to maintain, the existing AAC. This confidence influenced foresters to use optimistic assumptions in their calculation of the province's available timber supply.

Operability: All Sites, All Types, All Access

During the early years of logging in British Columbia, very little of the province's vast forest was physically or economically accessible for harvesting. Advancements in harvesting technology, improved access to export markets, and increasing prices for wood products combined to expand the extensive economic margin of supply or the limit of profitability (Williams 1993: 12). The economics of scarcity provided the impetus: as supply became relatively scarce, the price increased and new technology was developed so that the presently marginal or uneconomic wood became operable. It was believed this expansionary trend would continue until eventually 90 percent of the forest would be harvestable (Sloan 1945: 29-31). As a result, areas that were not economically harvestable were routinely included in the AAC calculation. This practice became known as "all sites, all types, and all access."

MacMillan Bloedel's Chief Forester, J.H. MacFarlane, recently explained this practice: if an area was expected to become harvestable within the length of one rotation, in this case eighty years or so, then it was included in the determination of the AAC, whether it was currently operable or not. He was "quite confident in all [his] experience that [MacMillan Bloedel] can log all the merchantable timber in that valley, from a physical logging standpoint" (Transcript of Proceedings 1992: 218).

The industry's optimism was based on experience: MacFarlane recounted how the extensive economic margin – what he defined as the "limit of profitability" – had expanded. As the accessible supply was exhausted, new supplies became operable with advancements in harvesting technology: the steam donkey, railway logging, truck hauling, portable spars, and grapple yarding have progressively increased the portion of forest accessible to harvesting. MacFarlane believed that, "over time, log values will change, logging costs come down, and technology takes over" (Transcript of Proceedings 1992: 216).

A belief in the power of scarcity to stimulate price increases and development of new technology was used to justify keeping the AAC at its present level. By 1980, opportunities for expansion were limited, if not non-existent, and conditions in the forest made the assumption of all sites, all access harder to accept. Timber supply analyses conducted between 1979 and 1981 netted down the gross forest land base to include only areas that were considered to be reasonably feasible for harvesting. The net operable area was estimated to be 50 percent of the total forested area: a substantial difference from the 90 percent once believed possible (BC Ministry of Forests 1984: B5).

The inclusion of areas that were unlikely to ever be harvested had had the effect of concentrating an inflated rate of harvest in the most accessible areas, usually the valley bottoms. The practice of harvesting the best first had left the worst wood and most difficult areas for the latter part of the rotation. In many regions where the second-growth was not yet mature, there were inadequate stocks of operable mature timber to support the current AAC until the growth would be mature. Unsubstantiated optimistic assumptions regarding operability meant too much of the mature timber had been harvested too quickly to ensure a smooth transition to second-growth forestry. It was estimated that the present AAC would fall down by an average of 25 percent within forty to sixty years (BC Ministry of Forests 1984: D6).

Rotation Age

Operability was not the only factor treated optimistically in the AAC calculation. The choice of the rotation age, or the length of time it would take the regenerated crop to mature, strongly affected the choice of AAC. It established the period over which the old-growth timber had to be rationed to ensure a smooth transition to second-growth harvesting. For example, a rotation age of 50 would mean the mature forest could be harvested in half the time and the AAC could be twice as high than if the rotation age were 100.

The rotation age was derived from yield projections of the growth of the regenerated forest. The complexity of modelling forest growth and the projection period, from 50 to 150 years hence, made prediction technically challenging. Future yields were estimated by measuring characteristics of the existing forest and then by making statistical projections of yield through a scattergram of data points. One glance at a scattergram of volume measurements would lead even an expert to conclude that the growth and yield projections were far from precise.

In giving evidence to the 1955 Royal Commission into the Forest Resources of British Columbia, Chief Forester C.D. Orchard estimated it would take 100 or 120 years to grow high-quality Douglas-fir saw logs and 60 years to grow pulp logs.[2] Chairman and Chief Forester of MacMillan Bloedel H.R. MacMillan proposed a rotation of 80 or 90 years (Sloan 1956: 236). The Commissioner considered these professional opinions and rejected both of them. Sloan recommended a rotation age of 60 years (1956: 241).

The Commissioner acknowledged that the harvest, sixty years hence, would be significantly different after the large, old trees were gone. This time seemed far away and he was optimistic about future prospects: the "adaptation of that growing stock to the uses of that period can safely be left to the ingenuity of future operators in the confident expectation that they will not miss the old-age trees of large growth they have never known" (1956: 236-7). The Commissioner encouraged a speedy liquidation of the old forest: "It does not seem necessary that industry of this generation should be sacrificed beyond that point in order to reproduce the kind of big-tree forest we value now, when even now industry is beginning to find profitable use for smaller trees." (1956: 236-7).

The professional opinions of two prominent Chief Foresters, H.R. MacMillan and C.D. Orchard, were dismissed by the Commission as being too pessimistic. In the face of inconclusive scientific evidence, the economic efficiency goal of sustained yield and the belief in the power of economic scarcity to drive technological innovation were used to select a rotation age. That age was not an objective technical calculation based on forestry expertise and scientific data.

Second-Growth Yield

Another way in which growth and yield data affected the AAC was by providing estimations of second-growth yield. Foresters were aware that, during the first rotation, volumes would be higher because of the extra inventory held in the mature and overmature natural forest. Because the second-growth forest would be harvested at a much younger age, most likely in the range of 60 to 100 years, regenerated stands would have lower volumes. This fact would mean that the AAC would have to reduced by approximately 25 percent on average across the province in response to this falldown in volume (BC Ministry of Forests 1984: D6).

Foresters believed the falldown could be averted through intensive silviculture or enhanced stewardship. By intensively managing the second crop, an inefficient Mother Nature could be improved upon and commercial timber yields increased – by as much as 100 percent (Science Council of BC 1989). Intensive silvicultural techniques included genetic improvement, planting to achieve full stocking of a site, prompt regeneration of commercial species, brush control to reduce competition, fertilization, pruning, and thinning.

Silviculturalist claims of yield gains were influenced by comparisons made to the agricultural sector where tremendous increases in yield had been experienced and to other forest regions – especially Sweden – where intensive management had produced substantial yield gains (Science Council of BC 1989). However, the magnitude of gains claimed for British Columbia could be neither substantiated nor refuted with hard data or research: reliable managed stand yield tables were not available at that time.

Even though yield gains from intensive silviculture could not be scientifically proven, the AAC remained at its 1980 level for over ten years. During this period, foresters lobbied for increased silviculture funding and researchers worked to develop managed stand yield tables. It was argued that there was no need to reduce the AAC for the falldown in second-growth volumes, because the difference could be captured through intensive silviculture.

Chief Forester W. Young was not convinced the first rotation old-growth AAC could be maintained into the second rotation. In a speech at the 1980 Canadian Forest Congress, he expressed concern with the "intensive forestry dream" of an new era of economic development supported by substantial increases in wood. First, massive amounts of funding would be required, and second, other demands on the forest land base were exerting downward pressures on timber supply. While some gains were possible, he believed the realistic potential was vastly overstated and ignored many of the other emerging environmental issues. Young called it a "blinker philosophy" (1981: 23-25). The expectation that intensive silviculture would ameliorate the timber supply falldown was unrealistic. Chief Forester Young challenged the government and the industry to demonstrate its commitment to intensive silviculture and its claims that the current AAC could be maintained. He declared his intention to reduce the AAC at the next five-year review unless improved performance could be demonstrated.[3]

Young's challenge was not taken. By 1990, the intensive forestry dream had proven to be just that, a dream. Secure funding for silviculture did not materialize from the public sector[4] and no significant investment had been made by the private sector.[5] Newly available managed stand yield tables forecast gains in second growth yield to be in the range of 5 percent[6] (BC Ministry of Forests 1991b: 1), a far cry from the unsubstantiated claims of 100 percent increases. Optimistic data assumptions were used to justify having kept the old-growth AAC too high for too long. This delay has made the transition to second-growth harvest levels more severe; reduced the options available for minimizing economic impacts; and contributed to environmental conflicts.

Under the sustained yield policy of maximizing economic efficiency, the onus was on the Chief Forester to scientifically prove the AAC must be reduced before he did so (Supreme Court 1993). This obligation has proven to be an onerous, if not impossible, task. Better inventory data, research, and more sophisticated modelling have reduced uncertainty, but, given the inherent complexity and unpredictability of projecting long-term timber supply, it is all but impossible to conclusively prove future impacts of higher short-term harvest rates. Uncertainty remains and the setting of the AAC is vulnerable to the influence of the economic values embedded in sustained yield. As long as the primary goal of harvest regulation is sustained yield, the Chief Forester will continue to find it difficult to take a conservationist approach in setting the AAC.

Scientific regulation failed because foresters relied far too much on science to mediate against the influence of economics and politics. Science could not be used to prove conclusively that short-term economic benefits should be forgone for the benefit of future generations. The aura of objectivity provided by portraying harvest regulation as a technical matter simply served to obscure the underlying influence of the economic values embedded in sustained yield.

The objectivity of science may also have contributed to the failure of many policy analysts to critique the conceptual model on which sustained yield is predicated. Under sustained yield, it was assumed that it was possible to reorder the forest to achieve perpetual equilibrium of supply and demand with growth and harvest. Unfortunately, the pursuit of equilibrium through reordering the forest and controlling the economy has had disastrous consequences for British Columbia's forests and communities.

Pursuit of Equilibrium: Control, Order, and Surprise

The ideal promised by the sustained yield model was perpetual equilibrium of supply and demand in perfect balance with the growth and harvest of the forest. It was a crop rotation model that required a so-called normal forest condition of an approximately equal distribution of all ages up to maturity. Each year, the mature trees would be harvested and the area regenerated with a new crop. The next-to-oldest trees would be left to mature and be harvested the following year. This cycle of growth, harvest, and regeneration would be repeated again and again to produce a yearly crop of mature trees.

The problem facing foresters in implementing sustained yield in British Columbia at the turn of the century was that its unmanaged, unharvested forests were far from the ideal structure needed. The forests were predominantly old and slow-growing. There was a storehouse of mature timber available for harvesting, but the soil was not growing new wood efficiently nor was the distribution of ages suitable for self-regulating crop rotation. The policy challenge was how to convert the province's old-growth natural forest into a normal forest as quickly as possible.

It was theorized that the conversion of a unmanaged new forest, such as was the case in British Columbia at the turn of the last century, to a sustained-yield normal forest would occur through a series of progressive stages. At each stage, there would be a different requirement for regulation depending on the relative balance between supply and demand.

Dr. Fernow, the first Dean of Forestry at the University of Toronto explained that "forestry is an art born of necessity, as opposed to arts of convenience and of pleasure. Only when a reduction in the natural supplies of forest products, under the demands of civilization, necessitates a husbanding of supplies ... does the art of forestry make its appearance" (1913: 2). Until resources became scarce, forests would not be valued, and therefore would be exploited and not managed. With increasing scarcity, forests would become more valuable, and scarcity would trigger the transition from one stage of development to the next.

At the first stage, policies would be limited to inventory, collection of royalties, and protection against fire. At the next stage, silviculture would be applied to ensure regrowth of the second-growth forest and harvest regulation implemented to ration the old-growth over the remainder of the rotation. At the last stage, a sustained yield forest economy would be reached (Fernow 1913: 6).

At this final stage, the forest structure would be normalized and the policies implemented to re-balance supply and demand would no longer be needed. The AAC would be limited by the physical structure of the forest, in which an approximately equal proportion of the forest would reach maturity each year. The desire to maximize economic and biological efficiency would act as an incentive to harvest only the mature trees: if cut too early, the average yield would not yet have been maximized; and if cut too late, the older, slow-growing trees would not be replaced soon enough with young, fast-growing trees. Once the supply and demand were balanced, annual growth would equal harvest and supply and demand would be balanced in perpetual equilibrium.

A review of the history of forest policy in British Columbia reveals parallels between this model of stages and the evolution of forest policy. It appears that the province has progressed through stages one and two, and is now in transition to stage three.

Stage One: Stimulate Demand to Reduce Supply

When Europeans first encountered British Columbia, the landscape was dominated by extensive forests. However, rough terrain, difficult operating conditions, and the large size of the trees restricted economic and physical operability. There was little domestic demand for timber and British Columbia's remoteness and relatively late development of its forests – compared to other regions in the United States and Canada – made entry into a volatile export market problematic. The harsh physical and economic environment made logging a risky venture: the forest economy of British Columbia in the early years was "boom and bust," with many more failures than successes (Williams 1989; MacKay 1985; Taylor 1975).

With stocks of timber supply in excess of demand, there was little danger of over-harvesting: the crucial issue in British Columbia at the turn of the century was under-harvesting. As such, the early objective of sustained yield policy was not supply-side regulation to restrict harvesting, as in eighteenth- and nineteenth-century Western Europe, but rather demand-side stimulation to encourage harvesting. In other words, to decrease the stocks of over-mature forest by increasing the demand for wood.

To stimulate demand, the government provided economic incentives through its tenure and pricing policies. Industrial development and investment were encouraged by assigning private property rights to timber without requiring the private sector to risk capital investment to purchase timber or land;[7] by making timber licences transferable; by not imposing restrictive and costly regulations on the industry; by absorbing management costs; and by implementing a system of administered pricing to minimize price fluctuations of a volatile export market.

The rate of harvest was not regulated or controlled during this period. It was believed that restrictive and costly regulation would negatively affect the development of the industry. Without an industry, the province would not be able to capture the economic potential of its forests.

Not regulating the harvest and providing economic incentives to the industry were means of initiating the harvest of the overabundant supply of old and slow-growing timber. These policies were consistent with stage one and marked the initiation of the conversion to a sustained-yield normal forest.

Stage Two: Regulation to Ration Supply of Old Growth

By 1940, the amount harvested had increased several-fold and a forest industry had been established. Harvesting remained unregulated. Douglas-fir, the most desired species, was being seriously overcut in coastal regions and localized scarcities were emerging (Mulholland 1937). It was feared that further unregulated harvesting of the over-mature forest would lead to future timber famines or age class gaps where there would be insufficient stocks of mature timber to provide a continuous supply. Foresters[8] were convinced that stage two in the development of sustained yield had been reached, and that harvest regulation was needed to ensure the old growth was carefully rationed over the remainder of the rotation period.

Harvest regulation became law in 1947. The Chief Forester was given the legal authority to set an allowable annual cut (AAC) of timber from Crown land. The objective of this policy was to control the conversion of the old-growth forest in an orderly manner. The AAC was an upper limit intended to ensure the old-growth forest was rationed over the remainder of the first rotation.

The AAC was also a lower limit. The goal of sustained yield harvest regulation was to convert the old slow-growing forest to a thrifty fast-growing normal forest as quickly as possible. The industry was not permitted to hoard wood or stockpile timber harvesting rights. To do so would have slowed down the conversion and reduced the overall benefits to society. A companion administrative policy – cut control – was instituted to penalize companies who exceeded their AAC (overcut) as well as those who did not harvest the full amount allotted to them (undercut).

Regulation of the harvest did not restrict economic growth in the era after the Second World War. Timber supply had proven not to be as scarce as thought. Previously inoperable timber became available with the opening of new frontiers, improvements in utilization, and the development of new technology. The AAC increased progressively until the 1980s (Williams 1993: 12).

By the 1990s, the extensive margin of economic supply had been reached and there were limited opportunities for further expansion. Many of the highest quality and most accessible stands of timber had been harvested and the end of the old-growth forest was in sight. Supply had finally become scarce. The AAC levelled off and foresters started planning for the transition to stage three.

Stage Three: Surprise

The transition to stage three was not going as forecast. The forest structure had not been normalized, the rate of harvest was declining in some regions by as much as 25 percent, employment rates had declined, rural communities were at risk with the threat of mill closures, and much of the old-growth forest planned for harvest later in the rotation was protected from logging or was subject to restrictive environmental regulation.

Too much state regulation has been blamed. Environmental regulation, administered prices, and a tenure system that favoured a few large companies have led to an erosion of private property rights and all but eliminated competition, two factors essential for market-based regulation (Vincent and Binkley 1992; Haley and Luckert 1989, 1993). Increased scarcity did not trigger a re-balancing of supply and demand, because the market was not free to respond by increasing prices and thus decreasing demand.

One aspect of this analysis has merit: rigidities created by regulation have prevented the achievement of sustainability. However, the assertion that entrenching private property rights and increasing competition will lead to perpetual equilibrium of supply and demand by way of market-based regulation is highly questionable. Among other assumptions, it presupposes that the model of steady state is valid. New theories in ecology challenge the myth of equilibrium and ecologists caution against attempting to control and reorder systems in the way prescribed by sustained yield (Holling 1973; Botkin 1990).

Holling has shown in his study of spruce budworm in New Brunswick that managing for stability by suppressing or controlling change can actually alter the system's functioning in such a way that the system becomes more and more vulnerable and less able to recover from disturbance (1973). He differentiates the concept of stability from that of resiliency: stability is "the propensity of a system to attain or retain an equilibrium condition of steady state or stable oscillation," while resiliency is "the ability of a system to maintain its structure and patterns of behaviour in the face of disturbance" (1986: 296).

Suppressing change and managing for stability reduces resiliency. As the system becomes less and less resilient, increasing levels of managerial control are required to prevent catastrophic or irreversible change. This control reduces resiliency even more. At some point, the control of change becomes impossible and the system collapses or reverts to another type of system. Holling calls this "surprise" (1986). Ecologists tells us that change is normal, essential to system functioning, and should not be eliminated or controlled (Holling 1973; Botkin 1990).

The reordering of the chaotic and inefficient natural forest to create a sustained-yield normal forest with evenly distributed ages producing timber on a crop rotation basis required a great deal of a managerial control. The structure of the forest was controlled by regulating the annual cut; decision-making was controlled by transferring the authority for setting the rate of harvest to the Chief Forester; and industry was controlled by tenure and pricing policies that all but eliminated competition and favoured a few large integrated companies (Marchak 1983). The outcome has been the evolution of highly rigid ecological, political, and economic systems. These systems have lost their resiliency or capacity to absorb change.

The natural forest has been simplified and restructured: there is less old-growth and more commercial species; and the soil has been degraded, nutrients removed, and wildfire suppressed. This strategy may have been well intentioned as a way to bring about economic and community stability; however, the policies have had negative ecological consequences. A sustained-yield normal forest is highly simplified, less resilient, and more vulnerable to widespread and catastrophic change than a natural forest. Biological diversity and the functioning of many ecological systems have been threatened.

Recent policy reforms have been introduced to address these ecological concerns. Amendments in 1988 to s. 7 of the *Forest Act* require the Chief Forester to consider the impact on the AAC of integrated resource guidelines and the forest practices now applied under the Forest Practices Code. A string of land use planning processes[9] have attempted to protect forest ecosystems through zoning areas for production, integrated, or conservation uses. Those areas designated for ecological conservation have been excluded from consideration in the determination of the AAC. The outcome has been increased protection of ecological and other resources values, but this improvement has been accompanied with a reduction in the AAC, increased harvesting costs, and more administration. It is not surprising that there has been significant backlash from industry, labour, and communities.

The government is attempting to manage the backlash through initiatives such as reducing the red tape associated with the Forest Practices Code and by allocating additional timber rights to those companies that create new jobs (e.g., the Jobs and Timber Accord). These initiatives in turn have generated a backlash from environmental groups who view them as an erosion of hard-fought policy reforms. There appears to be no middle ground. Positions are polarized and strongly held by diverse groups. The "win-lose" nature of forest policy reform can be attributed to the pursuit of stability through control, which has generated political and economic rigidities.

The pursuit of stability has spawned a conflict-oriented policy-making process. Non-forestry voices were intentionally excluded and only those considered insiders were given direct influence in AAC decision-making. They included Forest Service bureaucrats and industrial foresters who worked for companies with tenure. With such strong internal control and resistance to change, policy reform was largely driven by external pressure and protest (Dellert 1994). This type of policy community has been characterized by political scientists as "contested concertation" (Wilson 1990; Coleman and Skogstad 1990).

The policy process associated with such a closed institutional structure is not flexible or resilient. Change is resisted for fear of unacceptable or adverse consequences. Under these circumstances, it is extremely difficult to reform forest policies in response to an ever-changing social, economic, ecological, and political context. The result has been over two decades of conflict, protest, and polarization.

The political difficulties experienced in reforming forest policy have been exacerbated by economic rigidities whose creation can be linked to sustained yield policies. The AAC was fully allocated to licensees to ensure the rapid conversion of the old forest; mill capacity was overbuilt on the optimistic assumptions that old-growth harvest rates would be sustainable and that environmental concerns could be handled without AAC decreases; timber harvesting rights were linked to manufacturing; and the public forest was treated as a corporate asset to secure financing. Tenure policies explicitly favoured large companies: the advantages of corporate concentration – stability, generation of employment, increased efficiency, and improved forest management – were believed to outweigh the disadvantages of monopolistic control (Sloan 1956: 62-96).

Today's reality demonstrates that what was once thought of as an advantage has in fact turned out to be just the opposite. Sustained yield policies that pursued stability and control left the province with an over-capitalized and vertically integrated forest sector that relied too much on the large-scale production of low-value commodities and the expansion of the economic margin of supply for economic growth. The industry is having difficulty absorbing or adapting to reductions in the AAC, to changes to forest practices, and to shifts in global markets. The sector is rigid and fragile.

The equilibrium model presented by foresters and economists led policymakers to believe steady state was possible. It was not. The pursuit of stability has reduced the capacity of the forest, economic, and policy systems to respond and adapt to change without catastrophic impacts. The attempt to reorder and control the forest has not only failed, it has actually contributed in large measure to the current crisis in British Columbia's forests and communities.

Thus far, I have discussed the failure of scientific harvest regulation to achieve sustainability. In the next sections, I critique the substance of the policy itself. What was the policy intended to sustain? Did the policy regulate the correct indicator of sustainability?

Sustain What?

Sustainability is not a concrete technical concept with a single scientific definition. According to Romm, there are as many possible definitions as there are value perspectives. A specific meaning emerges as part of an adaptive social process (1993: 280). Gale and Cordray present at least eight possible meanings of sustainable forestry, each one based on an alternative mix of values and objectives, and each one requiring a substantially different policy framework (1991: 31-36).

The meaning of sustainability entrenched in British Columbia forest policy was sustained yield or the continuous supply of timber on a crop rotation basis. This meaning originated in seventeenth-century Western Europe where harvest regulation was first implemented in France to secure a steady supply of naval timbers (Pincetl 1993: 81). The concept was later refined and applied in Germany where it was used as an economic conservation policy and has been attributed as having "showed the way out" of the timber famine in the late eighteenth century (Fernow 1910: 33). In 1795, the chief forester for Prussia, Georg Hartig, declared: "From the State Forest not more and not less may be taken annually than is possible on the basis of good management by permanent sustained yield" (Rubner 1984: 171). In 1812, Hartig's colleague, Heinrich Cotta, the chief surveyor of Saxony, defined sustained yield as "highest yield, lowest costs and best covering of demand" (Rubner 1984: 171).

The interpretation of sustainability as sustained yield was imported to Canada by European-trained foresters in the early twentieth century. One of these was German-born and -trained Bernard Fernow. He was the founding dean of Canada's first forestry program at the University of Toronto and a prominent member of the Forestry Committee of the Commission of Conservation of Canada. In 1910, Fernow told the Commission it would only be a matter of time before unregulated harvesting would generate timber famines. He argued that if Canada implemented sustained yield, it could avoid the economically disastrous timber famines experienced by Western Europe in the previous century (1910: 33).

The meaning of "sustained" as sustained yield took root in British Columbia. In his report of the 1945 Royal Inquiry into Forest Resources, Chief Justice Sloan defined sustained yield as "a perpetual yield of wood of commercially usable quality from regional areas in yearly or periodic quantities of equal or increasing volume" (1945: 127). Sloan noted how dependent British Columbia was on the forest: if the forests were not perpetuated for future generations, the economic future would present a "very dark and dismal picture" (1945: 10). Forest use had to move from a "descending spiral" of unmanaged liquidation to an "ascending spiral" of sustained yield management – from mining to crop renewal (1945: 125). The meaning of sustained as sustained yield was entrenched in forest policy in the 1947 BC Forest Act. This meaning has persisted to the present day.

In 1993, the Sierra Club initiated a Judicial Review of the Chief Forester's application of the Forest Act (1979) in setting an AAC for Tree Forest Licence No. 44. It claimed the Chief Forester wrongly interpreted the meaning of sustained and illegally set the AAC at an unsustainable level. In his defence, the Chief Forester explained to the Court that the meaning of sustained intended by the Act was to maximize production and to ensure a continuous supply of wood. He substantiated his interpretation of the meaning by submitting an expert opinion paper.

In the opinion of the expert, the province needed to convert the unregulated, old, slow-growing forest to a regulated, young, fast-growing forest, equally distributed in all ages. Once this so-called normal forest structure was achieved, the annual growth could be harvested and a supply of wood maintained in perpetuity at the long run sustained yield level (Pearse 1993: 5).

The Supreme Court of British Columbia ruled that the Chief Forester had correctly interpreted the meaning of sustained as sustained yield and dismissed the Sierra Club petition (Supreme Court 1993: 35). According to the Court, the meaning of sustained intended by the Forest Act is the efficient production of timber on a crop rotation basis.

Regulate What?

Ultimately the purpose of sustained yield was community and economic sustainability. It was assumed that social and economic benefits would be sustained by sustaining timber production. As such, the sole indicator or measure of a sustainably managed forest estate was the production of an even flow of timber at the most efficient rate possible. The Chief Forester set the AAC, and then performance was monitored by regularly reporting on harvest rates, growth rates, regeneration success, and timber inventories. The sole regulatory mechanism used was harvest regulation.

Sustained yield was considered successful in the decades following the implementation of harvest regulation. Investment flowed into British Columbia, mills were built, and jobs created. The public was reassured that the forest was being managed sustainably at the same time that production was expanding and the economy growing. By the 1980s, the situation was changing.

Timber supply was no longer an automatic guarantee of employment. A recession in the early 1980s and stiff competition for export markets stimulated the sector to increase its efficiency through mechanization.

Employment levels dropped dramatically, even though the AAC did not. The ratio of direct jobs per cubic metre of wood harvested declined from about 1.7 in 1965 to about 1.0 in 1990 (BC, 1994: 4). Rural communities dependent on the forest sector were hardest hit.

Sustained yield failed to achieve community sustainability because it wrongly assumed a straight-line relationship between timber supply and employment.[10] By promoting sustained yield and harvest regulation as community stability policies, more direct indicators of community health were never regulated or reported. The unquestioned assumption that timber meant jobs blinded foresters to the inadequacy of harvest regulation as a social and economic policy. Timber supply has proven to be inadequate as an indicator of social and economic sustainability. It has also proven inadequate as an indicator of forest sustainability.

Environment: Constraints on Production

Until the 1970s, sustained yield forest management gave little if any direct consideration to other forest values. Forest inventory measured growing stock and categorized the forest as "over-mature," "mature," and "immature." Fire management was a priority to protect the forest, a valuable economic asset, from destruction. Silviculture was geared towards replacing the old slow-growing decadent forest and non-commercial species with a well-managed, high-yield second crop. Sustaining the forest was equated with sustaining timber supply. By the 1970s, the extent of timber operations was threatening other forest values and a Royal Commission of Inquiry on Forest Resources was directed to address this issue.

In the report of the 1975 Royal Commission, sustainability of other forest values was explained in economic terms: scarcity was generating controversy as a result of too many competing demands from the forest. The solution recommended was the implementation of coordinated land use planning (Pearse 1976: Ch.19). The forest was too scarce to afford the luxury of single use management – whether for timber production or for wildlife protection. Alienation of forest land to "single uses" such as agriculture, or wilderness preservation, was considered to be a threat to maximizing the total social net benefit. Through coordinated planning, the best pattern of uses could apparently be determined. Timber harvesting would be conducted within the context of multiple use planning and decisions would be made based on the capability and suitability of the land to produce the best mix of uses (Pearse 1976: Ch.19). This practice is known as integrated resource management (IRM).

IRM was a concept of the greatest good for the greatest number. According to IRM, coordination and integration would reduce conflict between competing uses and the overall benefits to society would be greater. In a keynote speech to a national meeting of professional foresters in 1979, Marion Clawson (1979: 228), a prominent forest economist, explained:

Forests in North America have a greatly unrealized potential. They are not now producing anything like as much wood as they can be made to produce economically. I am convinced that their output of recreation, wilderness, wildlife, and other non-wood outputs can be increased at the same time that wood growth and wood use are increased. Clearly, it will take managerial competence and often take capital as well, to achieve these increased outputs.

The "win-win" promise of IRM became policy with the 1979 Ministry of Forests Act. The industry participated in IRM with the expectation that other values could be accommodated without a decrease in the AAC or a substantial increase in harvesting costs. Throughout the 1980s, regional committees of resource professionals worked together to develop IRM harvesting guidelines that would protect other forest values by modifying harvesting schedules and cutblock design. The guidelines included setting limits on clearcut size, altering clearcut shape, maintaining riparian and wildlife buffers, and requiring longer green-up periods between the harvesting of adjacent areas.

Harvest practices were modified under IRM and there were many local successes. However, IRM failed to protect other values and failed to reduce conflict. Multiple use was criticized as a campaign against environmentalism and the so-called single use alienation of forest for non-timber production purposes such as parks. According to Wilson, multiple use, public involvement, and sustained yield were only "symbols" which "served as legitimating cornerstones" to counteract environmentalism and its critique of logging (1990: 155).

The basis for this critique was that, under IRM, the AAC had actually increased from 54 million cubic metres in 1975[11] to well over 70 million cubic metres by the early 1990s (BC Ministry of Forests Annual Reports: 1975, 1989/90, 1990/91, 1991/92, 1992/93). To environmentalists, this increase was inconsistent with the goal of protecting other forest values. To foresters, an increase in the AAC was not inconsistent with IRM, because it was believed that other forest values could be accommodated by altering how and when timber was harvested. By 1990, foresters were realizing this was not the case.

Forest Service field foresters complained that they could not design a twenty-year cutting plan at the present rate of harvest using the approved IRM guidelines. They advised the Chief Forester that either the AAC would have to be reduced or the IRM guidelines relaxed. The Ministry commissioned an internal review to investigate.

The review substantiated the concerns of field staff. It found that the IRM guidelines could not be accommodated without reducing the AAC (BC Ministry of Forests 1990). There was simply not enough mature forest to apply the IRM guidelines at the present rate of harvest: clearcuts were smaller, delay periods longer, rotation lengths extended, and reserve strips larger. It required the maintenance of a greater proportion of mature timber at any one time than the conversion to a normal forest allowed. Field staff estimated that AAC would have to be reduced by approximately 25 percent (BC Ministry of Forests 1990). It had been wrongly assumed that other forest values could be accommodated through rational planning and coordinated harvest scheduling. The AAC was too high to implement IRM or the new Forest Practices Code.

IRM failed to protect non-timber values because it was a modified version of sustained yield. It assumed that uses and values could be delineated as separated entities, that each one could be valued, and that a rational choice would be made that maximized utility to society. It cast other values as constraints on timber production and pitted one resource or value against another. It was not surprising that conflict increased.

From Sustained Yield to Sustainable Forests

With the raising of society's ecological consciousness and the failure of sustained yield to sustain communities and biodiversity, a new meaning of sustained is evolving. Ecological science has taught us that all plants and animals, living and non-living things are interconnected, and that each thing has a role in a complex web of life. Altering one component will affect the others, and if any one or more component is altered too dramatically, the entire system will be put at risk. The key to sustainability, according to ecology, is not the maximization of social net benefit from competing forest uses and values, but rather, the maintenance of a system's integrity. Ecological principles dictate that to sustain timber supply or any other forest value, the forest as a complex system must first be sustained. Then and only then can a flow of benefits, such as timber production, be sustained.

This new thinking is embodied in the 1992 National Forest Strategy: "Our goal is to maintain and enhance the long-term health of our forest ecosystems, for the benefit of all living things both nationally and globally, while providing environmental, economic, social and cultural opportunities for the benefit of present and future generations" (CCFM 1992: iv). Sustainability has been redefined to mean sustaining the forest ecosystem as a source of economic benefits – a substantively different interpretation from the timber production goal of sustainability yield. It implies that the forest and not timber production should be the primary focus of regulatory policy. This new meaning of sustainability has radical implications for reforming forest policy and harvest regulation.

First, it requires a reversal in the approach taken to uncertainty. When timber production was the primary goal, an optimistic approach was taken and the AAC was set at the high end of the range of uncertainty. The onus was on proving adverse impacts of a higher AAC before forgoing economic benefits. With forest sustainability as the primary goal, the approach should be refocused towards sustaining the forest. Production targets should be set at the low end of the range of uncertainty and the onus placed on proving that increasing or maintaining the AAC will have no adverse impacts on the environment. In other words, do not do anything that could harm the sustainability of the forest until it can be proven that harm will not result.

A second and more substantive policy reform is also required. Harvest regulation, regardless of whether the AAC is reduced, and regardless of whether state or market-based regulatory instruments are used, was designed to achieve sustained yield. It was an economic conservation policy that regulated the production or the output of a single resource, namely, timber. Sustaining the forest is an ecologically based policy that requires the regulation of biological diversity and ecosystem functioning. Harvest regulation will not do this. It is an outdated policy that is no longer appropriate and should be replaced.

Solutions

Much of the debate surrounding harvest regulation has involved differing opinions on whether the AAC was too high or too low, or whether market-based regulation would be a better policy instrument for achieving sustainability. Reducing the AAC, zoning the forest to separate high-yield timber production areas from conservation areas and then using market-based mechanisms to regulate timber supply may prove to be more successful at achieving sustained yield than current policies. These actions will not, however, prove to be any more successful than the current policies in achieving sustainable forests and sustainable communities. The achievement of sustainability is not just about limit setting, nor is it just about choosing between the state or the market to regulate timber supply. It is about a radically different notion of sustainability based on ecological principles of wholeness and the awareness of pervasive and unpredictable change.

Today's notion of sustainability requires sustaining the forest as a means of providing a mix of social, economic, and ecological goals. Timber can no longer be relied on as the single factor of sustainability nor its regulation as the primary policy for achieving sustainability. A new regulatory framework is needed, one that regulates the forest ecosystem and monitors a comprehensive set of social, economic, and ecological factors.[12] We must stop portraying the forest as a collection of single uses or resources which can be separated by resource use or land use designations. The forest ecosystem and social, economic, and political systems function as a whole and must be managed as such.

The forest policy framework must be reformed to embrace change rather than to exclude it. The quest for steady-state equilibrium should be discarded as an outdated and inappropriate model. The science of ecology tells us that systems are complex, dynamic, and unpredictable. Rather than attempt to reorder and control, policies should be designed with unpredictable change in mind. This reform means building responsiveness and flexibility into our policy and economic structures as well as in the forest. It means designing institutions and organizations that are mutable (Meidinger 1997: 361-80). It means letting go of control, and designing forest practices, policies, and economic strategies that are able to respond and adapt to our ever-changing world.

Until policymakers understand the substantive reasons for the failure of harvest regulation, policy reforms such as the transfer of regulation to the private sector is premature. At best, such reforms will be as ineffective as the current policies, and at worst, they will only intensify a bad situation. The core policy issue is not whether state or market-based regulation of timber supply is the way to achieve sustained yield or whether the AAC is too high or too low. Sustained yield is no longer an appropriate model and harvest regulation is no longer an appropriate policy for achieving sustainability.

Notes

This chapter is based on research conducted by the author for her master's thesis in Environmental Studies, *Sustained Yield Forestry in British Columbia: The Making and Breaking of a Policy (1900-1990)* (1994) and her PhD research on forest policy.

1 The report of the 1910 Royal Commission, which formed the basis the first Forest Act (1911), stated that the province's forests were too important to be left to politicians and required the "special talents" of foresters to take full advantage of the "great results" possible. The Commission recommended that a permanent organization, staffed with specialists, be established to oversee the implementation of conservation with its main purpose to "build up an exact knowledge" of the forests. The new organization would be headed by a Chief Forester, "a first-class scientific man," because "so much will depend upon the wise knowledge and sound judgment of this official" (Fulton 1910: 67).

2 Chief Forester Orchard based his estimations on a province-wide forest inventory conducted by the Forest Service between 1927 and 1935. (See Mulholland 1937.)

3 Chief Forester Young retired in 1982. Delays in reviewing the AAC meant the existing AACs were maintained throughout the 1980s.

4 Since 1979, several attempts have been made to secure long-term funding for silviculture. These efforts include the Forest and Range Resources Fund, the Forest Resource Development Agreement (FRDA), and the Forest Renewal Plan. Fiscal crises at both provincial and federal levels have gutted the funds or failed to renew funding agreements. Time will tell whether the current mechanism, Forest Renewal BC, will be successful. Its funding comes from stumpage increases, which, according to the industry are putting the industry at a competitive disadvantage globally.

5 In its 1984 Management and Working Plan, MacMillan Bloedel committed to an intensive silviculture program with an estimated increase in yield of 300,000 cubic metres. This program was dropped in its 1991 Management and Working Plan, because it was "found to be not currently practical or economic" (Appeal Board 1992: 37).

6 No financial analysis was conducted to determine the present net benefit from intensive silviculture. Analyses were done using several funding levels. The conclusion that a 5 percent gain was possible was based on what was believed to be a feasible level of public funding.

7 While ostensibly promoted as a conservation measure, the retention of public ownership

was an industrial development policy: the forest tenure system permitted industry to acquire long-term property rights to trees without having to purchase land or invest in forest management. The government absorbed the investment risk and bore the cost of reforestation in order to attract investment to British Columbia. It was only later, after the 1909 Royal Commission, that public ownership was regarded as a conservation policy.

8 In 1937, Chief Forester E.C. Manning urged the government to implement harvest regulation (see Manning 1937). In 1943, his successor, C.D. Orchard, presented a brief to cabinet that outlined sustained yield harvest regulation (see MacKay 1985).

9 Land use planning processes initiated by the government in the 1980s and 1990s include the Old-Growth Strategy, Parks and Wilderness for the 90s, the Protected Areas Strategy, and the Commission on Resources and the Environment.

10 Recent policy initiatives such as the 1997 Jobs and Timber Accord appear to be making the same mistaken assumption by linking job creation strategies directly to additional timber harvesting rights.

11 Based on a ten-year average for 1965 to 1975.

12 The *National State of the Forest Report* (Natural Resources Canada, 1996) with its fifteen social, economic, and ecological indicators may be an approach worth pursuing.

References

Ainscough, G.L. 1981. *The Designed Forest System of MacMillan Bloedel Limited*. H.R. MacMillan lecture, Vancouver

Appeal Board. 1992. *Appeal by MacMillan Bloedel Limited from a decision of the Chief Forester dated December 31, 1991 in respect of a determination of the allowable annual cut for the period January 1, 1991 to December 31, 1995*. September 14. Vancouver

BC Forest Service. 1976. *Annual Report 1975*. P. T 38. Victoria: Queen's Printer

BC Ministry of Forests. 1980. *Five Year Forest and Range Resource Program*. Victoria: Ministry of Forests

–. 1984. *Forest and Range Resource Analysis: 1984*. Victoria: Ministry of Forests

–. 1991a. *Annual Report 1989-90*. P. 85. Victoria: Queen's Printer

–. 1991b. *Enhancing Our Forest Resources: A Forest Renewal Plan (1991-1995)*. Victoria: Ministry of Forests

–. 1991c. *Review of the Timber Supply Analysis Process for BC Timber Supply Areas*. Final Report (Vol. I). Victoria: Ministry of Forests

–. 1992. *Annual Report 1990-91*. P. 51. Victoria: Queen's Printer

–. 1993. *Annual Report 1991-92*. P. 82. Victoria: Queen's Printer

–. 1994. *Annual Report 1992-93*. P. 98. Victoria: Queen's Printer

Botkin, D. 1990. *Discordant Harmonies: A New Ecology for the Twenty-First Century*. New York: Oxford University Press

Brundtland, Gro Marlem. 1987. *Our Common Future*. Oxford: Oxford University Press

British Columbia. 1994. *British Columbia's Forest Renewal Plan*. Premier's Office. Victoria, BC

Canadian Council of Forest Ministers (CCFM). 1992. *Sustainable Forests: A Canadian Commitment*. Ottawa: Canadian Council of Forest Ministers

Clawson, M. 1979. "Renewable Resources in the future of North America." *Forestry Chronicle* 55(6): 225-28

Coleman, W.D., and G. Skogstad. 1990. "Policy Communities and Policy Networks: A Structural Approach." In *Policy Communities and Public Policy in Canada: A Structural Approach*, ed. W.D. Coleman and G. Skogstad, 14-33. Mississauga, ON: Copp Clark Pitman

Dellert, L.H. 1994. *Sustained Yield Forestry in British Columbia: The Making and Breaking of a Policy (1900-1990)*. MES thesis, York University

Fernow, B. 1910. "Scientific Forestry in Europe: Its Value and Applicability in Canada." In *Report of the First Annual Meeting*. Ottawa: Commission of Conservation of Canada

–. 1913. *History of Forestry: In Europe, the United States and Other Countries*. Toronto and Washington, DC: Toronto University Press and American Forestry Association

Forestry Canada. 1992. *The Silviculture Conference: Stewardship in the New Forest*. Ottawa: Forestry Canada

Fulton, F.J. 1910. *Royal Commission of Inquiry on Timber and Forestry, 1909-1910*. Ottawa: Victoria: King's Printer

Gale, R.P., and S.M. Cordray. 1991. "What should forests sustain? Eight answers." *Journal of Forestry* 89(5): 31-36

Haley, D., and M.K. Luckert. 1989. "Forest tenure – Requirements, rights and responsibilities: An economic perspective." *Forestry Chronicle* 65(3): 180-2

–. 1993. "Institutional barriers to the sustainable development of British Columbia's forests." Paper presented at the 14th Commonwealth Forestry Conference, Kuala Lampur, Malaysia

Holling, C.S. 1973. "Resilience and stability of ecological systems." *Annual Review of Ecology and Systematics* 4: 1-33

–. 1986. "The Resilience of Terrestrial Ecosystems: Local Surprise and Global change." In *Sustainable Development of the Biosphere*, ed. W.C. Clark and R.E. Munn, 292-320. Cambridge: Cambridge University Press

MacKay, D. 1985. *Heritage Lost: The Crisis in Canada's Forests*. Toronto: Macmillan

MacMillan Bloedel. 1980, 1992, 1993. *Annual Report.* Vancouver: MacMillan Bloedel

–. 1991. *MacMillan Bloedel Limited Management and Working Plan No. 2 for Tree Farm Licence No. 44 and Forest Management Unit #74.* Final Draft. Prepared by B.O. Waatainen, RPF. Port Alberni, BC: MacMillan Bloedel

Manning, E.C. 1937. *Address by the Chief Forester to the Forestry Committee of the British Columbia Legislature.* November 2. Victoria

Marchak, P. 1983. *Green Gold: The Forest Industry in British Columbia.* Vancouver: UBC Press

Meidinger, E.E. 1997. "Organizational and Legal Challenges for Ecosystem Management." In *Creating a Forestry for the 21st Century*, ed. K.A. Kohn and J.F. Franklin, 361-80. Washington, DC: Island Press

Mulholland, F.D. 1937. *The Forest Resources of British Columbia.* Victoria: King's Printer

Natural Resources Canada. Canadian Forest Service. 1996. "Measuring Forest Sustainability: The Canada Approach." In *1995-96, The State of Canada's Forests*, 72-91. Ottawa: Queen's Printer

Pearse, P.H. 1976. *Timber Rights and Forest Policy in British Columbia.* Report of the British Columbia Royal Commission on Forest Resources. Victoria: Queen's Printer

–. 1993. *Determination of Harvest Rates in the Transition to Sustained Yield.* Background paper prepared for the Chief Forester. Vancouver

Pincetl, S. 1993. "Some origins of French environmentalism: An exploration." In *Forest and Conservation History* 37(2).

Romm, J. 1993. "Sustainable Forestry, An Adaptive Social Process." In *Defining Sustainable Forestry*, ed. G.H. Aplet, N. Johnson, J.T. Olson, and V.A. Sample, 280-93. Washington, DC: Island Press

Rubner, H. 1984. "Sustained-Yield Forestry in Europe and its Crisis During the Era of Nazi Dictatorship." In *History of Sustained-Yield Forestry: A Symposium*, ed. H.K. Steen, 170-5. Durham, NC: International Union of Forestry Research Organizations (IUFRO) Forestry History Group (S6.07)

Science Council of BC (SCBC). *Forestry Research and Development in British Columbia: "A Vision for the Future."* Prepared by the Forestry Planning Committee. Vancouver: SCBC

Sloan, G.M. 1945. *Report of the Honourable Gordon McG. Sloan, Chief Justice of British Columbia, Relating to the Forest Resources of British Columbia.* Victoria: Queen's Printer

–. 1956. *Report of the Honourable Gordon McG. Sloan, Chief Justice of British Columbia, Relating to the Forest Resources of British Columbia.* Victoria: Queen's Printer

Supreme Court of British Columbia. 1993. *Reasons for the judgment of the Honourable Mr. Justice Smith between the Sierra Club and the Chief Forester, Appeal Board (Vancouver, A930623) and between the Province of British Columbia and David S. Cohen, Gary Bowden, Charles Gairns.* (Vancouver, A9223847). December 22. Vancouver, British Columbia

Taylor, G.W. 1975. *Timber: History of the Forest Industry in BC.* Vancouver: J.J. Douglas

Transcript of Proceedings. 1992. *Re the Forest Act, R.S.B.C. 1979, c. 140 as amended, and in Re the appeal by MacMillan Bloedel Ltd. from a decision of the Chief Forester, dated December 31, 1991.* Volumes 1-6. August 10, 11, 12, 13, and 25. Vancouver, British Columbia

Vincent, J.R., and C.S. Binkley. 1992. "Forest-Based Industrialization: A Dynamic Perspective." In *Managing the World's Forests*, ed. N.P. Sharma, 93-137. Dubuque, IA: Kendall/Publishing

Williams, D.H. 1993. "Timber Supply in British Columbia: The Historical Context." In *Determining Timber Supply and Allowable Cuts in BC,* ed. Melissa J. Hadley, 9-15. Proceedings. Vancouver: Association of British Columbia Professional Foresters

Williams, M. 1989. *Americans and Their Forests: A Historical Geography.* Cambridge: Cambridge University Press

Wilson, J. 1990. "Wilderness Politics in BC: The Business Dominated State and the Containment of Environmentalism." In *Policy Communities and Public Policy in Canada: A Structural Approach,* 141-69

Young, W. 1981. "Increasing Demands on the Forest Resource – The Challenge of the '80s." In *The Forest Imperative,* proceedings, 1988 Canadian Forest Congress, 22-26. Montreal: Canadian Pulp and Paper Association

12
The Pitfalls and Potential of Eco-Certification as a Market Incentive for Sustainable Forest Management
Fred Gale and Cheri Burda

Introduction

Policymaking is subject to fads and fashions. In the 1970s, governments conceived the environmental problem largely as one of pollution abatement and prevention. They addressed industry's impact on the environment by legislating pollution regulations and production standards and by fining companies for non-compliance. Defenders of state regulation point to the success of this approach, citing policies that led to a reduction in chlorinated compounds in pulp and paper effluent, to a decrease in the use of chlorofluorocarbons (CFCs) in the manufacturing of refrigerators and car coolant systems, and to an increase in the area of land set aside for parks and wilderness protection. While these achievements were important, environmental progress was slow, implementation structures bureaucratic, and company compliance problematic.

In the 1980s, the emerging discipline of environmental economics launched an attack on the regulatory approach, claiming that it was inefficient and prevented companies from finding creative and novel solutions to environmental problems. Operating within a neoclassical economic paradigm and grounded in a theory of consumer preferences, environmental economists examined the feasibility of influencing social behaviour through tax policy (levying of "green" taxes on polluting activities such as fossil fuel combustion), cost-benefit analysis (pricing previously uncounted environmental costs and benefits in evaluating projects), market creation (establishing tradable permits allowing a certain volume of pollution to be released), green accounting (developing a system of national accounts to better reflect the environmental impact of production and consumption activities), and eco-certification (providing credible consumer information that certified products have been produced according to pre-determined standards of sustainability, including recyclability, recycled content, animal testing, toxicity, and biodegradability).

The jury is still out on whether this second, market-oriented wave of environmentalism will prove any more effective than the first one. The use of many of the above market instruments is still in its infancy. To the extent that a purely market-based approach ignores the need for a fundamental restructuring of our major social institutions, including that of the market itself, the market-based approach to environmentalism certainly warrants considerable scepticism. If coupled with significant institutional reforms, however, the use of market-based instruments could make a powerful contribution.

In this chapter, we examine the potential of one such market mechanism – eco-certification and eco-labelling – to contribute to the promotion of eco-forestry and a genuinely sustainable forest products industry in British Columbia. Specifically, we investigate the background to the eco-certification movement, examine its growing importance in the Canadian and British Columbian forest products industry, and assess the capacity of two different eco-certification schemes: the Canadian Standards Association's Sustainable Forest Management System (CSA), and the Forest Stewardship Council's Principles for Sustainable Forest Management (FSC) – to achieve genuinely sustainable forest management. This investigation reveals the struggle taking place between proponents of CSA and FSC certification schemes, pitting the claims of large, well-funded, mobile, but self-interested multinational corporations against those of a small, poorly funded, yet dedicated and credible environmental lobby. Notwithstanding its inequitable nature, we argue that the FSC can win this struggle in BC if it is well-organized, ensuring in the process that eco-certification fulfils its proper purpose.

Certification as a Market-Based Incentive

A market-based approach to the environment builds on the notion that corporations respond quickly and flexibly to changes in consumer tastes. A 1989 market survey, for example, indicated that 89 percent of Americans were concerned about the impact of their purchases on the environment and 79 percent indicated that they would be willing to pay more for a product packaged with recyclable or biodegradable materials.[1] Building on this concern, the environment began to emerge as a new business opportunity, sparking a significant increase in the number of new products labelled "green."[2]

Consumer awareness of, and demand for, less environmentally damaging products, if translated into actual consumer behaviour, creates an economic incentive for companies to produce and sell these products. It also results in the making of misleading claims. A well-known "green product" controversy in Canada was the Loblaws' marketing campaign in the late 1980s, endorsed by Pollution Probe. Labels were attached to such items as acid-free coffee and disposable diapers made from unbleached paper fibres.[3] McDonald's, too, took advantage of the opportunity to improve its environmental reputation by introducing CFC-free Styrofoam packaging (merely a less potent member of the CFC family).[4]

Dadd and Carothers investigated bogus claims of "greenness" and remain sceptical about the role that environmental marketing might play in changing corporate behaviour. They draw attention to the paradox of "green consumption," which continues to endorse the practice of consumerism. They warn that green consuming may divert action away from the real need for government regulation and institutional change.[5] Furthermore, the proliferation of company-specific green labels in the marketplace could generate consumer confusion and undermine the credibility of the entire approach.[6]

Consumer scepticism over individual company claims prompted industry associations, governments, environmentalists, and market consultants to set up different and competing eco-certification schemes in designated market sectors. Schemes promoted by industry associations and governments reflect a more arm's-length approach than the self-certification and labelling schemes adopted by individual companies. West Germany established the first government-sponsored scheme in 1978.[7] Canada initiated its own eco-labelling program ten years later when it set up the Environmental Choice Program (ECP). By 1995, ECP had certified 1,500 products that satisfied any of the following requirements: improves energy efficiency, reduces hazardous or toxic waste, uses recycled material, prolongs product life, can be reused, or in some other way is environmentally responsible. The criteria differ for each product or category and there are no overarching criteria or certification principles. Although applicants are encouraged to provide a description of the environmental benefits that occur at the various stages of production (where relevant and available), the process is not based on life-cycle analysis (LCA).[8] It is evident, therefore, that the ECP permits claims of environmental acceptability to be made about some otherwise very dubious products.

The failure to employ life-cycle analysis, the close relationship that often exists between government and business, the dependence of government on tax revenues from business activity, and governance by one of two or three broad-based, centrist political parties in most democracies all tend to weaken a government's resolve to implement tough environmental measures. The unwillingness of governments to act has prompted environmental civil society organizations (ECSOs) to support eco-certification schemes based on life-cycle analysis and the adoption of principles that put the environment, and not business activity, first. This type of eco-certification involves the full and active participation of environmentalists and ecologists in the establishment of the principles, criteria, and indicators to be applied, and limits and controls the power of business to weaken these standards in the interests of profits. An example of this approach is found in the work of the Forest Stewardship Council.

Green Forestry

The movement towards forest products certification began in the mid-

1980s, following widespread public concern over tropical deforestation. ECSOs, including the Rainforest Action Network in the United States and Friends of the Earth in the United Kingdom, launched boycott campaigns against tropical timber imports and sought intergovernmental agreement at the International Tropical Timber Organization on a certification and labelling scheme for tropical timber products.[9] This initiative was blocked by opposition from producing and consuming country governments, responding to pressure exerted by the tropical timber industry in the North and the South. The tropical timber industry feared that eco-certification and eco-labelling could quickly become a technical barrier to trade, encouraging European and North American consumers to switch to temperate and boreal timber and non-timber substitutes.

For most of the 1980s, ECSOs assumed that temperate and boreal forests were well managed in comparison to tropical forests and that the global forestry goal was to bring Third World countries up to Western standards.[10] This assumption began to be questioned in response to growing evidence of the negative impact of sustained yield forest management policies in Canada, Sweden, and Australia on biodiversity, riparian ecosystems, and soil stability. The emergence of the disciplines of conservation biology and landscape ecology provided an alternative framework within which to situate a critique of industrial forestry and promote an alternative, ecosystem-based approach.[11] In the 1990s, therefore, ECSOs expanded their critique of forest management to include temperate and boreal forests, and demanded the establishment of a credible certification and labelling scheme to guarantee that timber products from all countries come from genuinely sustainable forests.

Continued failure by the international community to take action on the establishment of a meaningful certification and labelling scheme obliged ECSOs to establish their own process of consultation and discussion. This development took place between 1991 and 1993 and culminated in the establishment of a new global forest organization, the Forest Stewardship Council.[12] The FSC's mandate is to encourage the institutionalization of a globally credible certification and labelling system in the forest products industry to assure consumers that timber carrying the FSC logo comes from forests stewarded in accordance with the key principles of ecosystem-based forestry.

Certification Schemes

In response to FSC's initiative, and out of concern to protect their domestic industries' access to foreign markets, many countries are establishing their own government- and industry-sponsored certification processes. The process of certifying and labelling products in general, and timber products in particular, is gathering momentum, and those most affected by it – corporations and governments – are manoeuvring to ensure that the system that eventually gets established serves their interests. Once almost exclusively an ECSO-sponsored initiative in the forest products industry, eco-certification and labelling has now become a matter of immense importance to governments and industry leaders around the world. Competing efforts to participate in or control the direction of forest certification has resulted in the emergence of a variety of certification schemes representing opposing interests, different standards, and a range of objectives.

First-, Second-, and Third-Party Schemes

First-party schemes are initiated by the timber companies themselves and involve an internal assessment of the company's systems and practices in regard to internally established guidelines or environmental objectives. A 1991 study in the United Kingdom investigated the "sustainability" claims of tropical timber companies. When these companies were challenged on the validity of the claims made, they quickly withdrew them rather than risk a legal action over false advertising. The study found that only three of the eighty companies in the sample were ultimately willing and able to substantiate their claims.[13] Although corporations in BC have not attempted to attach "eco-labels" to their forest products, certain claims have been made through advertising, information and public awareness campaigns sponsored by the individual companies, corporate associations such as the Forest Alliance, and the provincial government.[14] The content of this information is designed to convince purchasers that current forest management, planning, and practice in British Columbia are "sustainable."

Second-party certification schemes are promoted by governments, industry associations, and/or government-funded certification bodies. In several countries, the government is taking responsibility for the establishment of forest certification schemes. The Indonesian government, for example, has established a new organization: the Indonesian Ecolabel Agency (Lembaga Ekolabel Indonesia). In Canada, the Canadian Standards Association (CSA), with funding from the Canadian Pulp and Paper Association (CPPA), has developed a set of guidelines for the establishment of a Sustainable Forest Management System. The CSA approach is designed to ensure its compatibility with the quality management approach used by the International Organization for Standardization (ISO), an international agency comprising most of the world's national standards-making bodies.[15] The ISO's mission is "to promote the development of standardisation to improve the international exchange of goods and services, and to develop intellectual, scientific, technological and economic activity."[16]

Third-party eco-certification schemes are developed by parties that are knowledgeable about the environment, life-cycle analysis, and the ecosystem-based approach to resource stewardship. Such third-party schemes develop standards that put the needs of ecosystems first, and include both business and social objectives. The FSC, for example, an international environmental civil society organization, promotes third-party eco-certification through its Principles and Criteria of Forest Management. The FSC's principles and criteria have been developed with the full participation of environmental organizations, and the FSC is structured to ensure that environmental and social organizations retain control over industry interests. The FSC's principles and criteria are adopted by accredited, regional third-party certifiers. Accredited certifiers develop specific standards to certify operations in their region. A third-party certifier, the Silva Forest Foundation (SFF), conducted the first Canadian certification of a forestry operation in Vernon, BC.[17] SFF, which has developed a rigorous and specific set of standards for ecologically responsible forest use,[18] is a member of the Forest Stewardship Council.

Performance-Based and Systems-Based Schemes

FSC and CSA are the two main forest certification schemes competing internationally, in Canada, and in BC. The CSA, a second-party certification scheme that has received significant funding from the forest industry, and which has been unable to ensure the substantive participation of environmental and social organizations, is developing a management-systems approach to forest certification. This management-systems approach contrasts markedly with the performance-based approach of FSC. Unsurprisingly, the management-systems approach provides corporations with much more flexibility in the choice of technology to be used in the forest, the types of forest practices to be adopted, the volume of timber to be removed, and the degree to which biodiversity and other forest values are protected.

The FSC's performance-based certification scheme requires certifiers to assess the degree to which a forest operation is managed in accordance with FSC's principles and the set of detailed pre-determined social, environmental, and economic standards. These detailed standards are developed by national and regional working groups, such as the Pacific Certification Council (PCC) in the Pacific Northwest.[19] The FSC process not only certifies the forest practices of an individual or a company, but it also certifies the resulting *product* by tracing its "chain of custody" from the forest operation to log transport, through all stages of processing and shipping, to its receipt by the retailer. This process enables products to be labelled or stamped as "eco-certified" and to carry the FSC logo. Indeed, FSC was established to accredit third-party certifiers under a single, recognizable, and credible label to minimize consumer confusion resulting from the proliferation of different labels.[20]

The CSA's Sustainable Forest Management System involves the evaluation of the management *system* in place to conduct a forest operation. A certificate will be issued for a "Defined Forest Area" for which a Sustainable Forest Management System is in place and operating satisfactorily.[21] The forest stewardship performance requirements are set by the forest manager using broad-based principles established by the Canadian Council of Forest Ministers and following a process of public consultation with local stake-holders.[22] No predetermined set of performance standards is used by the CSA. Rather, a system of management and planning is established, through which a company sets its own standards in compliance with, but not nec-essarily beyond, government regulations and legislation.

The CSA approach is based on the ISO's 14001 Environmental Manage-ment System and suffers from the same defects.[23] The ISO objective is to facilitate international trade by standardizing and harmonizing product specifications. While in theory this objective does not necessarily lead to the adoption of the "lowest common denominator" standards, in practice it often does, given the power of industry lobbies to resist tough standards that threaten existing production strategies. Moreover, the CSA management-system approach does not involve product certification and the tracing of a timber product through the chain of custody from the forestry operation to the consumer. Confusion exists over the claims that a company, a part of whose operation is registered with the CSA, will be able to make.[24] While the company can advertise itself as having a "sustainable" forest management system in place, this claim does not mean that all, or even the majority, of the wood in its products comes from that Defined Forest Area.

The claims made by companies working under the CSA and FSC processes are fundamentally different. Under the CSA process, the company claims to have a Sustainable Forest Management System in place. The working assumption of the CSA scheme is that the adoption of this management sys-tem will ensure the sustainability of Canada's forest ecosystems. This assump-tion is unlikely to be tenable, however, given the weakness of the principles, criteria, and indicators developed by the Canadian Council of Forest Minis-ters and through the Montreal Process.[25] Although the CSA scheme allows for public input in the setting of management objectives, the process can be manipulated easily as a procedural requirement, because the company maintains complete control over all stages of the consultative process.

These defects in the CSA scheme can be counter-posed to the FSC scheme. The FSC approach makes substantive claims about the practices taking place in any individually or company-managed forest. Furthermore, the FSC logo assures consumers that the wood contained in the products comes from well-managed forests and that the product and its parts have been traced through the chain of custody to ensure that there has been no mixing of unsustainably produced wood.

Objectives of Eco-Certification Schemes

Forestry eco-certification has two primary objectives: the ecological objective of improving forest management and the economic objective of securing access to markets.[26] While both objectives are potentially com-plementary, there is an ever-present danger of the profit- and revenue-driven motives of industry and government eclipsing the ecological imperatives of sustainable forestry. The CSA process is deeply compromised. It is heavily funded by the Canadian timber industry and excludes the participation of the Canadian environmental movement. Moreover, the structure of its technical committee reflects predominantly government and business interests and the views of the industrial forestry establishment.[27] The CSA initiated its development of a sustainable forest management system *in response* to international and domestic criticism of Canadian corporate forest practices. The motive was defensive, the goal being to maintain Canadian company market share in threatened foreign, especially European, markets.

In contrast to the CSA, the FSC process originated *as a means* to promote better forest practices. The FSC is a unique international organization, both in the fact that it was established through the initiative of non-governmental organisations, and in its management structure. The highest authority of the FSC is its General Assembly, which comprises three chambers, representing economic, environmental, and social interests.[28] Social and environmental interests always weigh more heavily in the organization's decision structure than do economic interests, a unique and important model for the structuring of future global and national organizations. Nevertheless, the FSC's ability to adhere to its original mandate depends on the active participation of its social and environmental membership, and on the members' full understanding of the principles and practices of an ecosystem-based approach to forestry. There is the ever-present danger that the desire to appeal more broadly to a larger sector of the existing forestry establishment, coupled with an incomplete comprehension of the principles and practices of ecoforestry, could lead the FSC into compromises that undermine its credibility.[29]

The Pitfalls of Eco-Certification

The eco-certification movement encounters three major pitfalls, which threaten to undermine its capacity to achieve genuinely sustainable forest stewardship. First, all certification schemes reflect the underlying interests of those involved in their negotiation, and the bargains that are struck are reflected in the certification processes agreed to and in the principles and standards adopted. If industrial and governmental groups have too much power in negotiating the principles, standards, and processes of eco-certification, we should not be surprised that the resulting schemes "tend to amount to a marketing tool to sell wood" and "reflect short-term, profit-driven priorities."[30] There is an ever-present danger of industry gaining control of the eco-certification agenda and, in collaboration with government officials, manipulating the process to their own ends. This aim appears to be a central objective of the CSA process. Indeed, the dominant view in the industry appears to perceive eco-certification in precisely these terms. In a study conducted by the World Forest Institute, only a few industry participants believed that certification would actually improve forest management in the Pacific Northwest. Most viewed certification as an opportunity for forest producers to improve public relations and regain lost credibility.[31]

The second pitfall is the proliferation of eco-certification labels in the marketplace, each one competing with the other for consumer allegiance. While competition among labels is inevitable in the short term, given the various eco-certification processes currently under way both nationally and internationally, it is not desirable over the longer term. The proliferation of eco-certification schemes benefits status quo forestry, because consumers may be so confused about certification labels and claims that they are unable to discriminate between bogus and genuine logos in the marketplace. It is vital, therefore, that FSC win the logo competition, and it is here that the environmental network can help. Environmental organizations around the world can support the FSC process and work to discredit alternative processes.

The third pitfall, however, concerns the role that might be played by environmental organizations. The social forces gathered together in FSC are currently structurally weighted in favour of the social and environmental chambers. Compromises must, however, necessarily be made with those in the economic chamber if FSC is to move forward. There is a danger that these compromises will go too far, and the support of many environmental organizations will be lost. Simon Counsell, who represented Reforest the Earth at a recent general meeting of the Forest Stewardship Council, notes:

> Whilst none of the individual decisions taken by the meeting, with the possible exception of the change in chamber structure and voting power, will have major short term implications for the FSC, there seems to be evidence of a general trend towards marginalisation of NGO concerns – and those of small scale producers and forest managers – in the development of the organisation and its decision-making process ... It was also notable that many of the issues on the table were related to the requirements of large scale industry, whereas there was virtually no discussion of particular concerns ... that face smaller, community-based operations.[32]

Although all parties must recognize the need for compromises, FSC staff, executive, and council members, and particularly those in the environmental and social chambers, must be careful to ensure that the process of negotiation does not compromise the FSC's credibility and lead to the absurd, if currently hypothetical, situation where disaffected ECSOs publish pictures of FSC-certified operations where environmentally destructive forestry is being practised.

Eco-Certification as a Market Instrument for Ecoforestry in BC

As an ECSO initiative, certification is a market-based incentive with the objective of ensuring that logging and timber management activities protect the integrity of forest ecosystems.[33] Certification is the obverse of consumer boycotts, providing consumers the option of purchasing wood products from responsibly managed forests. In theory, consumer preference for eco-certified forest products makes its way back through the chain of custody through the retailer, wholesaler, and manufacturer to the logging company and the forest owner/manager, providing all with an incentive to practise ecologically responsible forestry.

Market Demand

Eco-certification influences forest practices and forest use only if there is an effective demand for sustainably produced wood products. The findings of market surveys suggest that there is a modest demand for eco-certified timber. A poll conducted in Great Britain on behalf of the World Wide Fund for Nature (WWF-UK) in 1990 found that 25 percent of people would "stop buying wood products made from trees such as teak or mahogany unless it could be guaranteed that they come from countries that were protecting their forests."[34] In a study by Winterhalter and Cassens, 68 percent of a sample of 12,000 consumers with annual household incomes of greater than US$50,000 indicated that they would be "willing to pay more for furniture whose construction material originated from a sustainably managed North American forest."[35] Ozanne and Smith report that approximately 10 percent of Americans (25 million) would be likely to seek out environmentally certified wood products.[36]

Taken together, these studies demonstrate the existence of a substantial group of consumers purchasing decisions on environmental considerations as well as on price, quality, and availability. If even only half of this group actually translates words into deeds, a market niche for certified timber products is assured. If "consumer sovereignty" is to function effectively, however, the desire of consumers to purchase certified forest products must be transmitted back through the chain of custody, providing an incentive for retailers to stock certified timber products, for the secondary and remanufacturing sector to buy eco-certified logs and lumber, and for logging companies and forest managers to switch from industrial to ecoforestry stewardship practices.

Chain of Custody

The retailer is the link between the consumer and the industry. Lober and Eisen argue that retailers have been reluctant to promote certification out of fear of destroying their reputation by marketing products with bogus claims, and because they did not have the expertise themselves to verify such claims.[37] This explanation, however, sits uneasily with the results of the study by Read and Associates, which notes the existence of numerous unsupported labelling claims made by companies in the United Kingdom. A better explanation for retailer reluctance is a lack of consumer pressure (and a lack of direct pressure from environmental organizations). Retail support for eco-certified products has developed in Europe as a result of major lobbying efforts by the environmental movement. BandQ, the largest "do-it-yourself" chain, and Sainsburys, a major retail outlet for forest products, are both now members of WWF's 95 Group, which consists of over fifty UK companies that have pledged to purchase only FSC-certified wood products by the year 2000.[38]

In the United States, Home Depot, the country's largest home building and improvement industry, has developed an Environmental Greenprint Program to promote and demonstrate the use of environmentally benign products.[39] The Greenprint Program is a partnership with Scientific Certification Systems (SCS), a certification agency based in California, which is responsible for auditing claims and for certifying participating companies. In addition, the Good Wood Alliance in the United States produces a list of sources, retailers, and distributors of certified "good wood" worldwide. The summer 1995 issue listed 35 retailers and wholesalers of "good wood," three of which were in Canada, but none in BC.[40]

Retailers in other parts of the world are thus becoming more interested in eco-certification and labelling schemes. There is some difficulty, however, in translating this retail interest along the chain of custody. The World Forest Institute, in a 1993 survey to assess stakeholder attitudes to eco-certification conducted in Washington, Oregon, and British Columbia,[41] noted that stakeholders closest to the consumers (retailers, architects, builders) had the most positive responses to FSC's initiatives and indicated they would have the least difficulties in implementing certification.[42] Manufacturers, on the other hand, did not believe that enough consumer demand existed, or that the customer would pay a premium for certified wood.[43] Manufacturers also doubted the feasibility of tracking wood from the harvest site to the consumer.

A survey by the Institute for Sustainable Forestry (ISF) confirms the findings of the WFI study. ISF assessed the attitudes of the wood products industry in Washington, Oregon, and California to determine whether companies would be willing to pay a premium for, and be willing to handle, certified wood products. This survey targeted small- to medium-sized manufacturers and retailers that were already selling certified and ecologically sustainable products. An overall objective of the study was to assess the feasibility of establishing a chain of custody link that would run from small landowners to small primary and secondary manufacturers, and from there to lumber retailers, custom furniture manufacturers, and high-end consumers who had been identified as willing to pay a 10 percent premium for certified wood.[44] The results indicated that consumer demand for certified wood products was not making its way back through the chain as effectively as it could. Although overall awareness of certification was high at 62 percent, awareness of certification and client requests for, and sales of, certified products increased with proximity to the consumer.[45]

Until recently, very little information existed concerning industry attitudes to eco-certification in British Columbia. Two 1996 surveys conducted in the province cast doubt on the industry's preparedness to exploit emerging North American and European markets for eco-certified wood products. The first survey, conducted by the Silva Forest Foundation, targeted small woodland owners and tenure holders, and small to medium producers and manufacturers. The objective of the study was to determine if a chain of custody was possible among the smaller producers and manufacturers, given that the former were more likely to participate in certification. Although the response rate for the study was low, the authors concluded, somewhat pessimistically:

> The market demand is not yet strong enough to prompt [small to medium producers and value-added manufacturers] to seek out and pay for third party certification. A major obstacle for producers and value-added manufacturers is the lack of interest on the part of primary and secondary manufacturers (i.e. sawmills, remanufacturers, etc.), therefore making the chain of custody certification difficult ... While certification is likely to become increasingly important over the next few years, until pressure is placed on the industry – either from consumers or government – certification will probably remain small scale.[46]

A second study that we carried out surveyed 1,144 primary and secondary manufacturers and retailers, representing all forest product primary and secondary manufacturers and retailers in British Columbia according to then current directories.[47] With a response rate of 18 percent, findings were that industry perceived little consumer demand for eco-certified forest products; that there was negligible consumer willingness to pay a price premium for eco-certified wood products; and that the CSA process was more credible than that of the FSC. The study validated a major finding of earlier studies in the United States, the tendency of attitudes towards eco-certification in general and to FSC eco-certification in particular to grow more positive the closer one gets to the consumer end of the chain of custody. Thus, for example, while only one primary manufacturer felt that the FSC process was the most credible in the eyes of clients (3 percent of the total primary manufacturing sample), ten secondary manufacturers and fourteen retailers (representing 16 percent and 15 percent, respectively) regarded FSC certification as the most credible in the eyes of clients.

Although there is now an awareness of certification in the industry, the major reasons that companies are reluctant to participate are a perceived lack of consumer and client demand, an uncertain supply of certified timber, an inability to separate certified timber from non-certified timber, and the variability of certified wood in terms of species and quality.

The Future of Eco-Certification in Canada

While industry participation in eco-certification is a necessary condition for the success of this market instrument, it is not a sufficient one. Studies indicate that many in the industry are leaning heavily towards supporting CSA's and ISO's management-systems approaches. Such approaches mislead consumers into thinking that the world's forests are being managed sustainably, while permitting industrial forestry practices such as clearcutting to continue. While it should not prove too difficult to undermine these approaches by attaching photographs of large clearcuts, such publicity risks harming all approaches to eco-certification unless carefully handled.

At one level, it is easy to appreciate why forest corporations prefer management-system to performance-based certification. The management-system approach avoids the difficult and costly requirements of following a chain of custody, conducting extensive field audits, documenting and addressing plant and animal numbers and air and water quality, and following prescriptive principles that conflict with industrial forestry and state-federal regulations and programs such as reforestation.[48] In particular, tracing timber products through the chain of custody is difficult and costly for large-scale operations when wood arrives from literally hundreds of different sources for manufacturing into composite materials such as plywood and engineered wood products.

Management-system certification appeals to corporations precisely because it does not challenge the underlying assumptions of industrial forestry and permits the continuation of sustained yield forestry with only the most modest restructuring. FSC's performance-based approach to certification is challenged by forest corporations because it does demand substantial changes in the technology employed, in the volume extracted, and in the practices used. A battle is brewing, therefore, between ECSOs and powerful national and global multinational corporations. The latter have tremendous financial power, the support of most of the world's governments, and a massive public relations capacity to communicate their message. They constitute a formidable foe.

On the other side stands the Forest Stewardship Council, promoting its performance-based eco-certification principles and standards. FSC is struggling to find financial resources to open offices in major timber-producing regions, and is treated suspiciously at best, and often with outright hostility, by national and provincial governments, which are deeply threatened by the revenue and economic implications of genuinely sustainable forest management.[49] The FSC currently lacks an active and effective public relations arm to ensure that its message is communicated to producing- and consuming-country consumers around the world.

As an alternative to the looming struggle between the two approaches, a recent WWF report by Chris Elliott and Arlin Hackman proposes a compromise.[50] Elliott and Hackman note that FSC's performance-based approach and CSA's management-system approach "both have enough support in Canada for continued disagreements between them to be extensive and damaging."[51] They argue that three options exist for eco-certification in Canada: continued conflict, mutual recognition, or some combination. Continued conflict is reckoned to be the worst case scenario, leading to a weakening of the credibility of both FSC and CSA approaches. In making the case for either mutual recognition or a combination, Elliott and Hackman compare the FSC and CSA systems across several criteria including level of NGO, industry, and government support; ISO compatibility; market credibility; and operational status. This comparison reveals that the two schemes have complementary qualities. Where FSC is high, CSA is low, and vice versa. That is, FSC rates well in terms of NGO support, but it is not welcomed by industry, while the CSA scheme rates high in terms of industry support, but is mistrusted by the NGO community. Elliott and Hackman suggest that this difference bodes well for some mix of the two schemes, either mutual recognition or a combination.

WWF's argument ignores the underlying ecological political economy of the development of the CSA and FSC approaches. Because the FSC approach was established before the CSA process, the real question is: why did industry not join the FSC process rather than establishing its own management-system approach? The answer lies in the substantially different functions that the two processes are designed to serve. The FSC process is designed to substantively alter the practices occurring in Canada's forests so that they conform to those mandated by an ecosystem-based approach to forest stewardship. The CSA is attempting to gain control of the certification agenda, prevent the establishment of substantive standards, protect Canada's export markets, maintain the country's volume-based export-industry, and safeguard the future of its large industrial forest corporations. In such circumstances, whatever the merits in theory of combining performance-based and management-systems approaches, the practical result of this combination, given the power of industry and government, would be to lower to an unacceptable level the principles, standards, criteria, and practices of the FSC approach.

A continuation of the struggle between FSC and CSA approaches thus seems inevitable in Canada. Such a struggle is unfortunate, because it will undoubtedly slow down the movement towards genuinely sustainable forest management, and creates an opening for more ecologically minded governments elsewhere to gain first-mover advantage in the emerging market for FSC-certified timber products.[52] On the other hand, environmentalists should be wary of any sudden agreement between FSC and CSA to join forces and combine their schemes. As the comments of Simon Counsell at the meeting of the FSC General Assembly cited earlier indicate, there are elements within the FSC itself that would like to see the organization move towards a more middle of the road, pro-industry, position. These elements could become increasingly powerful in the FSC if civil society members constituting its social and environmental chambers do not attend meetings and make their views known.

If the immediate future for eco-certification in Canada is a struggle between the FSC and the CSA processes, then environmentalists should consider a three-pronged strategy to promote the establishment of FSC certification in Canada. First, monitor vigilantly the work of the CSA and subject all corporate claims concerning the sustainability of their management systems to rigorous critical analysis. Resources must be spent to highlight the relationship between the corporate claim that it is implementing a sustainable-management system in a designated forest area, and the actual forest practices taking place. All corporations making such claims must, moreover, immediately become a focus for the most detailed scrutiny, so that alternative profiles can be communicated that challenge the corporate and government PR campaigns that are bound to be launched.

Second, strengthen the FSC's presence in Canada. The recent appointment of an FSC coordinator for Canada by the Canadian Working Group of the Forest Stewardship Council is a welcome development, but more must be done. In particular, the FSC needs to attract more Canadian members[53] to actively promote the establishment of FSC-accredited certifiers in all regions of Canada (and particularly British Columbia, where currently there is only one certifying body for the entire province), and to devote more resources to publicity, getting out the message about the potential advantages of eco-certification.

Third, establish the equivalent of the UK's WWF-'95 Group in other major European, North American, and Japanese markets. Any examination of the Canadian and British Columbia forest industry reveals its overweening dependence on export markets for its softwood lumber, pulp, paper, and newsprint products.[54] Ultimately, therefore, a Canadian strategy to promote FSC-certification in Canada must also have a substantial international component. The Canadian forest industry has a history of not responding to pressure from inside its borders, as the struggles over Clayoquot Sound and Temagami show. It is usually persuaded, as the graphic example of the softwood lumber dispute with the United States and the Coalition for Fair Lumber Imports illustrates, to alter its practices in response to real and perceived threats to its export markets in the United States, Europe, and Japan.[55] The recent announcement by MacMillan Bloedel to phase out clearcut logging in old-growth forests by 2003 constitutes the most dramatic example to date of the power environmentalists can wield by applying pressure in foreign markets.[56] The establishment of more and larger buyer groups, particularly in the United States, committed to purchasing FSC eco-certified timber products by the year 2000, would do more to alter industry perceptions of the desirability and feasibility of FSC eco-certification than any other measure. While there are encouraging signs that the Canadian forest industry is finally getting the environmental message and beginning to mend its ways,[57] the political reality is that only continued massive, unrelenting, and substantive pressure is going to bring industry to its senses.

Notes

1 D.L. Dadd and A. Carothers, "A Bill of Goods? Green Consuming in Perspective," in *Green Business: Hope or Hoax,* ed. C. Plant and J. Plant (Philadelphia: New Society Publishers 1991), 12.
2 Ibid.
3 For a detailed account of the Loblaws' saga, see G. Gallon, "The green product endorsement controversy: Lessons from the Pollution Probe/Loblaws experience," *Alternatives* 18(3) (1992): 16-25.
4 Dadd and Carothers, supra, note 1 at 16.
5 Ibid., 18.
6 C. Upton and S. Bass, *The Forest Certification Handbook* (Delray Beach, FL: St. Lucie Press), 45.
7 Dadd and Carothers, supra, note 1 at 15.
8 Environmental Choice Program, *Certification Overview* (Ottawa: Terra Choice Environmental Services 1995).
9 F. Gale, *The Tropical Timber Trade Regime* (London: Macmillan Press 1998).
10 This perspective underlies, for example, a very influential study of sustainable forest management in tropical forests carried out by Dr. Duncan Poore, a forest consultant with the International Institute for Environment and Development in London. See D. Poore, ed., *No Timber Without Trees* (London: Earthscan 1989).
11 An ecosystem-based approach to forestry is based a recognition that forests constitute complex ecosystems that have to be stewarded holistically to maintain ecosystem integrity and health. The ecosystem-based approach to forestry has important implications for forest practices, and replaces modern industrial clearcut forestry with selection logging, contour logging, horse logging, and patch logging. For further details on the ecosystem-based approach, see R.E. Grumbine, "What is ecosystem management?" *Conservation Biology* 8(1) (1994): 27-38; P. Alpert, "Incarnating Ecosystem Management," *Conservation Biology* 9(4) (1995): 952-5; and H. Hammond, "Forest Practices: Putting Wholistic Forest Use into Practice," in *Touch Wood: BC Forests at the Crossroads,* ed. K. Drushka, B. Nixon, and R. Travers, 96-136 (Madeira Park, BC: Harbour Publishing 1993).
12 For a good review of the history of FSC, see N. Dudley, J.-P. Jeanrenaud, and F. Sullivan, *Bad Harvest? The Timber Trade and the Degradation of the World's Forests* (London: Earthscan 1995).
13 M. Read, *An Assessment of Claims of "Sustainability" Applied to Tropical Wood Products and Timber Retailed in the U.K. July 1990-January 1991* (Godalming, Surrey: World Wide Fund for Nature 1991).
14 Examples of current publications include BC Ministry of Forests, *Providing for the Future: Sustainable Forest Management in British Columbia* (1996); and the Forest Alliance's *Choices* magazine at www.forest.org. These efforts reach beyond BC consumers. The Forest Alliance has communicated its message to European customers in a number of languages.
15 For a more detailed account of the history of eco-certification and labelling, see B. Ghezali and M. Simula, *Certification Schemes for All Timber and Timber Products* (Yokohama, Japan: International Tropical Timber Organization 1994).

16   Upton and Bass, supra, note 6 at 34.
17   For further details on Silva's certification of the Vernon forest operation, see "Good News! Silva Forest Foundation Conducts Canada's First Wood Certification," *Silva Forest Foundation Newsletter* (fall 1995): 1, 3.
18   The Silva Forest Foundation's primary standard for ecologically responsible timber management states: "All plans and activities must protect, maintain, and restore (where necessary) a fully functioning forest ecosystem at all temporal and spatial scales. Forest composition, structures, and functioning must be maintained, from the largest landscape to the smallest forest community, in both short and long terms." At the stand level, Silva's standards for ecologically responsible forest are detailed in extensive detail. They include: protecting and maintaining composition and structures to support fully functioning forests at all scales; using ecological tree-growing periods, for example, 150 to more than 250 years; using non-clearcut harvesting methods that maintain canopy structure, age distribution, and species mixtures found in healthy, natural forests; maintaining ecological succession to protect biological diversity; prohibiting pesticide use; minimizing soil degradation; protecting water and riparian ecosystems; and allowing the forest to regenerate trees through seeds from trees in and adjacent to the logged area. Cited in Silva Forest Foundation, *General Standards For Ecologically Responsible Forest Use*, Draft for Peer Review (March 1994).
19   For further details about its goals, structure, and operation, see W. Smith, "What is the Pacific Certification Council?" *International Journal of Ecoforestry* 11(4): 105-7; and D. Simpson, "Who is the Pacific Certification Council?" *International Journal of Ecoforestry* 11(4): 108-9.
20   Certification schemes currently in operation under their own labels are the Forest Conservation Program of Scientific Certification Systems (US, for-profit), Smart Wood Certification Program of Rainforest Alliance (US, non-profit), Woodmark of the Soil Association (UK, non-profit), SGS Silviconsult (UK, for-profit). Cited in C. Elliot and A. Hackman, *Current Issues in Forest Certification in Canada* (Toronto: WWF-Canada 1996), 13.
21   Ibid., 10.
22   The principles and a framework for action are presented in *Sustainable Forests: A Canadian Commitment* (Hull, QB: Canadian Council of Forest Ministers, 1992).
23   Standards are developed by the ISO in the following way: a particular industry sector expresses the need for a standard; technical committees determine the scope of standards; countries negotiate specific standards; and the ISO approves the national draft standards.
24   Discussion ensues over whether the CSA should "register" companies complying to CSA-SFM rather than actually certify forest operations, because certification requires a labelling scheme which is problematic without a chain of custody. Elliot and Hackman, supra, note 20 at 11.
25   See Upton and Bass, supra, note 6 at 127.
26   Elliot and Hackman, supra, note 20 at 10-12.
27   Ibid.
28   For further details on the structure of the FSC, see Upton and Bass, supra, note 6 at 186-90.
29   Two issues that illustrate the tensions that exist within the FSC have arisen in the past few years. The first concerns the certification by the Smart Wood Program of the Rainforest Alliance of the controversial forestry operation run by Flor y Fauna in Costa Rica. The issue concerns the claims made by Flor y Fauna about growth rates on its teak plantations and whether the certifying body is responsible for not only certifying the forest practices of the company, but also for ensuring that the claims it makes to potential investors are bona fide and supportable. For further details on this incident, see J.C. Centeno, "Forest certification as a tool for green washing," article posted on Infoterra listserv (infoterra@ pan.cedar.univie.ac.at), 17 November 1996; R. Donovan, "Flor y Fauna Certification: A Statement from Smart Wood" (New York: Rainforest Alliance, 1996). The second issue concerns the controversial recent 1997 agreement by FSC to add its principle number ten on forestry plantations. Although principle ten is carefully worded to ensure that plantations do not replace natural forests, and serve to "promote the protection, restoration, and conservation of natural forests," the principle does open the door to a much larger scale, industrial, and commercial form of forest operation than appears consistent with the principles of ecoforestry- and ecosystem-based forest management.

30 H. Hammond and S. Hammond, "What is certification?" *International Journal of Ecoforestry* 11(4) (1995): 102-3.

31 World Forest Institute, *Feasibility Study Regarding Forest Product Certification In Oregon, Washington and British Columbia* (Portland, OR: World Forest Institute 1993), 13.

32 S. Counsell, *Report on the First General Meeting of the Forest Stewardship Council, Oaxaca, Mexico, 27-28 June, 1996* (London: Reforest the Earth 1996).

33 Hammond and Hammond, supra, note 30 at 102.

34 Mike Read and Associates, *Truth or Trickery? Timber Labelling Past and Future* (Godalming, Surrey: WWF-UK 1994), 9.

35 D. Winterhalter and D. Cassens, "Telling the sustainable forest from the trees," *Furniture Design and Manufacturing* (August 1993): 101-6.

36 L. Ozanne and P.M. Smith, "Measuring the market: An opening for certified forest products," *Understory: Journal of the Woodworkers Alliance for Rainforest Protection* 5(4): 1, 5.

37 D. Lober and M. Eisen, "The greening of retailing," *Journal of Forestry* 93(4) (1995): 38-39.

38 See the speech by George White, in *The Business of Good Forestry,* ed. M. M'Gonigle, K. Stratford, and F. Gale (Victoria: Eco-Research Chair of Environmental Law and Policy, University of Victoria 1996).

39 Lober and Eisen, supra, note 37 at 40.

40 *Understory: Journal of the Woodworkers Alliance for Rainforest Protection* 5(3) (1995): 5-8.

41 World Forest Institute, supra, note 31.

42 Ibid., 1.

43 Ibid., 3.

44 Institute for Sustainable Forestry, *Marketing Assessment of Certified Sustainably Harvested Forest Products* (Redway, CA: Institute for Sustainable Forestry, 1995), 4-5

45 Ibid., A-1. A more recent study of US companies that sell FSC-certified wood and those that do not supports the contention that interest in certified products increases as one moves down the supply chain. See James Stevens, Mu Dariq Ahmad, and Steve Ruddell, "Forest Products Certification: A Survey of Manufacturers," Department of Forestry, East Lansing, MI, mimeo, 1998.

46 Silva Forest Foundation, *Directory of Wood Producers and Manufacturers for Potential Eco-Certification* (Slocan Park, BC: Silva Forest Foundation 1996), 7-8.

47 F. Gale and C. Burda, *Attitudes Towards Eco-Certification in the BC Forest Products Industry* (Victoria: Eco-Research Chair of Environmental Law and Policy, University of Victoria 1996). The report surveyed 213 primary lumber manufacturers using the *BC Forest Directory 1995 Edition* (Burnaby, BC: Independent Directories Inc. 1995); 381 secondary manufacturers included in B. Wilson and R. Ennis, *Directory of Secondary Manufacturing of Wood Products in British Columbia* (Victoria: Forestry Canada, Pacific Forestry Centre 1993); and 550 retailers listed in G.R. Tracey, ed., *1995 BSDA Directory* (New Westminster, BC: BSDA of BC 1995).

48 S. Berg, "Certification and labelling: A forest industry perspective," *Journal of Forestry* 93(4) (1995): 30-32.

49 Officials of the BC Ministry of Forests, for example, did not endorse the FSC initiative, did not believe in rapid changes in forest management practices, and did not consider the FSC's members to be important stakeholders. World Forest Institute, supra, note 31 at 2.

50 Elliot and Hackman, supra, note 20.

51 Ibid., 39.

52 Ibid. Elliot and Hackman report, for example, that major Swedish forestry companies have committed themselves to working towards FSC's performance-based certification standards.

53 FSC, "New coordinator at the Canadian Working Group of the Forest Stewardship Council" (Toronto, Ontario: Canadian Working Group of the Forest Stewardship Council n.d. [ca. late 1996]).

54 For a detailed account of BC's export dependence on low value-added commodity products, see C. Burda and F. Gale, *Trading in the Future: An Examination of British Columbia's Commodity-Export Strategy in Forest Products* (Victoria: Eco-Research Chair of Environmental Law and Policy, University of Victoria 1996).

55 For accounts of the softwood lumber dispute, see R. Hayter, "International trade relations

and regional industrial adjustment: The implications for the 1982-86 Canadian-US soft-wood lumber dispute for British Columbia," *Environment and Planning A* 24(1) (1992): 153-70; C. Yoder and W. Gilliland, "The Legal Context of Canada-U.S. Trade in Forest Products," in *Canada-United States Trade in Forest Products,* ed. R.S. Uhler (Vancouver: UBC Press 1991): 123-34; and L. Constantino and M. Percy, "The Political Economy of Canada-U.S. Trade in Forest Products," in *Canada-United States Trade in Forest Products,* 57-72.

56 Justine Hunter, "MacBlo to end clearcutting in old-growth forests," *Vancouver Sun* (10 June 1998), A1-A2.

57 For example, Western Forest Products Limited,a subsidiary of Doman Forest Products, which manages more than 850,000 hectares of public and private forest lands on the BC coast, recently announced its intention to become certified under the FSC process. See WFP, "WFP to Pursue Green Certification," Western Forest Products Limited, press release, Vancouver, 3 June 1998.

Part 4:
Legal Barriers to Sustainable Forestry

Preceding chapters have chronicled a variety of obstacles that tend to impede the transition to sustainable forestry, including political opposition, bureaucratic inertia, and daunting problems of institutional design and policy implementation. The purpose of this part is not to revisit these obstacles. Rather it is to consider whether and, if so, to what extent domestic and international legal regimes foreclose the adoption of sustainable forest policies of the type identified earlier in this collection.

In the United States, sustainability-oriented law and policy reforms are significantly constrained by constitutional protection of private property, in particular, by the requirement that government compensate private property interests adversely affected by environmental regulation. In chapter 13, Cohen and Radnoff consider the emerging and as yet unresolved question of whether, and to what extent, Canadian governments will be subject to analogous requirements.

Another area of continuing uncertainty and debate concerns how existing and future international trade agreements may affect the ability of national governments to implement sustainability policies in the resource sector. In chapter 14, Gale tackles this difficult question by considering how ecosystem-based forest management might be constrained by operation of the North American Free Trade Agreement (NAFTA), the General Agreement on Tariffs and Trade (GATT), and the World Trade Organization (WTO).

13

Regulation, Takings, Compensation, and the Environment: An Economic Perspective

David Cohen and Brian Radnoff

The right of an owner to use his land is not absolute. He may not so use it as to create a public nuisance, and uses, once harmless, may, owing to changed conditions, seriously threaten the public welfare. Whenever they do, the legislature has power to prohibit such uses without paying compensation.[1]

Introduction

Since the eighteenth century, support for the protection of private property rights has been ubiquitous. This support has always been and continues to be premised on sound principles. Such protection has encouraged economic growth and material progress unimagined by previous societies. It has also fostered freedoms and acted as a check upon state intrusion into private life. Notions of the importance of private property have become imbedded in our collective consciousness, to the extent that it is generally assumed any activity that derogates from the institution of private property will have a deleterious effect upon society. Hence, there has been a long-standing tradition through the common law or explicitly through constitutions to protect private property from government intrusion.[2] Even where it is thought necessary and in the interest of all citizens, the state can take private property only if it pays fair compensation in return.

The past 200 years have seen not only a growth in our prosperity and freedom, but at the same time a growth in the power and size of government. It has not taken long to realize that the institutions, like markets and private property, that formed the basis of the economic organization of our society were not perfect. In liberal democracies, it has been natural to turn to the state to remedy these defects, while attempting to keep the core of these institutions intact. However, these imperfections often require complex and involved solutions, that in turn require a substantial quantity of regulation. Through time, this regulation has become much more sophisticated, and the state is able to use more subtle means to accomplish its goals.

The increased intervention of the state in liberal democracies has resulted in a recognition that regulation can affect private property in almost the same manner and to the same extent as if the property had been taken by the state. If the state has to compensate when it physically takes possession of property or effects a legal transfer of property rights, then compensation should also be necessary when the same is accomplished through regulation. The essential insight is that both regulation and physical takings cause a loss of wealth or welfare to the owners of property, shifting that wealth or welfare to others – in fact, a physical taking can be seen as a special case of regulation directed towards land.[3] However, this realization has left important questions unanswered. Exactly when should we decide that a regulation becomes a taking? The strongest defenders of private property might claim that any reduction in the value of land caused by regulation must be compensated. But such a policy could have a severe effect on the ability of government to regulate efficiently, or in some cases, to regulate at all.

At the same time, there has been a growing awareness for the past quarter century of the importance of environmental issues to our society. The material wealth and prosperity we have enjoyed has come at a cost, a cost we are frequently reluctant to consider when making decisions. It has become patently obvious that protection of the environment may necessitate increasing levels of regulation. This is not to say that all progress must be checked – merely that individuals who make decisions that affect others must bear the full cost of those decisions. Unfortunately, it is precisely in this environmental context that we observe the greatest conflict between private property rights and the need to regulate. Regulation that protects the environment often involves some restriction on how private property owners might use their land. The question we set out to answer in this chapter is when should private property owners be compensated for regulations that, while having a positive impact on the environment, lower the value of their property.

This question is important for several reasons. The need to pay compensation has an obvious effect on the policy choices that governments make regarding the type and quantity of environmental regulation. The ability to receive compensation will have a corresponding effect on the choices that the owners of property threatened by regulation will make. In turn, our society as a whole will be affected through the quality of the environment and the level of investment and economic activity. The law in this area is still developing and we must be careful that it develops in the correct direction.

In "Legal Rights to Compensation for Takings," the first section of this chapter, we examine the current law in Canada regarding those rights. We argue that the current uncertainty of the law in this area only results in lengthy and costly determinations in an increased number of disputes, and must be remedied immediately. The following section, "Alternative Frameworks for Analyzing Compensation Policy," examines two of the more common frameworks used to analyze the regulatory takings problem. These frameworks will be rejected for lack of certainty and for creating perverse incentives in both the public and private sectors. "The Economic Framework" then provides an economic analysis of compensation in the case of environmental regulatory takings. This framework informs the discussion for the rest of chapter. We argue that the economic framework allows us to conclude that, in the case of regulatory takings, a presumption against compensation should prevail over a policy of compensation. A presumption against compensation provides certainty and credibility that the current law does not. Although such a presumption is not perfect, it works better than government compensation to preserve efficiency and environmental norms. "Considerations Favouring Compensation" expands the issue to include other concerns that have been raised to justify the payment of compensation. This section will conclude that although many of these concerns are valid, they do not lead to the rejection of a presumption against compensation. In fact, when considered in an environmental context, many of these concerns have less impact than they might in other contexts. Finally, "Compensation Concerns in an Environmental Context" examines other considerations that have importance in an environmental context. These considerations reinforce the general presumption against compensation, and also illuminate the contexts when specific adjustment policies for those affected by environmental regulation are justified.

This issue is especially critical given the current debate in British Columbia regarding the creation of a sustainable forest industry. Several of the contributions in this collection discuss the necessity of reforming the tenure system in British Columbia if there is to be any hope for sustainable forestry. Other new policies, furthering the end of sustainability, might have a dramatic effect on the manner in which forestry is carried out in British Columbia. The probable result of either tenure reform or other changes to the forest industry will be significant compensation claims by those parties adversely affected, at least partly on the basis of regulatory expropriation. The law regarding compensation for regulatory expropriations is too uncertain to be able to predict what the outcome of such claims will be. We hope to clarify this issue by making a strong normative argument, using economic analysis, that, presumptively, no compensation should result from regulatory expropriation claims arising from reforms leading to sustainable forestry. However, the conclusions of this chapter are much more far-reaching than the forestry industry in British Columbia. Economic and environmental norms support a presumption against compensation in the case of any environmental regulation. Given the current importance of environmental concerns, it is imperative the law and public policy move in the direction suggested here.

Legal Rights to Compensation for Takings[4]

Until 1978, property holders in Canada adversely affected by state action could obtain compensation in only two narrowly defined circumstances: first, if the regulatory action resulted in a formal expropriation of their ownership interest;[5] or second, if the property was injuriously affected.[6] Other forms of regulation resulting in a devaluation of property, such as municipal rezoning, were not compensable,[7] and the doctrine of injurious affection would only provide compensation where the construction (and not the operation) of a government undertaking resulted in physical interference with property. However, since 1978, Canadian courts have begun to expand the circumstances in which compensation will be payable to private property holders adversely affected by regulatory action. These decisions, far from relying on any implicit constitutional protection of property rights, have been grounded in common law doctrines.[8]

In the first of these cases, the Supreme Court of Canada, relying on nineteenth-century doctrines of statutory interpretation, awarded compensation in a case in which the federal government established a Crown corporation and created a monopoly in favour of that entity to regulate the trans-border shipment and sale of fish in Manitoba.[9] The importance of this decision hinges on the fact that the federal government did not formally expropriate the plaintiff's property but rather enacted legislation – the Fresh Water Fish Marketing Act,[10] – which used a Crown corporation to regulate the relevant industrial activity. The Court, interpreting the federal government's action as depriving the plaintiff of the goodwill of its business and effectively transferring it to the Crown through the creation of a monopoly, held that such an action was an expropriation of property.[11] Hence, the precedent was set that regulation could result in a compensable taking; moreover, the property need not be land, or even a tangible asset, as goodwill qualified as a type of compensable property.[12]

A further expansion occurred in 1984 when the Supreme Court of Canada, in *British Columbia* v. *Tener*,[13] awarded compensation to a plaintiff who claimed that economic losses were incurred as a result of the implementation of a provincial parks program by the government of British Columbia. In 1973, Wells Gray Park was reclassified by the provincial government in order to preserve its natural resources unless and until a park use permit was issued by the government. The plaintiff, David Tener, had invested considerable sums in preliminary development of certain mineral resources which he and others owned in the park. From 1973 until 1977, he engaged in several unsuccessful attempts to obtain the necessary government, authorizations to mine and exploit the area. Finally, frustrated with five years of what he considered to be bureaucratic obduracy and having received notice that a permit would not be issued, he sued the British Columbia government claiming compensation for the capital value of the mineral resources, his wasted expenditures, and his anticipated profits from exploitation of the mining claims.

At the Supreme Court of Canada, the majority judgment was delivered by Mr. Justice Estey. He supported Tener's proposition that the right to minerals, including the right of access necessary for their development, was a property right. Estey J. also held, as he was forced to by the logic of his reasoning regarding the "property structure," that the denial of access to the mineral resources constituted a recovery of part of the rights granted by the Crown in 1937 and that a taking occurred because the value of a state asset – the park – was thereby enhanced. Defining Tener's right in this manner then allowed him to invoke the statutory construction that "a statute is not to be construed so as to take away the property of a subject without compensation."[14] Madam Justice Wilson followed a similar course of argument and stated that "the vice aimed at is expropriation without compensation."[15] In her view, the denial of the access permit deprived Tener of his *profit à prendre* and constituted a taking insofar as the deprivation effectively resulted in the Crown removing an encumbrance from its own property. As Wilson J. said, "It would be quite unconscionable to say that this cannot constitute an expropriation in some technical, legalistic sense."[16]

A recent British Columbia decision has confirmed the principle enunciated in *Tener*. In *Casamiro Resources Corp.* v. *British Columbia (Attorney-General)*,[17] the British Columbia Court of Appeal applied *Tener* to a case where the holder of mining rights had acquired them after the creation of a park, which was subsequently reclassified, and where the holder of those rights did not own the surface mining rights. Southin J.A. held that the reclassification had the effect of "reducing the Crown grants to meaningless pieces of paper."[18] Therefore, there is authority to suggest that regulatory takings by the state require compensation to property owners adversely affected by those takings. The Supreme Court also seems to have taken an expansive view of the types of compensable property.

However, there are several exceptions to this general principle. Cases like *Genevieve Holdings* v. *Kamloops*,[19] and *Steer Holdings Ltd.* v. *The Government of Manitoba and the City of Winnipeg*,[20] suggest that compensation will not be payable where governmental regulation leaves the landowner with property which retains some value, albeit limited. Even if the potential value of an unregulated landowner's interest – recognized in negotiations by a municipal government – is drastically reduced by provincial legislation prohibiting the pursuit of development plans, compensation will not be payable absent a formal taking.[21]

The courts have also tended to maintain a purely formalistic view with regard to whether the property right affected is one recognized at common law – where compensation would be payable[22] – or whether it is entirely a creation of statute – where, in the absence of express wording requiring it, compensation would not be payable. The Manitoba Court of Appeal, in *Home Orderly Services Ltd.* v. *Government of Manitoba*,[23] held that the decision of the Manitoba government to replace the plaintiff's home care services with direct provision of the same service by the government itself did not constitute an effective expropriation of private property, although the plaintiff's business was, in any meaningful sense, terminated by the decision. The Court found that the plaintiff's entitlement was enjoyed at the sufferance of the public and did not exist independently of the government's option to contract out for the service in the first place. The distinction this decision highlights is that between a "property entitlement" recognized at common law and a contractual or statutory entitlement derived through legislation.[24] In *Manitoba Fisheries* v. *The Queen*, the Court was able to find that it was protecting rights originating in the common law and grant compensatory relief. In *Home Orderly Services*, the decision was not to protect and not compensate for what the Court found to be rights derived simply from governmental action.[25]

Recently, this distinction was applied by the British Columbia Court of Appeal in *Cream Silver Mines Ltd.* v. *British Columbia*,[26] in which the Court of Appeal overturned a decision of the British Columbia Supreme Court which had held in favour of the plaintiff's claim to compensation.[27] The Court of Appeal noted that the rights in question (that is, recorded and located claims) differed from the rights inherent in the Crown-granted mineral claims of *Tener* and *Casamiro* in that the former, unlike the latter, were solely creations of statute and were not recognized at common law. Nor were they capable of registration under the British Columbia land registry system. Southin J.A., writing for a unanimous Court, upheld the appeal of the British Columbia government on the basis that the legislature had evinced no intention to compensate holders of this type of claim when exercising its regulatory powers.[28]

Finally, even if there is no statutory mention of whether compensation will be payable upon an expropriation or a regulatory taking, the courts may still deny compensation if it can be determined that the purpose of the statute implicitly denies it. In *B.C.M.A.* v. *British Columbia*,[29] the British Columbia Court of Appeal denied compensation to doctors who were not permitted to balance bill.[30] Lambert J.A. stated that the rule of statutory interpretation requiring compensation in the absence of express wording to the contrary "is not a device by which the courts can enable a claimant to outwit the Legislature."[31] Based on the language of the Medical Service Plan Act, the fact that this was a legislative taking and the context in which the Act was passed, Lambert J.A. was satisfied that the legislature intended to regulate the practice of balance billing without permitting any compensation. In essence, Lambert J.A.'s judgment demonstrates that where regulation is designed precisely to remove a particular welfare gain, it is absurd to argue that the gain should be recovered or reinstated through compensation from the regulating authority.[32]

This discussion of the recent Canadian case law demonstrates the potential availability of compensation to the owners of property and interests closely identified with property when the government, in a specific regulatory action, devalues the ownership interest to such an extent that it is effectively rendered worthless. This is true only in the absence of statutory wording to the contrary. In this area, the law seems to be approaching the position taken in the United States with regard to regulatory takings.[33] However, compensation might not be available when the regulation leaves the landowner with some limited use, even if the property is drastically reduced in value. Nor might compensation be available if the regulated right is a creation of statute or merely the granting of a discretionary privilege, like a quota or licence. And the courts may still deny compensation where there is no statutory wording to the contrary when it can be established that the statute implicitly denies compensation.

This rather complicated position reflects the complexity of the regulatory takings issue and judicial inability to resolve the fundamental concerns involved. The court is aware of the significant economic costs that are incurred by individuals, communities, and local economies when the state regulates. There is also a "fairness" sentiment, reflecting distributive justice ideals, that individual property holders should not be obliged to bear the entire cost of a government program.[34] But there is also a recognition that requiring compensation can impose substantial transaction and settlement costs on governments, thereby hindering necessary regulation. The record of recent cases indicates that the judiciary recognizes many of the problems but has not been able to develop a general framework to resolve the issue. Decision-making seems to have occurred on a case-by-case basis.[35] As the courts realize that perhaps they have gone too far in expanding the ambit of compensation, they seem to have retreated to deny compensation in very similar circumstances when it was previously available.

This deliberate retreat might be the only logical way to explain the different results in, for example, the *Casamiro* and *Cream Silver* decisions. The distinction between a right recognized at common law and a right created by statute is untenable. A real property right can be created by statute and an individual would treat such right as equivalent to any other property right. Why (apart from a judicial bias in favour of judicially created rights) should such a right be treated differently for the purposes of compensation?[36] Moreover, the argument that only a registrable right is compensable begs the question. There is nothing magical about a registrable right; it can be registered merely because the legislature has determined that it can be. The court is not determining why certain property rights should be compensable as compared to others, but engaging in an exercise in judicial line-drawing. The results of these two cases can be better explained by their timing than by any resort to formal reasoning.[37] The court's exception for regulations that do not destroy all the value of the property also makes little sense. Such an exception could result in few compensated regulatory takings, because rarely does regulation prohibit all potential land uses, while at the same time resulting in a large transfer of wealth to the state from owners affected by regulations. Although these distinctions and exceptions allow much flexibility in deciding individual cases, they provide no guidance and result in great uncertainty.[38]

The result of this uncertainty is higher costs for all the parties involved. There are an increased number of disputes that must be settled through a very expensive litigation process. Both the government and landowners face increased risks and difficulty in making efficient future plans. Such uncertainty is a function of the complexity and novelty of compensation for regulatory takings. As the law becomes more mature, one might expect more definite rules from the courts, although this process can be long and costly. The law in the United States has gradually developed over several decades, and is still far from settled.[39] It is possible that the law in Canada might continue to develop parallel to that of the United States, resulting in an increasing expansion of the availability of compensation in the event of regulatory takings.[40] A recent report for the government of British Columbia recommended an expanded right to compensation in the case of regulatory takings.[41] Moreover, the increasing conformity of Canadian law to American law might be accelerated by the provisions of the North American Free Trade Agreement.[42] Given the importance of this issue to the development and interpretation of environmental policy, and the impact of the resolution of the policy debate on economic investment and environmental quality, it is important to create a framework that might help to answer if and when compensation should be available in the event of regulatory takings. It is to the creation of such a framework that we will next turn.

Alternative Frameworks for Analyzing Compensation Policy

In the preceding section, we argued that the courts have been unable to arrive at a satisfactory framework for analyzing the availability of compensation in the case of regulatory takings. Their solution, it seems, has been to take a flexible case-by-case analysis that will continue to result in increased uncertainty and higher costs. In this section, we will examine some other potential frameworks for analyzing the compensation issue. We will discuss the "police power" regulatory framework which is employed by the American courts. Then we will go on to discuss generally the wisdom of using a framework based on either the type of regulation or the type of property regulated. Both frameworks will be found wanting, leading to the next section, in which we develop an economic framework that we argue should be used to formulate compensation policy in the case of regulatory takings in an environmental context.

Compensation and the Police Power

It is generally accepted that the government has the power to engage in the regulation of activity that is causing harm to others. This use of the "police power" by the government to prohibit or otherwise constrain harmful activity does not give rise to compensation. The Supreme Court of the United States has indicated that compensation is not payable where the government regulates activity that would have constituted a private nuisance at common law if left unregulated.[43] The reverse of this argument, also generally accepted, is that when the government regulates in such a way so as to provide benefits to others, compensation should be required.[44]

This analysis is to some extent useful. It does seem morally justifiable to allow the state to regulate without compensation in order to prevent harmful activity.[45] However, the problem with the police power framework is that it is impossible to distinguish government action designed to regulate harm-causing behaviour from government action that creates benefits for other members of the community.[46] It is for this reason that the distinction has been continually criticized.[47] The attempt to use a formal nuisance test to determine if a regulatory action will result in compensation is similarly impractical. Such a test can be defended only if one assumes that determining whether there is a nuisance is a formal exercise, and that legal definitions of nuisance do not have political, economic, and normative content themselves. Such a test would mean that most regulatory measures aimed at addressing a wide variety of environmentally harmful activity will give rise to compensatory rights. Measures to protect ecosystem biodiversity, to regulate carbon dioxide emissions, to preserve old-growth forests, and so on, would all constitute takings rather than the use of the police power. Only regulatory measures that protect the possession and use and enjoyment of property – interests protected through the law of private nuisance – would constitute non-compensable regulation.

From an economic prospective, the distinction between causing harm and creating benefits is also unhelpful. If one assumes that the benefits from regulation outweigh the costs to those regulated, all regulation is indistinguishable because it all results in a net benefit to society. At that point the only question becomes the effects of different distributions of those net benefits on the actors involved. This question, of course, is only another way of asking when and how much compensation should be available in the event of a regulatory taking. If the benefits do not outweigh the costs, no regulation should be undertaken and so the compensation question should not arise.[48] Therefore, the police power framework results in an unworkable distinction that is also problematic from an economic perspective.

Frameworks Based on the Type of Regulation or Property Interest
The police power framework can be seen as a specific case of a framework based on the type of regulation enacted. If the regulation prevents harm-causing behaviour, no compensation is warranted. Hence, it is an example of using a framework based on the creation of a category to determine when compensation should be available. The two main categories that one

might rely on in formulating a framework of analysis are those based on the type of regulation or those based on the type of property interest.

Basing compensation decisions on the type of regulation results in an analysis that necessitates the use of distinctions. Some distinctions, although they might make sense in theory, like the police power distinction, are impossible to apply in practice. Conversely, other distinctions that can be applied in practice have no underlying justification. For example, for a long time, the government was required to formally expropriate land before compensation would be payable. The move towards acceptance of the concept of regulatory takings has been a recognition that the distinction between a regulation that destroys the value of land and a formal taking makes no sense in practice. However, it equally makes no sense to recognize only the regulatory takings that destroy all the value of land, because as long as the state regulates so as to leave some value, there would be no requirement for compensation. The result of such a policy would be to countenance large-scale redistributions of wealth from property owners to the state without recourse to the property owners. All regulation affecting land has the same effect and the difference is a matter of degree.

Moreover, distinctions based on the form of regulation are not justified because of their effect on the owners of property. When individuals undertake investment decisions, they are unconcerned about the form of regulation. Any regulation that adversely affects their property will lower the value of that property. Consequently, when making their investment decisions, the possibility of government regulation will be a cost that owners will take into account.[49] The precise form of government regulation that happens to increase costs, devalue assets, or reduce expected income will be irrelevant. Investment decisions will factor in the expected reduction in the profitability of investments without regard to the formal instrument the government employs to achieve its regulatory objective.

Distinctions based on form can create certainty insofar as all interested parties know under exactly what conditions compensation is available. But governments will have an incentive to avoid regulatory decisions that will lead to compensation because these decisions will impose extra costs on the use of regulation. The array of regulatory instruments from which governments can choose a means to implement policy is limited only by the imagination of the regulator.[50] It is not difficult to predict that, when firm lines are drawn, the state will only regulate, as far as it is possible, so that it is unnecessary to pay compensation. This outcome might result in less than optimal regulatory policies if the most efficient regulatory policies would trigger compensation; alternatively, less efficient regulatory instruments may be selected where they minimize the costs associated with compensation. Another result might be a less than optimal quantity of regulation if most forms of regulation required compensation. Whether such effects exist and their magnitude depends on where the distinction is made and what the costs and availability are of alternative regulatory schemes. Regardless of the potential adverse effects of determining compensation based on the form of regulation,[51] the fact remains either there will be no independent justification for doing so, or the distinction employed will be unworkable in practice.

Determinations based on the type of property right being expropriated also lead to similar problems. Knowledge that some ownership interest is compensable while other interests are not would result in the non-compensable property having less value. Economic distortions would result if there were no economic justification for the difference in value. Moreover, there would be an incentive to regulate only non-compensable interests, even if more efficient results could be achieved through regulation of compensable interests. The strength of such incentives once again depends on the costs involved.

The forestry industry in British Columbia provides an instructive example of the problem of focusing on the type of regulation or the type of interest regulated. For example, a regulator might be able to reduce output levels by increasing royalty payments a licensee must make, or by imposing stricter biodiversity guidelines. If both raise costs equivalently for licence holders, they will be indifferent to the policy chosen. However, if the increase in royalty payments is not a compensable regulation, while the new biodiversity guidelines are, the regulator will have an incentive to choose the former instrument. If the new biodiversity guidelines have a more salutary effect towards promoting sustainability, focusing on the type of regulation leads to less optimal regulation. Similarly, the provincial government could determine that private landowners deserved compensation for regulatory expropriations, while licensees did not. If the primary value of the land in either context is the value of the old-growth timber resources, licensees would face a greater risk of a loss compared to private owners. In addition, regulators would have an incentive to regulate licensees, but not private landowners. Due to the focus on the type of interest regulated, timber resources on the private land would be more valuable than the same resources on the licensed land, and more intensive investment would take place on the private land, even though there is no economic justification for this difference in value or investment.

The point is that all regulation has the same effect, a reduction in the value of the regulated property. There is no justification for focusing on the form of the regulation or the type of the interest regulated. While this approach might provide increased certainty, it would not produce the optimal quantity or quality of environmental regulation, and it could cause economic distortions. We should find a framework that does not depend on creating distinctions in order to draw lines, but one that provides a strong normative basis for making decisions that result in positive consequences.

The Economic Framework

In this section, we outline the framework that will serve as the basis for the remainder of this chapter. We begin by introducing an economic analysis of compensation for environmental regulatory takings, and conclude that the trade-off between risk and investment incentives is the fundamental insight into the compensation issue which economic analysis provides to policymakers. If it is possible to solve the problem of risk created by the threat of uncompensated takings, either through diversification or insurance, the efficient solution is to not compensate for regulatory takings. This leads into a general discussion of the possibility of insurance for this type of risk. Although such insurance is not generally provided through the market, it is our position that there are few impediments to the provision of such insurance. To the extent that difficulties in providing insurance cannot be solved through the market, it is unlikely that relying on the government to mitigate the risk of a regulatory taking will improve the situation. Finally, we examine some issues relating to transaction costs that are raised by a no-compensation presumption. In general, such a policy should lead to lower costs, thereby providing further support for such a presumption.

Our adoption of an economic framework is not meant to indicate that economics provides the only or most important means of analyzing regulatory takings. Economic analysis is burdened with assumptions that often make conclusions difficult to apply to the real world. Moreover, efficiency is certainly not the only norm through which to judge policy decisions. We use an economic framework as a starting point, and then go on to examine several other important issues implicated in this question. In the end, however, we are able to conclude that the efficient solution to the compensation problem results in a policy that provides few impediments to effective environmental protection.

The Trade-Off Between Risk and Investment Incentives[52]

It is common to find the terms "reliance" and "reasonable expectations" used in legal analyses of the question of compensation for regulatory takings.[53] These terms have definite economic connotations, but they should play no part in any economic analysis of compensation for takings.[54] Arguments justifying the payment of compensation based on reliance are inherently circular; certainly, the government could announce tomorrow that it would no longer compensate for takings and no property owner could argue from then on that he or she was relying on compensation in making investment decisions. Arguments concerning expectations are similar to reliance in that credible changes in policy can change expectations. There is a further question: what constitutes a reasonable expectation?[55] One might argue that a property owner has a reasonable expectation of the ability to derive maximum income from his or her property; however, one could also argue a property owner has no reasonable expectation to be able to visit harms on third parties without their consent. Perhaps the most important point to be made against reliance and expectation arguments is that they represent appeals for the continuation of the status quo. In an ever-changing world, such appeals make little sense. In an environmental context, with constantly improving information on what constitutes optimal protection, such appeals are nonsense.

The real economic effect of regulatory takings can be illustrated through an example. Consider a firm that owns an interest in a piece of land rich in timber resources. The government requires all firms that cut down trees to obtain a licence, and as part of the terms of that licence a fixed royalty must be paid to the government based on the value of the trees cut down. However, the piece of land lies in an environmentally sensitive area, one in which the current royalty rate does not fully account for the cost of extracting those timber resources. At some future date, the government might either enact legislation to regulate strictly the quantity of resource extraction allowed on the land or raise the royalty rate to account for all the costs of the timber extraction.[56]

The result of the threat of government regulation is to impose a risk of a loss on the owners of this land.[57] The imposition of risk is generally undesirable because it creates disutilities[58] and can result in inefficient investment.[59] Hence, the owners of the firm will want to adopt some strategy for mitigating the risk of regulation. Insurance and diversification are common strategies for mitigating risk.[60] When the government provides compensation for takings, this compensation represents a form of insurance because the government is protecting owners from any loss caused by regulation. With full compensation for the value of the loss caused by the regulation, any risk the firm faces is completely mitigated. Thus, government compensation of losses experienced by the owners of land caused by regulation serves to solve the problems caused by the risk created through the threat of regulation.

However, compensation acts like full insurance in that the owners of the firm no longer will bear any of the risk of regulation, and this fact will be reflected in the investment decisions of the firm. Compensation externalizes the cost of investing in resource sectors subject to the risk of government regulation to all taxpayers.[61] The result is that investment in the forestry venture will become more attractive; but if such investment is harmful to the environment, generating costs in excess of the value of the extracted resources, it should not be encouraged.[62] When the cost of regulation is not internalized, investments might be made that would not otherwise have occurred and these investments will be destroyed if the government must later regulate.[63] Government compensation results in over-investment, because investors do not bear the real costs of their decisions.[64]

To solve the compensation question, it is necessary to examine both risk and incentives simultaneously.[65] The problem can be solved if one assumes that government risks and risks faced by participants in the market are essentially similar.[66] This assumption seems a fairly accurate reflection of the real world. The source of a risk should have no bearing on investment decisions; investors should be indifferent whether their investment is destroyed by government action, by an act of God, or by the action of competitors. For a rational investor, it is only the probability of the risk actually occurring that should affect the level of investment. The market has developed a variety of mechanisms to deal with a plethora of risks. If there is a market solution to the problem of the risk created through the threat of uncompensated takings, compensation becomes unnecessary. Investment decisions will accurately reflect the costs involved, meaning that over-investment will not occur, while investors will mitigate risk through the market. If the market can only partially or imperfectly solve the problem of risk, one might also analyze whether government action can improve the market response.[67]

In the forestry example, the threat of an uncompensated taking can be dealt with through a strategy of diversification.[68] If the firm is owned by a large group of shareholders, the individual shareholders can diversify their portfolios. A large firm can also invest in other projects whose risks are unrelated to the venture in question.[69] Another method of mitigating risk is to purchase insurance. Unlike government-provided compensation, private insurance would have to be paid for by the firm that received it. Insurance companies would set premiums and deductibles to reflect the risk of a taking, and through the payment of these costs, the firm would internalize all the risk of investment. Private insurance currently exists for a variety of risks, but apparently it does not exist for the risk of uncompensated takings. It is necessary to determine why that is, if such an insurance market could develop, and if not what the government could do, if anything, to mitigate the risks of regulation.

The Availability of Insurance for Regulatory Takings

To the extent that diversification is not possible,[70] firms that wish to mitigate the risk of an uncompensated regulatory taking would need to purchase insurance. This chapter will not attempt to determine to what extent diversification is not possible by companies involved in projects that are potentially subject to environmental regulation. Many firms involved in resource extraction are quite large and would have little difficulty in diversifying to avoid risk.[71] However, environmental regulation affects not only large firms involved in resource extraction, but also smaller-sized companies and developers that are not publicly owned and whose ability to diversify is limited. For those firms who cannot diversify, the development of insurance to mitigate the risk of regulatory takings is critical.

Currently, there appears to be no private market for insurance against the risk of regulatory takings. This situation might reflect the fact that the government, to some extent, already offers such insurance in the form of compensation. However, the compensation offered, because of uncertainties with the law, would seem far from complete.[72] The question is whether an insurance market would develop if the government stopped providing compensation.

Commentators who argue that private insurance would not be provided to mitigate the risk of regulatory takings identify two main impediments to the formation of such a market.[73] The first is the problem of moral hazard; that is, to the extent that insurance covers losses, firms might engage in riskier activities because they have less incentive to avoid losses.[74] This impediment is merely a restatement of the fact that full insurance distorts what would otherwise be an efficient decision-making process. This problem does not exist only in the context of regulatory takings, but with insurance generally. In a takings context, where the loss is often directly proportional to the size of the investment, there should be little difficulty in setting an insurance premium that re-internalizes the cost of the risk.[75] If this step cannot be taken, another solution is to provide partial coverage by requiring deductibles or co-insurance so that the insured bears some portion of the loss.[76] Yet another solution is to monitor the behaviour of the insured and base full insurance on specified behaviour.[77] The market cannot always solve the incentive problem while spreading all the risk, but there is no reason to believe that the market will not provide the best trade-off between risk and incentives in the regulatory takings context, as it does in other contexts.

The other impediment to the provision of insurance is the problem of adverse selection. Adverse selection occurs because often the manner in which the probability of loss varies among individuals, although known by those individuals, is not ascertainable by the insurance company. This informational asymmetry means the insurance company either will not offer insurance or must set premiums in such a way so that a significant number of individuals might purchase less insurance than they otherwise would, or none at all.[78] This problem assumes that insurance companies will be unable to detect, at sufficiently low cost, differences in the probability of losses across those who might wish to purchase insurance. This outcome might be true in some contexts; however, it is doubtful whether it is true in the context of regulatory takings. In almost every instance, insurance companies would have equal access to information that individual owners possess to ascertain the probability of loss on an individual piece of land.[79] So long as insurance companies have equal access to information regarding the possibility of future regulation on a particular piece of land, the adverse selection problem should not arise.[80]

It is doubtful whether problems of moral hazard and adverse selection will keep private insurance markets from forming in a no-compensation regime. Over time, one should expect insurance to become available. However, because moral hazard cannot always be solved perfectly, full insurance might not always be provided. Government action could provide more complete coverage, but this point does not suggest that government compensation is warranted. It is likely that private arrangements can balance risk spreading and incentives in the most efficient manner.[81] Private parties would pay a higher price for more complete coverage if the distortions of incentives that would result were outweighed by the benefits of additional risk spreading. Hence, further risk spreading by the government beyond what would be provided by the market is undesirable; if more compensation was desirable, it would have been provided by the market.[82]

Transaction Costs

If, as we argued, insurance should be available to mitigate the risks of uncompensated regulatory takings, such insurance will be provided at a cost. All insurers experience administrative costs, including gathering information, writing contracts, and determining the amount payable in the event of a loss.[83] These costs might be substantial, especially in the case of low probability events.[84] This fact does not mean that insurance will not be offered, although administrative costs might increase premiums sufficiently to discourage some individuals from purchasing insurance.[85]

The costs, though, must be balanced against the costs of any government compensation policy. Current compensation policy requires administrative expense only in the event of a loss.[86] However, this process requires a very expensive determination through judicial proceedings. The reduction in the value of the land in question is also very difficult to calculate, and owners have incentive to argue for very high losses.[87] Furthermore, if a firm's value has been diminished in a minor way as a result of government regulation, it will be expensive if not administratively impossible to distinguish the impact of government action from other factors.[88] In the current regime, costs before the loss are lower than in a no-compensation regime, but costs after a loss are much higher under current compensation policy.[89] Moreover, current policy is by no means certain, and seems very much to rely on case-by-case determinations.[90] The result is an increase in the risks borne by those affected by regulation, because it is impossible to determine in advance if compensation is available.

The greatest benefits can be reaped through a credible and certain compensation policy. Certainty will ensure that no extra risk is created through the process of determining the availability of compensation while a credible policy can have the desired *ex ante* effects on incentives.[91] The current system of judicial determinations in an adversarial process delivers neither of these benefits and does so at a high cost. A much better system, if firms are able to diversify and insure, would be a presumption against compensation in the event of environmental regulatory takings.[92] Such a policy would involve administrative costs, but it is doubtful they would be higher than under the current system of compensation. An alternative would be a system of compensation that did not rely on the courts. Such a system might be very similar to actual government-provided insurance and could be run in much the same manner as private insurance in order to preserve incentives.[93] However, government-provided insurance might not provide the optimal quantity and quality of environmental protection.[94] We conclude that allowing the market to mitigate risk under a no-compensation policy is the most efficient solution to the general problem of environmental regulatory takings.[95]

Considerations Favouring Compensation

The economic framework described above suggests that there should be a presumption against paying compensation in the event of regulatory takings. However, other factors that support the payment of compensation might affect or alter that conclusion. In this section, we examine the most common arguments used to support compensation in the event of takings. We begin by examining the potential social costs of no compensation: Michelman's "demoralization costs" and Dana's accelerated development thesis. The former, although potentially a factor in some situations, should not be important in the case of environmental regulation. The latter thesis depends on the inability of individuals to mitigate risks, an assumption which we find unrealistic. We then discuss problems specifically related to government in the context of regulatory takings, including the concept of fiscal illusion and considerations of fairness. It is unclear whether these problems exist at all, but it is clear that they do not exist in an environmental context. As a whole, this section serves to reinforce the general conclusion that the best policy is one of a presumption against compensation in the event of environmental regulatory takings.

Social Costs in a No-Compensation Regime

Demoralization Costs
Without compensation, regulatory takings have the effect of causing losses to a small number of people in order to benefit the community as a whole. Those who lose from this process might experience costs caused by the apparent unfairness of a majority of citizens using their power, through both legal and political institutions, to exploit and injure without redress. Michelman has described these as demoralization costs and he defines them in the following manner:

> the total of (1) the dollar value necessary to offset the disutilities which accrue to losers and their sympathizers specifically from the realization that no compensation is offered, and (2) the present capitalized dollar value of

lost future production (reflecting either impaired incentives or social unrest) caused by demoralization of uncompensated losers, their sympathizers, and other observers disturbed by the thought that they themselves may be subjected to similar treatment on some other occasion.[96]

Michelman's rule is that compensation should only be paid when demoralization costs exceed settlement costs. Settlement costs "are measured by the dollar value of the time, effort, and resources which would be required in order to reach compensation settlements adequate to avoid demoralization costs."[97]

The concept and existence of demoralization costs has been much discussed in the literature regarding regulatory takings.[98] Some commentators have argued that demoralization costs are merely another formulation for the risk of an uncompensated taking.[99] If this argument is true, the general economic framework deals with demoralization costs in its treatment of risk and incentives. A more plausible interpretation is to treat demoralization costs as distinct from other risks, caused by majoritarian exploitation which is "strategically determined" and therefore more costly than random losses.[100] In this view, even if insurance were to exist to mitigate the risk of regulatory takings, it could not replace or restore lost property rights taken by the majority, and hence would not relieve demoralization costs.[101]

One should question how much demoralization costs depend on preconceived notions of property rights. Those who hold very strong views about the importance of private property would seem to be the most affected by demoralization costs. Those who believe that rights to private property should be absolute will advocate compensation because such a policy restricts the state's ability to inflict a loss on an owner of property and preserves freedom.[102] If conceptions of property were to change to reflect growing knowledge about the environment and economy, it is likely that demoralization costs would have much less importance.[103] This question – whether property rights deserve special protection in a democracy[104] because of the link between property and liberty – is one of political philosophy and is beyond the scope of this chapter.[105] However, we advocate an instrumentalist justification for the protection of property rights, and it follows that when protecting property rights reduces welfare, there is no justification for protecting those rights.[106] In the case of environmental harm, protecting property rights only serves to encourage investment in the very activities that are generating harm. In our opinion, it is necessary to achieve some balance between the protection of private property and the protection of the environment.[107]

More significantly, it is unlikely that demoralization costs are particularly important in an environmental context. Property owners affected by environmental regulation will normally wield sufficient power to avoid suffering majoritarian exploitation.[108] Moreover, those affected who are concerned with the environment and are aware the regulations in question serve to protect the environment are unlikely to be as concerned with majoritarian exploitation.[109] People who care about other people and their community may not feel "exploited" if they know that what is being done is for the good of the community.[110] So long as awareness of environmental issues continues to grow, demoralization costs are likely to have a diminishing impact on individuals subject to environmental regulatory takings.

The Race to Develop

In a recent paper, Dana has argued that the absence of compensation would encourage property owners to accelerate development to avoid regulatory losses from future preservation legislation.[111] Compensation would serve to reduce this development and may facilitate preservation. Accelerated development occurs because owners of land, with potentially profitable investments, are concerned that future regulation will prohibit them from making these developments. By developing before regulation occurs, owners can make their investments and other land will be affected by regulation.[112] So long as their investments remain sufficiently profitable to justify developing early, and the risk of regulation is sufficiently high, this accelerated development will occur. Accelerated development creates costs in terms of allocative efficiency, in that investments are made before it is optimal to make them, and in terms of preservation because the quantity of undeveloped land is reduced.[113]

Although this thesis is interesting, it is highly unlikely such a race to develop would ever occur. Dana assumes in his model that markets for insurance against regulatory takings will not develop and that many firms will be unable to diversify.[114] As we have argued, it is quite probable that a private insurance market for regulatory takings would develop and that diversification would be an available strategy for many firms involved in resource extraction.[115] So long as investors can protect themselves against the risk of uncompensated regulation, there is no threat of accelerated development.[116] Moreover, any system of *ex ante* compensation would still distort investment incentives, involve large administration costs, and could have a negative effect on the quantity and quality of environmental legislation.[117]

Problems Associated with Government in a No-Compensation Regime

The analysis presented so far has proceeded under the assumption that government regulates efficiently so that the benefits derived from the environmental legislation in question exceed the losses imposed on individuals whose land is affected. There is nothing to guarantee this result, especially if it is assumed that governments, like private actors, will ignore externalities or discount the cost of activities that will not be faced directly.[118] This problem has been called "fiscal illusion." The problem of fiscal illusion is thought to arise particularly in the context of uncompensated takings because private individuals bear all the costs of the taking while the government bears none.[119] Through compensation, the cost of takings is internalized back to the government, resulting in more efficient regulatory decisions.[120]

The existence of fiscal illusion relies on the assumption that decision makers have a bias towards discounting costs more than they discount benefits.[121] Neither benefits or costs are borne by decision makers and it is unclear why one should be discounted in greater proportion than the other. It is likely that the direction of bias towards benefits or costs will depend on how dispersed or concentrated the benefits and costs of the regulation are, and the composition of the affected groups.[122] In the context of environmental regulation, fiscal illusion should not be problematic because the costs imposed are visited on small and often powerful groups while the benefits of such regulation are normally widely dispersed. Furthermore, regulation to aid the environment often results in long-term benefits that might only accrue when the current government is no longer in power.[123] Moreover, even if fiscal illusion exists, it is not clear that internalizing the cost of regulation would result in efficient regulation. Governments unconstrained by capital markets, labour markets, and product markets would not be affected by cost internalization.[124] Finally, policies of compensation could lead to attempts by rent-seeking individuals to manipulate programs resulting in a quantity of regulation and compensation greater than the optimal level.[125]

Related concerns regarding government in a no-compensation regime involve possible abuses of power.[126] It is unlikely that concerns with government behaviour warrant compensation; even if it did, the costs associated with that behaviour would have to be balanced against the costs associated with compensation before one could advocate a policy of compensation. If fiscal illusion or abuse of power do present real difficulties, instead of solving the problem through compensation, environmental regulatory decisions in a no-compensation regime should be subject to review.[127] Individuals affected by regulation could challenge it on the basis that it is not for a bona fide environmental purpose, or that the benefits of the proposed regulation are too small to warrant the losses that would be created.[128] The result of such an inquiry would be to uphold the regulation and the no-compensation result, or for the government to be forced to provide compensation or withdraw the regulation. The onus of proof in such an inquiry should rest with the government, because it would have the best access to the information that led to the decision to regulate. This process would increase administrative costs to a limited extent,[129] but those costs should be balanced against the benefits of increasing the likelihood of efficient regulation.

Compensation Concerns in an Environmental Context

In this final section, we outline some important considerations concerning compensation for regulatory takings in an environmental context. Although these considerations might arise in other contexts, examination of their importance in those areas is beyond the scope of this chapter. We begin by examining the negative effect a requirement for compensation can have on both the quantity and quality of environmental legislation. A presumption against compensation allows legislators the ability to achieve optimal environmental regulation. Regulation may represent public attitudes and environmental values in advance of the market. This being the case, it allows one to regard losses caused by environmental regulation as pecuniary externalities, which, like all such externalities, should not be compensated. Finally, we consider the effect of environmental regulation on those who are least able to protect themselves against the risk of uncompensated legislation.

The first part of the section also examines the effect on the small firm while the second part discusses the impact on affected industry workers and resource communities. It is here that compensation might be warranted out of considerations of fairness, although only in certain specific instances. Overall, examination of the environmental considerations serves only to strengthen the thesis that, generally, there should be a presumption against paying compensation in the event of environmental regulatory takings.

The Effect of Compensation on Regulatory Decisions

Problems of fiscal illusion and abuse of power are frequently considered in the regulatory takings literature.[130] We have argued that these concerns are not important in the environmental context, or can be solved if they do exist. Another important consideration concerning the environment is the impact of compensation on regulatory decisions.[131] Regulators who must make decisions as to how to regulate must also be aware of the fact that in certain cases compensation will need to be paid. As compensation becomes available in a greater number of regulatory takings, the need to pay compensation will become an increasingly important consideration for the regulator. It is likely that the requirement of compensation will have a negative effect on environmental regulation.

Compensation requirements will have an adverse effect on the quantity of regulation. Regulation that requires compensation can prove very costly to the government, which must pay the compensation upfront, but reaps the benefits in the long term or not at all.[132] Regulation that prohibits development can exacerbate the problem by creating a loss in tax revenue. Most governments today are under financial constraints and the ability to raise large sums of revenue to pay for environmental regulation is extremely limited. The result is that regulators will be very cautious in where they decide to regulate. Often, the societal benefits of environmental regulation will exceed the costs, but because the costs to the government will exceed any benefits it might receive, it will be reluctant to enact such regulation. Only in the most egregious cases, where there is a public outcry for action, will regulation be enacted, and even then it is likely budget-minded bureaucrats will be hesitant to do so. The result is an insufficient quantity of regulation due to the requirement of compensation, because it does not benefit the government to enact necessary environmental regulation.[133]

Compensation requirements will also have an adverse effect on the quality of environmental regulation. Compensation is currently available in only certain instances of regulatory takings, although the availability of compensation continues to increase.[134] For the reasons discussed above, regulators will avoid enacting regulation that requires compensation and will prefer less costly regulation that does not require compensation. Regulators will make choices based on the costs and benefits to government instead of the costs and benefits to society. One form of regulation might be optimal insofar as it delivers the greatest net benefits to society. However, another form of regulation that involves lower net benefits but lower costs to government might be preferred by the regulator. The result will be less efficient regulation being enacted in order to reduce government costs, providing less than the optimal quality of environmental protection.[135]

For regulators, the environment presents an important and complex problem. New regulation is constantly needed as new scientific information becomes available. Hampering regulators by requiring compensation will only serve to slow or perhaps even reverse the process of protecting the environment. Fortunately, a policy of compensation for environmental regulatory takings is unnecessary and economically inefficient. Moving to a presumption against compensation would help ensure that regulators are not hindered from enacting the optimal quantity and quality of environmental regulation.[136]

Environmental Regulation as a Mimic for Market Competition

Losses experienced by private actors as a result of changes in the demand for a product or due to legal competition by other firms are referred to as pecuniary externalities. These losses are never compensated, because they do not represent a change in the productive capacity of the firm but are, in fact, transfers of wealth that are necessary for an efficient market.[137] If losses due to regulatory change can be considered as pecuniary externalities resulting from collective decisions, compensation is not justified for the same reasons as it would be for pecuniary externalities caused by market exchanges.[138] This is the case when regulatory action is properly perceived as a reflection or manifestation of the community's preferences for and attitudes towards a firm's products. When the government regulates in an effort to force the internalization of social costs, or to reflect changes in private tastes for a particular product which, for reasons of market dysfunction, have not been reflected in the market, regulation can be seen as directly analogous to competition in the private sector, and, as a pecuniary externality, should not give rise to compensation.

This argument assumes that a collective decision to take a particular course of action is necessarily welfare-maximizing.[139] This assumption is incorrect if for any reason government decisions do not represent the desires of the society as a whole. For example, if government is not a conduit for citizen preferences, but is simply a vehicle for private interests to use the authority of the state to their personal advantage, government regulation would not represent a pecuniary externality.[140] However, in the case of environmental regulation, it is probable that the collective action (i.e., environmental regulation) does represent citizen preferences. The market failure that necessitates government action is an inability of the government to disseminate information it possesses regarding the environment.[141] Because the public is unaware of the new information, tastes do not change to reflect it, but they will as soon as the information is properly assimilated. The government, aware of this delay and the danger that in the time period before tastes change much environmental damage is possible, can decide to enact regulation now that will have the same effect as the eventual change in tastes.[142]

So long as the proposed environmental regulation maximizes social welfare, the payment of compensation will be a wealth transfer with no allocative effects. We do not believe it is unrealistic to suggest that most environmental regulations are welfare-maximizing. If new regulations are subject to some *ex post* review to ensure their efficiency,[143] one can be fairly certain that resulting losses represent a pecuniary externality and should not be subject to compensation. In general, then, a compensation policy for environmental regulatory takings would seem to be unwise.

Environmental Regulation and Small Producers/Resource Communities

The Small Producer[144]

Small producers are more susceptible to loss from environmental regulation because they are limited in their ability to mitigate risk. Small producers might not have sufficient assets to follow a diversification strategy; if insurance is not available or is too expensive, the small producers might have no ability to mitigate risk at all. Consider the previous example of the land containing valuable timber resources which might be subject to regulation.[145] If the firm is owned by "Owner," a single individual with total assets of $1.5 million, of which $1 million is represented by the value of the land containing the timber resources, Owner will have inadequate assets to diversify into other projects.[146] Owner can try to sell part of his or her interest to others, perhaps in the form of equity, in order to mitigate some of the risk. If this sale is not possible, the small firm still has the ability to mitigate risk by switching portfolios – in other words, selling the land entirely and reinvesting assets in less risky or more diversified investments. In a no-compensation regime, selling the risky asset and reinvesting in a more diversified portfolio is a realistic strategy if there is no other method to mitigate risk.[147]

The problem for the small producer is not mitigation of risk, which can be accomplished, but the loss in wealth associated with the change to a no-compensation policy. In the above example, before the change in policy Owner's land was worth $1 million, and now assume it is worth $750,000. If Owner sold the land to mitigate the risk of the investment, no one would pay more than $750,000 for it, which means a personal loss of $250,000 or one-sixth of his or her net worth. This reduction in net worth is, of course, merely a correction for the over-investment caused by previous compensation policy which distorted incentives. However, this fact is probably of little solace for an individual who has just experienced such a large loss relative to total wealth.[148] Considerations of fairness suggest that perhaps the owner ought to be compensated to some degree for the massive loss.

The need for compensation results from the switch to a no-compensation regime causing huge one-time reductions in wealth. Any compensation policy to correct for this problem would exist to remedy the one time reduction in wealth – other compensation would be unnecessary because from then on the risk of uncompensated regulation would be capitalized into the value of the land. Whether such losses would occur sufficiently often to warrant a general policy of compensation depends on the number of small firms with large investments relative to their total assets that might be subject to the risks of environmental regulation. It also depends on the extent to which the threat of uncompensated regulation was not previously capitalized into the price of the land.[149] It would seem to us that the number of firms that would suffer such a large relative decline in total wealth would be small and might justify compensation in certain "hard luck" cases, but they would certainly not warrant a general compensation policy.

The Resource Community

The small firm, limited in its ability to mitigate risk, is still able to do so because it can sell its assets and reinvest the proceeds. Individuals employed by the industry or firm affected by regulation are more constrained in their ability to bear the risk of legal change and more likely to absorb large losses relative to their total wealth as a result of regulatory reform.[150] The main asset that these individuals possess is their human capital and it is difficult to diversify investments in human capital; unlike the owner in the previous example, it is not possible to sell one's human capital and reinvest the proceeds elsewhere. Where regulations affect employees with specialized skills or those in isolated or economically depressed regions, it is likely these constraints will result in large losses to the individual employees.[151] The existence of large unions might provide some protection for employees affected by environmental regulation.[152] But not all employees are unionized and unions can only protect their employees to a certain extent.

The government already provides forms of assistance for workers in transition with policies like employment insurance. Such policies are only partial and will be of limited help to employees who, because they are isolated or have very specific investments in human capital, must undergo substantial relocation or retraining costs. A more specific government policy of providing compensation for workers displaced by environmental regulation would be problematic. It would be necessary to determine who qualified for the program and the amount of compensation,[153] which might be an administratively difficult and expensive process. Such a compensation scheme of cash payments might also not serve to solve the underlying constraints that created the problem in the first place, but might only delay the necessary displacement.[154]

The most sensible transition policy for this type of situation is a government-sponsored training program that can correct the underlying constraints causing unemployed workers the greatest losses. Such policies can retrain workers and relocate them to areas with better opportunities. Although in theory such a program makes sense, the empirical evidence on the effectiveness of such programs seems mixed.[155] Such programs can also be expensive to run and administer. If such policies are thought desirable, clear guidelines are necessary to determine where and when they would be available. They should only be available where there is a sufficiently large dislocation caused by environmental regulation to justify the intervention; and such programs should only be run where one can determine that there is a good chance for their success.[156]

Conclusion

Determining the correct compensation policy for environmental regulatory takings raises many complex issues that are impossible to solve perfectly. Nonetheless, it is important for the sake of our environment and for those who own land subject to the risk of an environmental regulatory taking that compensation policy be known, credible, and certain. Without such a policy, regulators face a much more difficult time enacting optimal regulations to protect the environment. Property owners face uncertainties and increased risks which will have an effect on the economy as a whole. The current law does not serve us well in terms of giving us a coherent framework with which to address all the issues involved. The case-by-case reasoning that seems to predominate judicial thinking on this issue is a very poor solution to a very important problem. We cannot be concerned with the form of the regulation or the property interest involved. We must recognize what the true effect of regulation on property is, and not be concerned with formal requirements like the existence of an actual taking. We must avoid using distinctions, like creating harm or providing benefits, that are impossible to apply in the real world.

We advocate a presumption against compensation in the event of environmental regulatory takings. Such a policy is economically efficient in that it best solves the trade-off between investment incentives and risk spreading. This policy is not a perfect solution because it relies on the market to mitigate risk for those subject to uncompensated regulation. We are confident that the market can achieve the best trade-off between incentives and risk spreading. But in the event the market fails to provide adequate risk spreading, government-provided insurance is preferable to compensation. In an environmental context, compensation is always a problematic policy because it encourages over-investment that causes harm to the environment. Only in certain specific contexts, for example, when dealing with small firms that suffer large losses in wealth, or employees who are constrained in their ability to deal with risk, might compensation ever be justified. Even then, other policies, like retraining programs for displaced workers, are preferable to direct compensation.

Efficiency and economic growth are not the only important considerations when forming compensation policy. The best policy solves the economic problems and encourages an optimal level of environmental protection. A presumption against compensation is especially effective here, because it relieves regulators from the necessity of considering compensation costs when making regulatory decisions. A presumption against compensation combined with an *ex post* review can help ensure the efficiency of environmental regulation while at the same time controlling problems of fiscal illusion and abuse of authority. Finally, when environmental regulation is regarded, as it must be the majority of the time, as a mimic for market preferences, it only creates pecuniary externalities that, on efficiency grounds, do not justify compensation. Thus, a presumption against compensation works best to preserve efficiency and to preserve the environment.

The greatest impediment hindering a presumption against compensation are current notions of property rights. Property rights played a central role in the formation of Western democracies and are still considered fundamental to the health of democracy and the economy. The idea of allowing regulatory takings without compensation would seem a great intrusion and interference into property rights. We advocate an instrumental notion of property, in which the protection of private property is only justified when it advances societal welfare. If private property is regarded as an end in itself, there is the potential for great harm to be visited on the environment when land use conflicts with protection of the environment. It is unlikely in a modern liberal democracy that the refusal to pay compensation when an environmental regulatory taking occurs will damage the institution of private property or the economy. Such a policy is merely a recognition that a balance must be struck between property rights and the environment.

More immediately, in the context of British Columbia, the drive for a sustainable forestry industry is intimately connected with the potential for compensation claims arising from tenure and/or other types of reforms. Whatever losses flow from these changes, it is important that great caution be exercised in providing compensation. Such compensation will be an impediment to making the necessary reforms to achieve sustainability. The fact that firms have been allowed, and perhaps even supported by the government, to practise forestry in a non-sustainable manner in the past should not create rights of compensation when new knowledge sheds light on destructive practices. The fact is that economic and environmental analysis does not support awarding such compensation, and it is unlikely that advocates of compensation can find an alternative justification that is convincing.

Notes

1 Brandeis, J., dissenting, in *Pennsylvania Coal Co.* v. *Mahon*, 260 U.S. 393 (1922) at 417.
2 The analysis of this chapter will focus on the law in Canada, where there is a common law statutory presumption that property is not to be taken without compensation. The United States, like Australia, has a constitutional provision (the Fifth Amendment) that requires compensation in the event of a taking. The Fifth Amendment reads: "nor shall private property be taken for public use, without just compensation."
  Section 51 of the Australian Constitution reads: "The Parliament shall, subject to this Constitution, have power to make laws for the peace, order, and good government of the Commonwealth with respect to: ... (xxxi) The acquisition of property on just terms from any State or person for any purpose in respect of which Parliament has the power to make laws."
3 Alternatively, a regulation can be seen as a special case of a taking of land. The effect of both is the same: to lower the value of the land to the owner.
4 Resource compensation schemes in Canada and many other jurisdictions (including the United States, Australia, and New Zealand) are canvassed and discussed in the Schwindt Report, a recent British Columbia commission on this issue. In general, all the jurisdictions canvassed affirm the principle that formal takings of property should result in compensation. See R. Schwindt, Commissioner, *Report of the Commission of Inquiry into Compensation for the Taking of Resource Interests* (Vancouver: Province of British Columbia 1992), A-1.
5 Unlike the United States, where there are constitutional guarantees at both the federal and state levels, there is no guarantee of compensation upon the expropriation of property in either the United Kingdom or Canada. Furthermore, there is no common law right to compensation. As Lord Parmoor stated in *Sisters of Charity of Rockingham* v. *R.*, [1922] 2 A.C. 315 (P.C.) at 322, "Compensation claims are statutory provisions. No owner of lands expropriated by statute for public purposes is entitled to compensation, either for the value of the land taken, or for damage, on the ground that his land is 'injuriously affected,' unless he can establish a statutory right." However, courts in Canada, in the absence of clear wording to the contrary, will apply a principle of statutory interpretation which raises a presumption in favour of the payment of compensation where property has been expropriated through government regulation. See E. Todd, *The Law of Expropriation and Compensation in Canada*, 2nd ed. (Carswell: Toronto 1992), 31-38. For a discussion suggesting the statutory presumption might actually be a common law right,

see R.J. Bauman, "Exotic expropriations: Government action and compensation," 52 *Advocate* 561. The principle that there is no common law or constitutional right compensation for expropriation was recently confirmed in *Cream Silver*, see infra, note 26 and in *Bell Canada* v. *Unitel Communications Inc.* (1992), 99 D.L.R. (4th) 533 (F.C.A.), in which the Court made clear it is strictly a question of the construction of the authorizing statute that will determine whether compensation will be payable and the amount that will be due. For a recent confirmation of the statutory presumption, see *Toronto Area Transit Operating Authority* v. *Dell Holdings Ltd.*, [1997] S.C.J. No. 6 (QL) at paras. 20-23.

6   Injurious affection denotes damage done to land through negligence, trespass, or nuisance, or in the course of the exercise of statutory powers. Examples of this latter category may include partial expropriation – where the remaining land suffers damage or is devalued due to its severance from the expropriated portion – or damage suffered due to actions upon neighbouring property – such as the construction and use of schools, parking lots, firehalls, and so on. See Todd, supra, note 5 at 328-93, and Bauman, supra, note 5 at 575-8.

7   For a discussion of the so-called re-zoning exception, see I.M. Rogers, *Canadian Law of Planning and Zoning* (Toronto: Carswell 1973), 123. See also, *Hartel Holdings Co.* v. *Calgary (City)*, [1984] S.C.R. 337, 8 D.L.R. (4th) 321; and *Vancouver (City)* v. *Simpson*, [1977] 1 S.C.R. 71.

8   See supra, note 5. See generally, P.W. Augustine, "Protection of the Right to Property Under the Canadian Charter of Rights and Freedoms," *Ottawa Law Review* (1986): 55. The recent constitutional debates resurrected proposals to entrench property rights in the Canadian Constitution. See *Shaping Canada's Future Together: Proposals* (Ottawa: Supply and Services Canada 1991), 3.

9   *Manitoba Fisheries Ltd.* v. *The Queen*, [1979] 1 S.C.R. 101. Ritchie, J., speaking for the Court, referred to *Attorney General* v. *De Keyser's Royal Hotel Ltd.*, [1920] A.C. 508, in which Lord Atkinson quoted from Bowen L.J. in *London and North Western Ry. Co.* v. *Evans*, [1893] 1 Ch.16, 28: "The Legislature cannot fairly be supposed to intend, in the absence of clear words shewing such intention, that one man's property shall be confiscated for the benefit of others, or of the public, without any compensation being provided for him in respect of what is taken compulsorily from him."

10   R.S.C. 1970, c. F-139.

11   Compare this result to a 1980 case in which the Federal Court considered a compensation claim brought by a firm which had been licensed under federal legislation to raise rainbow trout. In *La Ferme Filiber Ltée.* v. *The Queen*, [1980] 1 F.C. 128, the Court denied compensation to the firm when, given a decision of the government to prohibit raising trout in the plaintiff's area, the firm's licence was not renewed. The two cases may be distinguished on the ground that in *Manitoba Fisheries* the government replaced an existing business, effectively transferring the private firm's goodwill to the government, while in *La Ferme Filiber* there was no replacement of the existing firm – that is, no goodwill was taken. Another perspective might see *La Ferme Filiber* decided on the ground that the licence was a discretionary privilege, which is not considered by the court to be a compensable property right. See discussion infra, note 24.

12   One might say that this outcome represents an expansive view of property; see Bauman, supra, note 5 at 568. Compare this to the Court's less expansive view of property in *Cream Silver* (see infra, note 26).

13   Reported in (1985), 17 D.L.R. (4th) 1 (S.C.C.), dismissing an appeal from the British Columbia Court of Appeal, (1982), 133 D.L.R. (3d) 168, [1982] 3 W.W.R. 214, which had allowed an appeal from the British Columbia Supreme Court (1980), 114 D.L.R. (3d) 728.

14   *Attorney-General* v. *De Keyser's Royal Hotel Ltd.*, [1920] A.C. 508 at 542; see also supra, note 9.

15   *British Columbia* v. *Tener*, supra, note 13 at 23.

16   Ibid., 25.

17   Reported at (1991), 91 D.L.R. (4th) 1 (B.C.C.A.).

18   Ibid. 10-11. The Crown grants in both *Tener* and *Casamiro* were similar in that they were both grants of real property rights. The difference is that the grant in *Casamiro* did not reserve any surface rights to the grantee. Thus, *Casamiro* can be seen as authority that mineral rights imply surface rights. Another plausible interpretation is that the grant of mineral

rights without surface rights is really only a contract entitlement generated by resource policy. Such an interpretation would result in *Casamiro* being an extension of *Tener*, insofar as not only real property rights but also contract entitlements are compensable.

19 Reported at (1988), 42 M.P.L.R. 171 (B.C. Co. Ct.).

20 Reported at [1992] 2 W.W.R. 558 (B.C.S.C.), aff'd [1993] 2 W.W.R. 146 (B.C.C.A.).

21 This was the situation in *Steer Holdings,* in which the province offered some $400,000 for the plaintiff's land, the offer being reduced to $48,000 after legislation prohibiting the development was enacted. For a recent confirmation of this exception, see *Purchase* v. *Terrace (City),* [1995] B.C.J. No. 247 (QL) (B.C.S.C.); and *Harvard Investments Ltd.* v. *Winnipeg (City)* (1995), 129 D.L.R. (4th) 557 (Man. C.A.) (designation of hotel as heritage site not giving rise to compensation because not all value was taken).

22 Although the property right is recognized at common law, as noted, compensation would only be required as a result of the statutory presumption. There are common law property rights, but there is no common law right to compensation when those property rights are affected; see supra, note 5.

23 Reported at (1987), 43 D.L.R. (4th) 300 (Man. C.A.), leave to appeal refused (1988), 54 Man. R. (2d) 160 (S.C.C.).

24 See supra, note 9. Another way of examining this distinction is to say that a discretionary privilege, like a quota or licence, is not a compensable property right. Such a privilege is the creation of a statute and entirely dependent upon the whim of the legislature; see Bauman, supra, note 5 at 569. In practice, such a distinction makes little sense, because at any time the legislature can end a common law right. Hence, those rights are also discretionary in nature. The distinction seems more an exercise in judicial line drawing; see infra, discussion at note 37 and accompanying text.

25 For an application of this distinction in the context of a marketing board quota, see *Sanders* v. *British Columbia (Milk Board)* (1991), 77 D.L.R. (4th) 603 (B.C.C.A.), in which a milk quota was held not to be compensable property. For an opposite result, see *Ackerman* v. *Nova Scotia* (1988), 47 D.L.R. (4th) 681 (N.S.S.C, T.D.).

26 Reported at (1993), 99 D.L.R. (4th) 199 (B.C.C.A.), leave to appeal to the Supreme Court of Canada denied, October 14, 1993.

27 Reported at (1992), 85 D.L.R. (4th) 269 (B.C.C.A.).

28 For an application of *Cream Silver* in another context, see *Stafford* v. *British Columbia,* [1996] B.C.J. No. 1010 (QL) (B.C.S.C.), in which a claim for compensation arising from the imposition of a quota on a licence to hunt bears was rejected.

29 Reported at [1985] 2 W.W.R. 327 (B.C.C.A.).

30 Balance billing was permitted under an agreement between the Medical Services Commission and the BC Medical Association. If a new annual fee schedule was not settled by March 31, doctors were allowed to bill the difference between the old and proposed new fee schedule. This practice was ended by the *Medical Service Plan Act,* S.B.C. 1981, c. 18.

31 *B.C.M.A.* v. *British Columbia,* supra, note 29 at 332.

32 This case demonstrates the difference between an intentional redistributive policy and a regulatory change that creates an incidental loss. It would defeat the purpose of the intended redistribution (e.g., a tax change) to compensate those who "lose" as a result of the change. See L. Kaplow, "An Economic Analysis of Legal Transitions," *Harvard Law Review* 99 (1986): 511 at 519.

33 The U.S. Supreme Court first recognized the possibility of regulatory takings over seventy years ago. In the leading case of *Pennsylvania Coal Co.* v. *Mahon,* 260 U.S. 393 (1922), the Supreme Court held that a statute prohibiting miners from exercising mining rights constituted a taking under the Fifth Amendment to the Constitution. See also *Goldblatt* v. *Hempstead,* 369 U.S. 590 (1962), in which the Court held that regulation of gravel excavation did not constitute a taking, but affirmed the position that regulatory action could constitute a taking requiring compensation.

34 See for example, the judgments in *Tener,* supra, note 13, and *Casamiro,* supra, note 17.

35 For a detailed discussion of regulatory compensation on a case-by-case basis, see Kaplow, supra, note 32 at 558-60 and infra, note 91 and accompanying text.

36 If it was true that statutorily created property rights were treated less favourably than

common law property rights, the holders of the statutory rights would value them less highly than the common law rights. There is, however, no independent economic justification for such valuation. In fact, such treatment, by leading to these different valuations in cases of very similar property rights, would probably be quite inefficient from an economic prospective.

37   The court, realizing it had ventured too far in *Casamiro*, responded by going in the other direction in *Cream Silver*. This reasoning might also help to explain the very different decisions in the similar cases of *Manitoba Fisheries* and *Home Orderly Services*.

38   For another criticism of the current Canadian law regarding regulatory takings, see the Schwindt Report, supra, note 4 at 50-51.

39   For over seventy years the US Supreme Court has recognized the possibility of compensation in the case of regulatory takings (supra, at note 33). The potential to receive compensation has increased dramatically in the last eight years. In *Nolan* v. *California Coastal Commission*, 483 U.S. 825 (1987), the Supreme Court held that the State of California was obliged to compensate owners of beachfront property whose development rights were conditioned on the granting of a public easement through a portion of their property.

   More recently, in *Lucas* v. *South Carolina Coastal Council*, 60 L.W. 4842 (1992), the Supreme Court held that an environmental protection law that prohibited construction of two homes on beachfront property constituted a taking where it rendered the plaintiff's property useless. In *Dolan* v. *City of Tigard*, 114 S. Ct. 2309 (1994), Dolan opposed conditions to prevent flooding and reduce traffic congestion the city put on her proposed development. The Supreme Court, in reversing the Oregon Supreme Court, ruled a taking had occurred even though the conditions did not deprive the owner of all the value of her land. For a complete discussion of this decision, see J.H. Freis and S.V. Reyniak, "Putting takings back into the Fifth Amendment: Land use planning after *Dolan* v. *City of Tigard*," *Columbia Journal of Environmental Law* 21 (1996): 103.

   Two recent Federal Court decisions have gone even further than the Supreme Court's latest position. See *FC Florida Rock Industries* v. *U.S.*, 18 F. 3d 1560 (1994), and *Loveladies Harbor* v. *U.S.*, 28 F. 3d 1171 (1994) (even a partial taking requires compensation). For a good review of the recent case law and its implications, see M. Blumm, "The end of environmental law? Libertarian property, natural law, and the just compensation clause in the federal circuit," *Environmental Law Journal* 25 (1995): 171.

   See generally, S.E. Loper-Friedman, "Constitutional rights as property? The Supreme Court's solution to the takings issue," *Columbia Journal of Environmental Law* 15 (1990): 31; D. Dana, "Natural preservation and the race to develop," *University of Pennsylvania Law Review* 143 (1995): 655 at 658-62; J. Sax, "Property rights and the economy of nature," *Stanford Law Review* 45 (1993): 1433 at 1434-6; and note, "Taking back takings: A Coasean approach to regulation," *Harvard Law Review* (1993): 914 at 915.

40   An example of this trend is the current expropriation law in Ontario, where it has been established that an owner is not required to provide land for public use, and if private land is to be designated for public use, it must be acquired within a reasonable time. See J. Mascarin, "Confiscation without expropriation – "Public" official plan designations of privately owned lands" (1992), 9 M.P.L.R. (2d) 43.

41   See Schwindt, supra, note 4. Before proposing a number of specific recommendations for legislative amendments to the applicable statutes in the areas of forestry and mining in British Columbia, the Schwindt Report makes two general recommendations in terms of compensation policy. First, the report states that by allocating private rights to public resources, the Crown has created expectations backed by investments, and, as a result, when these interests are taken for a public purpose, the owners should receive compensation. Second, there would be no useful purpose served by legislating a definition of "taking." The distinction between interference with a property right and a taking should be flexible enough to accommodate changed conditions and, therefore, the report recommends that the issue should be dealt with on a case-by-case basis by legislators or, in the final instance, by the courts. For an opposing view, see G. McDade, *Report on Compensation Issues Concerning Protected Areas: A Draft Discussion Paper* (Vancouver: Sierra Legal Defence Fund 1993).

42   See D. Schneiderman, "*NAFTA*'s taking rule: American constitutionalism comes to Canada," *University of Toronto Law Journal* 46 (1996): 499, who argues the taking rule in

Article 1110 of NAFTA will result in American constitutional law regarding expropriations being incorporated into Canadian law. Note that these provisions cannot be used by Canadian companies in Canada, only by an investor resident in one of the other party states. Note also that these provisions apply to investments (which are broadly defined in Article 1139).

43 This position was confirmed in *Lucas* v. *South Carolina Coastal Council*, supra, note 39. The Court held that compensation could be payable except where the law was designed to prevent a private nuisance. Moreover, one could not argue the nuisance exception was met by demonstrating that the regulatory measure was in the public interest. In *Keystone Bituminous Coal Association* v. *DeBenedictis*, 480 U.S. 470 (1986), the Supreme Court upheld legislation requiring coal mining companies to leave pillars of coal in order to prevent surface subsidence as a justified exercise of the police power to prevent property owners from harming others.

See generally E. Freund, *The Police Power* (1904). Professor Freund draws the distinction between regulations which prevent harm to others, for which compensation is not paid, and regulations which require the transfer of benefits to others, for which compensation is payable. For an excellent more recent treatment of this issue, see J. Sax, "Takings and the police power," *Yale Law Journal* 74 (1964): 36. Professor Sax draws the distinction between government acting as a mediator among competing interests, which is a non-compensable use of the police power, versus government acting to enhance its resource position, where compensation is required.

44 This argument seems to be the reasoning of Estey J. in *Tener*, supra, note 13. Regulations that add nothing to property are not compensable, while regulations that create a public park result in benefits for others and should be compensable. See also "Exotic expropriations," supra, note 5 at 570.

45 A good example is the attempt by the federal government to prohibit all advertising of cigarettes and other tobacco products in Canada through the enactment of the *Tobacco Products Control Act*, S.C. 1988, c. 20. This legislation was recently deemed unconstitutional by the Supreme Court of Canada (see *R.J.R. MacDonald Inc.* v. *Canada (Attorney General)*, [1995] 3 S.C.R. 199). However, while the advertising ban was in effect, there was no offer of compensation by the government, even though the prohibition on advertising may have had a substantially adverse impact on the economic position of firms engaged in tobacco manufacturing. For a discussion concerning how recent attempts by the federal government to legislate the plain packaging of cigarettes might attract successful compensation claims by US tobacco companies under NAFTA, see Schneiderman, supra, note 42 at 523-35. New and stricter legislation concerning tobacco advertising, Bill C-71, has received Royal Assent.

46 For example, the government might regulate a firm engaged in clearcut logging. Imagine that the government reduces the firm's cutting rights, that permits are required for individual cuts which must comply with rigorous aesthetic and biodiversity guidelines, that clearcutting practices are replaced by alternative forest practices, and that extremely intensive silviculture obligations are required of the firm. Is the state preventing harm to others who would have been worse off as a result of the firm's forest practices? Or is the state demanding that the firm benefit tourists and tourism operators, commercial fishing operations, hunters, and others who enjoy wildlife? The best position seems to be that the state is doing both.

47 See Dana, supra, note 39 at 664-6; D. Bromley, "Regulatory takings: Coherent concept or logical contradictions?" *Vermont Law Review* 17 (1993): 647 at 655-6; T. Miceli and K. Segerson, "Regulatory takings: When should compensation be paid?" *Journal of Legal Studies* 23 (1994): 749 at 754; and D. Kendall, "The limits of growth and the limits to the takings clause," *Virginia Environmental Law Journal* 11 (1992): 547 at 580-2. Sax's views are discussed and criticized in L. Blume and D. Rubinfeld, "Compensation for takings: An economic analysis," *California Law Review* 72 (1984): 569 at 577-8.

48 Moreover, the police power framework only examines the consequences of regulation to owners of property, without examining the effects on those employed by the industry affected. See the discussion above in "The Resource Community," pages 322-3.

49 To be precise, this cost represents the increase in risks that investors must face, as will be

discussed later in this chapter. All risk represents a cost that investors factor into their decisions. Here, as elsewhere in this chapter, it is assumed that market-created risks and government-created risks are equivalent and investors treat both in the same manner. See infra note 66 and accompanying text.

50 These instruments include, beside obvious regulation, economic subsidies, tax policy, direct government ownership or production of services, licensing and certification processes, standard setting, and information dissemination. See Stanbury and Vertinsky, Chapter 3 in this collection.

51 The "form" of regulation may also include delay strategies that would tend to significantly distort regulatory policy. For example, regulators faced with a decision to permit resource development in a particular area – or not – or to further regulate ongoing resource development activities – or not – may simply delay their decisions until investors, aware of these uncertainties, simply withdraw from the market. Where delay, of itself, is non-compensable, and other regulatory options will result in compensation, regulators may consciously or unconsciously develop this strategy in order to forgo the payment of compensation. To the extent that such delay is inappropriate, it should be addressed within the structure of the overall compensation policy. For an example of an award of compensation in an expropriation case involving a delay in the decision to expropriate, see *Toronto Area Transit Operating Authority* v. *Dell Holdings Ltd.*, supra, note 5. For an opposite result, see *British Columbia* v. *Granite Development Ltd* (1987), 44 D.L.R. (4th) 707 (B.C.C.A.).

52 Much of the analysis that follows is based on the economic framework developed by Kaplow in L. Kaplow, "An economic analysis of transition rules," *Harvard Law Review* 99 (1986): 511. Kaplow refers to investment incentives as *ex ante* incentives, and we use the terms interchangeably. For a simpler but similar analysis of this trade-off, see R. Cooter and T. Ulen, *Law and Economics* (Scott, Foresman 1988), 198-202.

53 The use of the term "reasonable investment backed expectations" is familiar in the American case law on regulatory takings; see articles cited supra, note 39. The Schwindt Report, supra, note 4, uses reliance and expectations as important variables in determining the availability of compensation.

54 See Kaplow, supra, note 52 at 522-7.

55 In other words, what is a legitimate expectation? Certain actions, legitimate at one time might become illegitimate or unreasonable with improved information. The maker of a hazardous product, previously thought not hazardous, should not have an expectation of being able to continue to sell that product; see Kaplow, ibid., 524. Kaplow also argues that expectations are a matter of degree in that investors act on probability estimates and do not take an all-or-nothing approach. Thus, whether legal change is expected depends very much on the individual context (Kaplow, ibid., 526). Two additional problems with expectations analysis involve the difficulty in observing expectations in advance, and that they are subject to strategic misrepresentation after the fact. See C.F. Runge, "Economic implications of wider compensation for 'takings' or what if agricultural policies ruled the world?" *Vermont Law Review* 17 (1993): 723 at 729-31.

56 This example is derived from events currently occurring in British Columbia. On 21 December 1995, Timberwest started action A954337 and on 22 December 1995, MacMillan Bloedel commenced action A954359. In both cases, these companies held Tree Farm Licences with fixed royalty rates. Through the *Forest Amendment Act*, S.B.C. 1995, c. 24, the provincial government has significantly raised these rates, which will effectively quadruple by 2001. As a considerably increased cost of extracting timber from the land subject to the Tree Farm Licences, these changes will substantially lower the value of the Tree Farm Licences in question. See "Forest firms sue NDP for $240 million," *Vancouver Sun* (29 December 1995): A1; *Statement of Claim, Timberwest*, in action A954337 at 6 para. 17; and *Statement of Claim, MacMillan Bloedel*, in action A954359 at 10 para. 21.

Note how this example demonstrates the poverty of analysis based on reliance or expectations. Any reliance on the ability to extract timber resources that the owner might possess is clearly inappropriate if there is a risk such extraction might be discovered to be harmful to the environment. Expectations that the government will refrain from regulating are also inappropriate if there is the risk that timber extraction will be discovered to be dangerous to the environment. This is especially so because the potential risk of regulation can be easily capitalized into the price of land (see infra at note 57).

Finally, this example also demonstrates the poverty of analysis based on the type of regulation or property interest. See the above section "Frameworks Based on the Type of Regulation or Property Interest," pages 307-9. It is of little consequence whether the provincial government significantly raises the cost of timber extraction or simply takes away the Tree Farm Licences. Both actions cause a loss to the owners of the Tree Farm Licences, and the question is merely the magnitude of the loss. Similarly, the owners of the Tree Farm Licences have a *profit à prendre*, but the value of this interest is based on the value of the timber resources on the land. If the firms in question simply owned the land, and therefore the timber resources, the value of the land would still be primarily based on the value of the timber resources on the land. If compensation was based on actual ownership as opposed to the holding of the tree farm license, tree farm licences would be worth less than actual ownership of the timber resources, even though there is no underlying economic justification for the difference in value.

57 It is easy to demonstrate how the risk of regulation will lower the value of the land. Assume the value of the land, if one can extract timber resources, is $1 million, and only $500,000 if no extraction is possible. If there is a 50 percent chance of regulation and the owners of the firm are risk neutral, the new value of the land would be $750,000.

58 Disutilities are created by risk if one assumes that individual utility curves exhibit a diminishing marginal utility of wealth. This would mean that individuals are risk averse to some degree, which is probably a realistic approximation. See Blume and Rubinfeld, supra, note 47 at 600-2; Kaplow, supra, note 52 at 527; and R. Posner, *Economic Analysis of Law*, 4th ed. (Boston: Little Brown and Company 1992), 57.

59 Inefficient investment might occur because individuals will vary in their aversion to risk. A highly risk averse individual will more heavily discount the expected value of a project (called the risk premium) than an individual with less risk aversion. If A has a project that has an expected value of $100,000 and B has a project on the same land with an expected value of $90,000, A's project is the most efficient and should be the one that proceeds. But if A is highly risk averse and calculates a risk premium of $25,000, A will not bid more than $75,000 for the land. If B is risk neutral or not very risk averse, B can outbid A for the land and proceed with a less efficient project. See Blume and Rubinfeld, ibid., 587-8.

60 See J. Quinn and M. Trebilcock, "Compensation, transition costs, and regulatory change," *University of Toronto Law Journal* 32 (1982): 150-1, where the authors point out that risk averse individuals facing uncompensated regulation might respond by insuring through various contractual arrangements, by diversifying or reducing their investment, or by bearing the risk and continuing as before.

61 See Kaplow, supra, note 52 at 531

62 It can be argued that not compensating for regulatory takings would result in a reduction in investment in a jurisdiction with such a policy. This is certainly so; using the example infra, note 68, investment would be reduced by 25 percent due to the risk of uncompensated regulation. This loss in investment, it must be emphasized, represents no real loss to the economy because it represents over-investment caused by government compensation policy. Removal of such over-investment will produce a more efficient and environmentally sensitive community. However, removal of over-investment can also result in hardships, discussed above in the section "Environmental Regulation and Small Producers/Resource Communities," pages 321-2, which might be dealt with some form of limited government compensation policy. Another potential difficulty could arise, in the BC forestry context, if tenure reform involves strengthening property rights in order to give owners more incentive to make environmentally beneficial investments, for example, investments in silviculture; see Haley and Luckert in this collection. The threat of an uncompensated taking might reduce incentives to make such beneficial investments. However, investments in sustainability would be likely to reduce the possibility of regulation as such investments would give the government little reason to regulate to promote sustainability. A strategy of sustainability investments could reduce the risk of regulation, lowering the value of other investments, and therefore a no-compensation strategy might actually *increase* the incentive for making such investments.

Bromley argues that the owners of land subject to compensation after regulation enjoy inflated income streams. Because the value of any piece of land is merely the present value of future income streams, his concept of inflated income streams seems equivalent to the concept of over-investment. See Bromley, supra, note 47 at 675.

63 In the forestry example, any investments made to the land to extract the timber would be wasted if the government does not permit forestry to continue. If the land was in a highly earthquake sensitive area, it would be necessary for the owners, when determining the optimal level of investment, to take into account the risk of an earthquake that could destroy the timber resources. The risk of the government regulating to disallow timber extraction would have a similar effect and should also be taken into account. If the government provides compensation for regulation but not for earthquakes, the firm would take one risk into account but not the other. Compensation distorts what should be an otherwise efficient decision-making process.

64 W. Fischel and P. Shapiro, in "Takings, insurance, and Michelman: Comments on economic interpretations of just compensation policy," *Journal of Legal Studies* 17 (1988): 269 at 275-7, examine takings from the perspective of property rules (which involve consent) versus liability rules (which involve compensation). They argue that compensation requires a liability rule (strict liability) which induces moral hazard and results in over-investment. This analysis, from a different perspective than that presented in the text, reaches essentially the same conclusion.

Some commentators argue that compensation results in a windfall for the owners of land; see Bromley, supra, note 62 at 674 (losses occurring when the regulatory status quo changes were really temporary windfalls, reaped at the expense of others who bore the external costs), and "Taking back takings," supra, note 39 at 921 (the risk of uncompensated regulation is taken into account in the price; hence, any compensation is a windfall). Compensation only results in a windfall to the extent the risk of an uncompensated taking is capitalized into the price of land. In the example supra, note 57, a price of $750,000 would fully capitalize the risk of a regulatory taking. Because regulatory takings are potentially compensable (see discussion of the law in the first section of this chapter), it is unclear to what extent the threat of uncompensated regulation is currently capitalized in the price. It is likely that if the government were to announce a policy of no compensation in the event of regulatory takings, land potentially subject to a taking would lose value, but only to the extent the risk of non-compensation had not been previously capitalized into the price. Therefore, current owners would probably suffer losses and compensation, although creating distortions on investment incentives, would not result in a windfall. From that point on, the risk of non-compensated takings would always be capitalized into the price for all future buyers and sellers. See Fischel and Shapiro, ibid., 288-91.

Note how capitalization of the risk of an uncompensated taking into the price of land weakens the argument for compensation. Such capitalization would occur if a no-compensation policy were credible and certain. In that event, anyone who bought property after the new rule was announced would not be hurt at all as the risk of an uncompensated taking would be reflected in the lower price of the property. The buyer would already be fully compensated. See Posner, supra, note 58 at 58.

65 This point is made by Kaplow, supra, note 52 at 532. Because the incentive and risk effects lead in opposite directions, examining them in isolation will lead to different conclusions without solving the underlying problem. For example, Blume and Rubinfeld, supra, note 47, discuss the problems of risk associated with regulatory takings and how government compensation can solve that problem. However, they do not discuss incentives. In L. Blume, D. Rubinfeld, and P. Shapiro, "The taking of land: When should compensation be paid?" *Queen's Journal of Economics* 99 (1984): 71, the same authors examine the incentive issue while the article does not focus on the problem of risk. They conclude that if compensation is based on full market value, compensation for a taking is inefficient because investors bear less than optimal risk and over-investment results (at 73). It is the different focus that allows the same authors to come to different conclusions, both of which cannot be right.

Dana, supra, note 39 at 677-81, also discusses the risk and investment incentives trade-off, but he greatly simplifies the issues involved as if the problem is merely one of counte-nancing over-investment or under-investment. The non-availability of risk mitigation mechanisms is central to his thesis, an argument we reject (see infra, note 114).

66 See Kaplow, ibid., 533-6.

67 This point is made by Kaplow in the context of all transitional relief (see Kaplow, ibid., 535).

68 Diversification is possible through investment in projects or stocks that have a low corre-lation (i.e., returns are not related). It is possible to almost completely mitigate all risk associated with an individual project (what is termed unique risk) through investment in about ten uncorrelated projects. See R.A. Brealey and S.C. Meyers, *Principles of Corporate Finance* (Toronto: McGraw-Hill Ryerson 1986), 133-6.

69 However, investment in about ten projects of equal size would be necessary to mitigate all unique risk. Only a very large company could afford such a strategy; diversification through the actions of shareholders is a more realistic option. Companies that cannot diversify would need to purchase insurance in order to mitigate the risk of a taking. See discussion in "The Small Producer," pages 321-2, for a discussion of the problems for firms that have difficulty mitigating risk.

70 Supra, note 69.

71 MacMillan Bloedel provides an excellent example of a firm that would be unlikely to experience substantial difficulties in following a strategy of diversification. In fact, the forestry industry in British Columbia is characterized by large firms. In 1990, the ten biggest firms held 69 percent of cutting rights; see M'Gonigle in this collection, Ch.7. It should be noted that the other contributions to this collection advocate reform of the forestry industry away from big-firm concentration. Ironically, these large firms, who have the most to lose under such proposals, are also likely to be in the best position to mitigate the risk associated with such changes.

72 See discussion of the current law on regulatory takings in the first part of this chapter. The current Canadian law (and American law to a much lesser extent) leaves many regulatory takings uncompensable. The uncertainty created might be another reason that insurance markets are slow to develop. With the outcome of many regulatory takings cases hard to predict, insurance companies do not know the characteristics of the market for insurance and when the government is obligated to pay compensation. Moreover, the increasing liberalization of the law in terms of compensation results in insurance companies not wishing to offer insurance that the government might soon offer, or be obligated to offer, in the form of compensation.

It should be noted that insurance companies already offer many types of political risk insurance. See M. Levy, *International Finance* (New York: McGraw Hill 1990), 401-8; and C. Johnson, "Ranking countries for mineral exploration," *Mining Journal* 314 (1990). Also see Posner, supra, note 58 at 58, who comments: "If the point is that the risk of a government taking would be less readily insurable than that of a natural disaster, because it would be less predictable, one is entitled to be sceptical. The government's eminent domain takings vary less from year to year than the losses from earthquakes; and insurance can be bought against expropriation of property by foreign governments."

73 These are discussed in Blume and Rubinfeld, supra, note 47 at 592-7; Kaplow, supra, note 52 at 537-45; and W. Fischel and P. Shapiro, supra, note 64 at 286-7.

74 This also includes having less incentive to participate in the political process that deter-mines regulatory changes. See Blume and Rubinfeld, ibid., 594. To the extent this might occur, insurers would then have incentive to replace landowners in attempting to influ-ence public policy; see Kaplow, ibid., 605.

Blume and Rubinfeld, ibid., 594, and Fischel and Shapiro, ibid., 286, go further and argue that, with private insurance, insured individuals might bribe officials for informa-tion to obtain an advantage over their insurer. If such a problem exists, it would also have an effect on the adverse selection problem, discussed, infra note 78 and accompanying text, which depends on the relative information the insurer and the insured possesses. Insofar as such arguments depend on individuals behaving illegally, one would think that criminal sanctions act as a deterrent against such behaviour. Furthermore, if one was to relax the assumption that individuals behave in a legal manner, it would be difficult to make policy recommendations on any subject.

75 A premium that equals the probability of a taking multiplied by the value of the investment, would preserve efficient incentives. If there is a 10 percent chance of a regulatory taking that will completely destroy the value of the land, then the premium will equal 10 percent of the value of any investment made to the land and the insurance policy would pay 100 percent of the value of the investment. Investors would be left with a net return of 90 percent, which exactly equals the net expected value and thus preserves correct investment incentives. See Kaplow, supra, note 52 at 539, who notes that the relationship between the loss and the size of investment need not be linear, only that the "relationship be a known function of the level of investment."

76 This compromise is quite a common solution to the moral hazard problem, although it means that not all the risk borne by the owner of land is mitigated. Auto insurance is an example where deductibles are quite common. The compromise might be completely acceptable because of the diminishing marginal utility of income. The costs of bearing risk are relatively large when losses are substantial but are much smaller when uninsured losses are small. See Kaplow, ibid., 538. Blume and Rubinfeld, supra, note 47, do not discuss the potential adequacy of partial insurance. Firms that can only partially insure can use financial markets to diversify the remainder of the risk. The use of partial insurance can make diversification available to a wider group of landowners, because less risk will need to be diversified through the market.

77 Kaplow, ibid., 538-9. Blume and Rubinfeld, ibid., 593-4, use the example of fire insurance to demonstrate the moral hazard problem; after fire insurance is purchased, an owner has little incentive to run a fire prevention program that will greatly reduce the possibility of fires. Surely, this circumstance is precisely the type of case where it is possible for the insurance company to monitor, at a relatively low cost, whether the fire prevention program is being run. Insurance contracts contingent on risk reduction arrangements are quite common.

78 See Kaplow, ibid., 543-5, and Blume and Rubinfeld, ibid., 595-7. Adverse selection occurs when insurers set premiums based on the average probability of loss. Individuals with a high probability of loss will purchase such coverage, but individuals with a below average probability of loss will be unlikely to purchase coverage, because it will cost more than the expected value of their losses. If enough low probability losers are selected out of the market, the average probability of loss will rise and insurers will have to raise their premiums, continuing the process until either no insurance is offered or few people are able to purchase insurance.

79 Rational investors must, to the best of their ability, determine the probabilities of loss when making investment decisions. There is no reason why rational insurance companies cannot make the same calculations with the same information. Insurers should have the same access to information about the legislative process as do owners of land. Ascertaining investments on land and the value of land is not like ascertaining driving skill or the health of the insured, where adverse selection is more likely to be a problem. Companies that suffer from some inadequacy of information can always make the sharing of information by the insured a prerequisite to the obtaining of insurance.

80 If owners of land can bribe officials or obtain other illegal advantages, adverse selection would be a problem; see supra, note 74. This problem should be solved through criminal sanctions. Governments can help cure adverse selection by ensuring interested parties have equal access to information about the legislative process.

81 See Kaplow, supra, note 52 at 541.

82 This argument assumes that the government is in a worse position than private parties to judge risk preferences and to assess incentive effects. In other words, the market has better information than the government and absent remediable imperfections, the market should be relied on (Kaplow, ibid., 540, and accompanying footnote). This assumption is not unrealistic.

   Reliance on the market to solve problems of risk associated with the risk of uncompensated regulation should, to some extent, weaken the concerns of those who argue that uncompensated takings threaten individual liberty through their effect on private property. See infra, note 102 and accompanying text. Reliance on the market is normally associated with a high degree of individual freedom.

83 Kaplow, ibid., 545.
84 See Kaplow, ibid., 546, and Blume and Rubinfeld, supra, note 47 at 598-9. In the case of most regulatory takings, the possibility of a taking will increase as time goes on, turning a low probability event into a high probability event. At some point it is likely that insurance will be offered, when the possibility of a taking occurring is sufficiently high. Because of the gradual change in the probability of a taking, Kaplow suggests that takings insurance would be similar to life insurance in that as time goes on additional coverage can be purchased but at higher premiums (see Kaplow, ibid., 595, and accompanying footnote).
85 If administrative costs reflect social costs and are unavoidable, the failure to purchase insurance would be efficient; see Kaplow, ibid., 546.
86 Kaplow, ibid., 547; see discussion of the law in the first section of this chapter.
87 Kaplow, ibid., 547. Calculating the quantity of compensation (or magnitude of loss) presents many difficulties under the current law and is a very difficult process. The Schwindt Report, supra, note 4 at 54-66 devotes a lengthy section to the problem of calculating compensation. For empirical work on problems of compensation and calculation, see J. Knetsch and J. Sinden, "Willingness to pay and compensation demanded: Experimental evidence of an unexpected disparity in measures of value," *Queen's Journal of Economics* 99 (1984): 503 (individuals might require a larger sum to forgo their rights of use or access than they would pay to keep to the entitlement). Also see Quinn and Trebilcock, supra, note 60 at 126-7, and Posner, supra, note 58 at 58 for discussions of the difficulty of calculating losses from regulation. For a thorough economic analysis of the calculation of compensation for the taking of property, see J.L. Knetsch, *Property Rights and Compensation* (Toronto: Butterworths 1983).
       Blume and Rubinfeld, supra, note 47 at 618-20 also discuss problems of measuring compensation. They suggest that using the market value is inappropriate and can lead to moral hazard problems, because over-investment on the land will lead to greater compensation. Their solution, a valuation based on the unique value of the land or the willingness to pay to avoid the taking, are, in their own words, "difficult to ascertain." Their suggestion seems impractical and is probably unworkable.
88 Other factors might include changes in technology, product markets, management strategies, productions costs, etc. In *Keystone Bituminous Coal Association* v. *DeBenedictis*, 480 U.S. 470 (1986), the Supreme Court held that legislation which did not make it "commercially impracticable" to mine coal, but merely affected some 2 percent of the coal in the field served by the coal mine, would not trigger compensatory obligations on the part of the state.
89 See Quinn and Trebilcock, supra, note 60 at 139-42 for a discussion of why transaction costs in a compensation regime might be particularly high, perhaps even exceeding the amount of compensation in question.
90 See discussion of the current law in the first section of this chapter.
91 See Kaplow, supra, note 52, who discusses the problems with case-by-case determinations at 558-60.
92 Our current system compensates for some losses but does not tax gains due to regulation. If gains and losses are symmetrical, as Kaplow argues, ibid., 553-5, advocacy of certain treatment for losses should be accompanied by advocacy of a corresponding position for gains. Our position, that losses should not be compensated and gains not be taxed, is consistent. Posner, supra, note 58 at 60, argues against compensation for regulatory takings because a policy based on efficiency considerations should tax windfalls as well as compensate losses. Such a policy would be administratively impossible. He comments: "Imagine the difficulties involved in the government's identifying, and then transacting with, everyone whose property values were raised or lowered by government regulation of the price of natural gas or heating oil."
93 The government could determine premiums and charge them to individual owners through instruments like land taxes and then pay compensation after the fact (see Kaplow, ibid., 603). This option might be desirable if private insurance was not offered or was too expensive. It is unclear whether such a system would be preferable or even operate as well as potential market arrangements. If market arrangements are available, we believe they should be relied on and that government action should be necessary only in

the event of market failure. See also "Taking back takings," supra, note 39 at 922-3 (the problem of risk is solved through insurance and government provided insurance is better than compensation if no private insurance exists).

94　See infra, "The Effect of Compensation on Regulatory Decisions," pages 319-20.

95　For a contrary view, see Blume and Rubinfeld, supra, note 47 at 610-11 who advocate compensation if two conditions are satisfied. First, the loss must be substantial, in that it significantly reduces the owner's net worth, and second, the party harmed must be substantially averse to risk. This solution ignores the *ex ante* effect of compensation on incentives. Moreover, their solution seems highly impractical: in practice it would be difficult to measure an individual's aversion to risk. Furthermore, basing compensation on risk aversion would give individuals incentive to structure their affairs so that they qualify for compensation. For a discussion of the difficulties in determining *ex ante* risk distribution, see Quinn and Trebilcock, supra, note 60 at 153-7.

Miceli and Segerson, supra, note 47, argue that two rules for compensation are efficient. The *ex ante* rule awards compensation if the regulated land use was efficient at the time it was first engaged in. The *ex post* rule awards compensation if the regulation was imposed inefficiently. The *ex ante* rule, which conditions compensation on the efficiency of the land use decision, would be very difficult to determine in practice and would likely change over time. The *ex post* rule, which conditions compensation on the efficiency of the regulatory decision, is similar to a no-compensation rule with *ex post* review to determine if the regulation is for a bona fide environmental purpose. This option is exactly what we propose infra, "Problems Associated with Government in a No-Compensation Regime," pages 317-8.

96　F. Michelman, "Property, utility, and fairness: Comments on the ethical foundations of just compensation law," *Harvard Law Review* 80 (1967): 1165 at 1214.

97　Michelman, ibid., 1214-15. In addition, the project or regulation in question should only go ahead if the expected gains exceed the lesser of the settlement or demoralization costs.

98　See Kaplow, supra, note 52 at 560-1; Blume and Rubinfeld, supra, note 47 at 578-9; Dana, supra, note 39 at 669-77; Fischel and Shapiro, supra, note 73 at 281-7; and Quinn and Trebilcock, supra, note 60 at 152-3.

99　Kaplow, ibid., 561. Kaplow argues the first component of demoralization relates directly to risk while the second component refers to the negative effects of uncompensated takings on future investment. Also Posner, supra, note 58 at 58.

100　Fischel and Shapiro, supra, note 73 at 281-3 (noting the causes of demoralization at 282). To illustrate demoralization costs, they note the difference between, "a watch that is stolen and a watch that is lost."

Another perspective is that demoralization costs involve risks over which individuals have little control. Individuals will respond to such risks more significantly than others if they are conscious of them in their decision-making process, if these risks are random, and if they may have catastrophic impacts. See M. Douglas and A. Wildavsky, *Risk and Culture* (Berkeley: University of California Press 1982), 1-48.

101　Fischel and Shapiro, ibid., 287.

102　This view is most commonly associated with Richard Epstein who argues that for the sake of protecting private property, all reductions in value caused by regulations should be compensated. See R. Epstein, *Private Property and the Power of Eminent Domain* (Cambridge: Harvard University Press 1985). For comment and criticism of Epstein, see Blumm, supra, note 39 at 192; and Miceli and Segerson, supra, note 95 at 764-7.

103　In a market economy, businesses constantly adapt to non-compensated regulation; non-compensated environmental takings force individuals to better adjust and adapt to changing circumstances and knowledge. See Sax, supra, note 39 at 1449, who comments: "The non-compensation norm in circumstances of social change reflects a decision to encourage adaptive behaviour by rewarding individuals who most adroitly adjust in the face of change," and at 1450, "Diversification and timely dis-investment of lands unsuitable for development are techniques of economic adaptation."

104　Bromley, supra, note 62 at 676, argues that the availability of compensation gives land an advantage over other factors of production which must bear windfall gains and losses. He comments at 655: "Why is it politically and socially acceptable to prevent the cultivation of marijuana or prostitution and child pornography, on a parcel of American real estate, while a law to enhance the habitat for endangered flora or fauna, or a ruling to influence the nature and scope of development consistent with locally articulated norms is seen as a fundamental invasion of someone's alleged 'rights'?"

105 Note, though, that slavish devotion to the protection of property bodes ill, not just for regulatory expropriations, but for any policy that redistributes wealth. If inequality in the distribution of wealth has an adverse effect on freedom, constricting the state's ability to redistribute wealth denies society an important tool for enhancing liberty.

106 Bromley, supra, note 62 at 682, comments, "The social role of property is as an inducement to undertake socially valued endeavors – nothing more."

107 For an example of a definition of property incorporating environmental norms, see Sax, supra, note 39 at 1442-6, who proposes a new property definition that is compatible with what he terms the economy of nature. He proposes a usufructory model with judicial safeguards in cases of government abuse.

108 Dana, supra, note 39 at 676. He comments at 674, "Those most often subjected to uncompensated natural preservation regulation – groups of wealthy individuals and businesses are likely to be taken very seriously by legislators." In the forestry context, the composition of the industry, which is characterized by large firms, makes feelings of majoritiarian exploitation unlikely (see supra, note 71).

109 Large corporations, which might only have profit motives at heart, can often benefit from being seen as environmentally friendly. Such companies must balance good publicity, which is very valuable, with the profit to be made in environmentally detrimental ventures (factoring in the bad publicity such ventures will create). Companies expecting benefits from good publicity at least as large as the profit from the foregone venture, would not experience demoralization costs.

110 In this regard, one should differentiate environmental legislation from, for example, the construction of an expressway. Both are government projects requiring takings; but those whose houses are taken for the expressway might enjoy none of the benefits while suffering the loss of their homes, as compared to suburban residents who have faster transportation and enjoy all the benefits at no cost. Environmental regulatory takings allow the owners to share in the benefits of a better environment. Most environmental takings are also only partial takings that allow owners uses other than those prohibited by the legislation.

However, some stakeholders in resource industries, like labour, might still suffer demoralization costs due to regulation that affects investment, and therefore employment. Due to the specific nature of human capital in resource industries, these costs might be quite high. These costs might be mitigated to the extent that employees do care about the quality of the environment. Nonetheless, there is more justification for compensation for labour displaced by environmental regulation than there is for the owners of property affected by such regulation; see infra, "The Resource Community," 322-3.

111 Dana, supra, note 39.

112 Dana is not referring to all environmental regulation but only that which preserves land, for example, preserving wetlands. Once a developer fills in a piece of wetland and builds on it, there is little likelihood that land will be subject to regulation, because it is much more difficult to restore wetlands than it is to preserve land that remains as wetland. Hence, already developed land will not be subject to regulation. This caveat obviously narrows the magnitude of any potential accelerated development.

Note that the race to develop has limited application to certain resource sectors, like forestry, where the quantity of extraction is limited through legislation. For example, in British Columbia, forestry firms are limited in how much timber can be extracted by annual allowable cuts. These limitations add another constraint on the ability of firms to develop before regulation occurs.

113 Dana, supra, note 39 at 692-3.

114 See Dana, ibid., 677-81 and 690.

115 See "The Availability of Insurance for Regulatory Takings," pages 312-4. Dana's analysis suffers from several additional difficulties. Dana assumes that individual investors will be able to predict relatively accurately the risks of regulation and the potential for

development. If individual developers can accomplish such calculations, there seems little reason why large insurers could not also collect the same information. See the discussion supra, note 79. Dana also contradicts himself at 674 when he comments that demoralization would not be a problem because property owners have sufficient power (they are "groups of wealthy individuals and businesses") to be taken seriously by legislators. He later comments at 690, "Some land developers have relatively limited assets and hence limited capacity to self-insure through diversification." If most property owners are groups of wealthy individuals and businesses, diversification should be of little difficulty. See the discussion on diversification, supra, note 80. The few who cannot diversify might engage in accelerated development to the extent insurance is not possible; but this is a much smaller number of developers than Dana contemplates, making the problem less important and perhaps inconsequential. Finally, Dana also points out that reducing the risk of regulation through investment in the political process is possible, but does not work in practice (at 688-90). However, to avoid accelerated development, it is only necessary for developers to prolong the process until the optimal time for investment. He notes (at 694 and accompanying footnote) that developers have the political clout to delay the legislative process.

116 The choice firms face is not diversification or insurance. So long as firms are able to engage in some combination of the two strategies, the risk of uncompensated takings can be mitigated.

117 Dana, ibid., 696, notes the potential large administrative costs of *ex ante* compensation and the likely negative effects of compensation requirements on legislators. See also the discussion infra ""The Effect of Compensation on Regulatory Decisions," pages 319-20. Dana suggests a system of partial *ex post* compensation and *ex ante* payments (696-706). This system assumes the government can calculate the risk of regulation and the potential value of investments. Even Dana doubts such measurement is possible by the government (at 707). Such a complex system of payments would be difficult to administer and would still result in *ex ante* incentive distortions.

118 See D. Mueller, *Public Choice* (New York: Cambridge Surveys of Economic Literature 1979), 153-67.

119 See Kaplow, supra, note 52 at 567-70; Blume and Rubinfeld, supra, note 47 at 620-22; the Schwindt Report, supra, note 4 at 31; Fischel and Shapiro, supra, note 73 at 285; and "Taking back takings," supra, note 39 at 923-5.

120 Another proposal to solve the problem of fiscal illusion is to have property owners pay to escape regulation instead of the state paying compensation; the state would sell regulatory exemptions. See "Taking back takings," ibid., 923-5. This "Coasean" solution to the problem of fiscal illusion is problematic as it violates two of the assumptions necessary for Coase theory to function. First, the initial distribution of property rights might affect demand for those rights and second there would be potentially substantial transaction costs (difficulties noted at 926-7). Note, though, how this idea is really another example of the use of a market instrument, like a "pollution permit," instead of regulation to protect the environment. Another potential solution has been suggested by Cooter and Ulen, supra, note 52 at 201. They argue the government could buy an option from the property owner to entitle the government to buy the property at any time within a specified interval at a price specified in the contract. This solution is a contractual method of solving the incentive-risk trade-off, while at the same time internalizing the cost of regulation and expropriation to government. Such a solution would be impractical where transaction costs are high, and this situation is likely to be the case if there is more than a small number of owners.

121 See Kaplow, supra, note 52 at 567-8.

122 See Kaplow, ibid., 568-9, and Posner, supra, note 58 at 60.

123 See the Schwindt Report, supra, note 4 at 31. The report comments that the unimportance of fiscal illusion "is particularly relevant to projects such as the creation of ecological reserves where benefits are thought to accrue not just at some future period, but in perpetuity."

124 See D. Cohen, "Regulating the regulators: The legal environment of the state," *University of Toronto Law Journal* 40 (1990): 213; and Quinn and Trebilcock, supra, note 60. Considering the current fiscal situations in the US and Canada, it is unlikely this assumption is true. Runge, supra, note 55 at 735, argues that public choice theory suggests a govern-

ment will not behave like a profit maximizing firm, so internalizing regulatory costs is unlikely to have a positive impact.

Also note that if the responsibility for revenue and taxation are separated from the decision of where and when to regulate, cost internalization might have little effect (see Kaplow, supra, note 52 at 569).

125 See Runge, supra, note 55 at 734-5, who uses the example of agricultural subsidies to demonstrate the behaviour of rent-seeking groups. Quinn and Trebilcock, supra, note 60 at 142, argue that rent-seeking considerations do not support a case for or against compensation.

126 See Kaplow, supra, note 52 at 572-5. This issue is the concern that the decisions of the government would reflect *mala fides*; for example, singling out certain individuals or groups for losses.

127 Although judicial review is a possibility, it is likely, because of the inexpertise and inexperience the courts have with environmental issues, that the courts would not be the most appropriate forum. An independent administrative agency with the necessary environmental expertise would be better equipped to judge the regulatory policy in question. Parties would also save time and expense in avoiding the court system and taking their dispute before a board specifically constituted to hear it.

128 This solution to the compensation question is similar to the *ex post* compensation rule articulated by Miceli and Segerson, supra, note 95, which they argue is an efficient solution. Compensation for environmental regulatory takings would be based on the efficiency of the regulatory decision. They note that compensation under this rule would generally not be paid (at 750).

The example of the actions commenced by Timberwest and MacMillan Bloedel, supra, note 56, are instructive. If these firms were legitimately concerned that the decision of the provincial government was not for an environmental purpose, but was, for example, merely an attempt to raise revenue, an inquiry would be justified. Such an inquiry, hopefully before a body with the requisite expertise, would replace the court actions.

129 Although it is likely challenges to environmental regulation would be common at first, in time the decisions of the courts or board would provide guidance on how to judge the efficiency and propriety of environmental regulation, resulting in fewer challenges over the course of time.

130 See infra, "Problems Associated with Government in a No-Compensation Regime," pages 317-8."

131 This concern is rarely mentioned in the literature; most commentators are solely concerned with the possibility of too much instead of inadequate legislation. For brief discussions of the problem, see Dana, supra, note 39 at 696 and "Taking back takings," supra, note 39 at 927.

132 Regulation that preserves the environment can lead to increased popularity, although most environmental regulation either has long-term benefits that might accrue for the next government or benefits that accrue in perpetuity (see supra, note 123 and accompanying text).

133 The magnitude of this effect will depend on the availability of compensation; it is likely this effect is, currently, not very strong in Canada because of the limited availability of compensation (see the discussion of Canadian law in the first part of this chapter). The trend though, has been towards increased availability, which will have an adverse impact on the quantity of legislation, ironically at the same time as environmental awareness continues to increase.

134 See the discussion of Canadian law in the first part of this chapter.

135 This effect probably has, at this point, a fairly large impact given the current law which requires compensation in some cases (complete takings) but not in others (partial takings). Regulators have an incentive to enact regulations that only are partial takings, regardless of their overall efficiency, because complete regulatory takings would result in huge compensation costs for the government. As compensation becomes more available for regulatory takings, this effect should lessen, because the ability of the regulator to regulate so as to avoid compensation will be reduced.

136 A process of *ex post* review, as suggested in "Problems Associated with Government in a No-Compensation Regime," pages 317-8, would help ensure that enacted environmental regulation was efficient.

137 Compensation would, in this case, reduce competitive incentives and distort market signals; these signals are necessary to induce a reallocation of resources to their most highly valued use.

138 See Quinn and Trebilcock, supra, note 60 at 142-3, and Knetsch, supra, note 87 at Ch.8.

139 Quinn and Trebilcock, ibid., 143-4.

140 This is an example of rent-seeking behaviour, which is behaviour by individuals and firms who use governments to extract wealth from others which they cannot obtain consensually through markets. See J. Buchanan, T. Tollison, and G. Tullock, *Towards a Theory of the Rent-Seeking Society* (College Station, TX: Texas A. and M. University Press 1980).

141 Either the government receives the information from the scientific community (and there is a time lag in conveying it to the general public), or the information is so complex that it is difficult to educate people to understand it. That public tastes take a long time to reflect new information is probably a result of some combination of these points. In addition, consumers probably take time to change their habits. Eco-certification is one method of ensuring consumers can differentiate between environmentally-friendly and -unfriendly products.

142 For example, imagine a firm that produces paper products made from virgin fibres. If it is known that cutting down old-growth forests has an adverse environmental impact, consumers will change to buying paper products made from only recycled fibres. However, if consumers are only just starting to assimilate the new scientific information and tastes are slow to change, the government would be justified in regulating the mill to produce paper products made only from recycled fibres and in not offering compensation if the regulatory change inflicted a loss on the firm.

143 See the discussion infra, "Problems Associated with Government in a No-Compensation Regime," pages 317-8.

144 Our thanks to Professor Mark Gillen for his helpful insights on this section.

145 Supra, note 56 and accompanying text.

146 See discussion of diversification, supra, at notes 68 and 69. With an investment of $1 million, Owner might need to invest as much as $10 million in ventures with uncorrelated returns in order to completely mitigate all unique risk associated with the mining venture.

147 If there are a large number of small producers in the same position as the owner in this example (that is, firms that have an overwhelming proportion of their assets tied to one high-risk investment), the threat of uncompensated regulation might help to more efficiently restructure the industry by causing these small firms to sell their assets or share them with larger firms. Such a restructuring would shift risky assets to those that were in the best position to diversify their investments, thus lowering the overall level of risk in the industry.

148 For a discussion of the importance of the relative size of the loss in determining the necessity of compensation, see Quinn and Trebilcock, supra, note 60 at 162-3. If an individual is risk averse (that is, has a declining marginal utility of wealth), huge losses relative to total wealth are likely to cause huge disutilities.

149 See supra, at note 64.

150 Quinn and Trebilcock, supra, note 60 at 158.

151 Quinn and Trebilcock, ibid., 159-60. For example, human capital in the forestry sector is highly specialized. Regulation in this sector, resulting in a decrease in investment and therefore a decrease in employment, would likely lead to significant losses for labour.

152 Large unions might be better able to negotiate contracts with higher wages in return for employees bearing more risk. Unions might also be able to provide forms of insurance to employees who lose positions as a result of environmental regulation.

153 For a discussion of the problems in measuring compensation, see supra, note 87 and accompanying text.

154 Former employees who received cash payments might have little incentive to retrain or relocate. The example of the Newfoundland cod fishery is instructive in this regard, because years of support payments have had little positive effect on the region. The result has been little relocation, retraining, and reduction of the number of fishers.

155 For a more complete discussion on government training programs and their effectiveness, see Quinn and Trebilcock, supra, note 60 at 135-9.

156 Forest Renewal BC was created in 1994 by the British Columbia government as a scheme
     to create a $2 billion fund through increased stumpage fees to forestry companies. Part of
     the fund is to be used for job skill training and job creation for forestry workers displaced
     as a result of the new Forest Practices Code. This is the type of transition policy that might
     be appropriate in the case of regulatory expropriation. See "BC forest plan puts big load on
     loggers," *Globe and Mail* (29 April 1994): A23; "Environmentalists hail new forest plan,"
     Vancouver Sun (15 April 1994): A1, A2; "Forest policy expected to make communities suf-
     fer," *Vancouver Sun* (26 January 1995): A18. The plan has come under attack as not being
     particularly effective; see "Forest renewal gets poor mark from Munro," *Vancouver Sun* (29
     September 1995): D1. Moreover, the provincial government gutted $400 million from this
     fund to help remedy the provincial debt. See "If a mill dies," *Vancouver Sun* (23 November
     1996): B1, B2.

14
Ecoforestry Bound: How International Trade Agreements Constrain the Adoption of an Ecosystem-Based Approach to Forest Management

Fred Gale

Sustainable forest management is not occurring in BC for want of policy. As contributors to this volume have clearly shown, a vast number of policies, both incentive-based and regulatory, are available to promote sustainable forest use. Community-based tenure reform, competitive pricing systems, eco-taxation and subsidies, ecologically based zonation, devolution of administration, and eco-certification can be used to restructure industry behaviour to encourage profitable and sustainable production. The problem is not a lack of policy, but a lack of what is euphemistically called "political will," a term that glosses over the underlying structures of power rooted in corporations, unions, and the state. This triangular configuration of social power appears set to resist badly needed reforms of the forest products industry until the last stand of old-growth forest is cut. As globalization proceeds, moreover, the triangular alliance gains strength from the web of international trade and investment laws woven by the World Trade Organization (WTO) and the North American Free Trade Agreement (NAFTA). Tied to a free trade and investment discourse, the Canadian state and its sub-national entities are ceding policy-making power to remain competitive in the emerging, single global market for unsustainably produced forest products (UPFPs).

Proponents of sustainable forest policy in BC not only contend with special interests outside and inside the state that lobby to maintain industrial forestry, but also with those pushing stronger, freer, international trade and investment regimes on a reluctant public. The two major aims of agreements such the General Agreement on Tariffs and Trade (GATT), the World Trade Organization (WTO), and the North American Free Trade Agreement (NAFTA) are straightforward and unambiguous. The first aim is to restrict the right of states to discriminate between foreign and domestic producers of the same product, creating in the process a level playing field for companies to compete internationally. The second aim is to liberalize trade and investment relations by reducing and removing tariff and nontariff barriers (NTBs). As the number of such trade agreements increases, their general effect is to integrate Canada and BC ever more tightly into a single global economic system, reducing and constraining their capacity to legislate in key policy areas. Indeed, the current negotiations taking place at the Organization for Economic Development and Cooperation (OECD) on a Multilateral Agreement on Investment (MAI) promise to deepen the globalization process dramatically, opening states up to further multi-national corporation (MNC) investment and formalizing corporate rights to sue governments over alleged expropriation of assets.

The steamroller of globalization raises critical questions about the international legality of many policies available to national and sub-national levels of government. The mere existence of international trade agreements can be used as a blanket rationale for avoiding needed policy reform when such policies might conflict with the provisions of GATT/WTO and NAFTA. In this chapter, I take a critical look at some of the key policy provisions of an eco-conversion strategy that might conflict with BC/Canada's international trade obligations. I argue that, although increasingly constrained, BC can still make a good case for policies that promote eco-forestry. That case, moreover, needs to be made now, before existing unsustainable trade and investment regimes become stronger, foreclosing future options.

An Eco-Conversion Strategy for the BC Forest Industry

Drawing on the literature of the ecosystem approach, six general principles can be identified.[1] These are the principles of holism, complexity, uncertainty, scale-dependence, diversity, and long-term planning. Applying these principles in practice makes an ecosystem-based approach to forestry a tall order.[2] Ecoforestry places the forest in a landscape context, requiring that all relevant features of the landscape be identified and considered. A narrow focus on trees is abandoned and much more information than the quantity, quality, and species of tree is required. The goal of ecoforestry is to protect and enhance the composition, structure, and function of a forest ecosystem, not to replace the natural forest with simplified, monocultural tree-farms. Large-scale clearcutting designed to create even-aged stands gives way to variable-retention and selection logging systems. Consequently, the annual volume of timber produced under an ecoforestry regime is significantly less than that produced under the current industrial model, because more values than fibre production are being optimized over a much longer period of time.[3]

In contrast to industrial forest practices, therefore, an alternative strategy grounded in an ecosystem-based approach to forest use is required. Such a strategy is emerging from the theory and practice of ecoforestry.[4] The essential elements of this strategy have been outlined elsewhere and involve three dimensions of change.[5] First, BC's commodity-based export strategy in forest products must be reversed, with small volumes of high value-added goods replacing large volumes of low value-added products such as softwood lumber and pulp and paper. Second, there must be a reorientation away from certain kinds of capital-intensive production methods, especially the use of such technology as massive feller-bunchers that are ecologically inappropriate. Third, there must be a shift away from corporate domination of the industry to empower rural communities to take much greater control over the resources in their jurisdictions. The eco-conversion strategy proposes to restructure fundamentally the BC forest sector in the interests of BC's forests and its forest communities. The strategy is, not unexpectedly, vigorously opposed by those forces currently benefiting from the existing set-up: large corporations, unionized labour, and government agencies. On the other hand, the strategy could be made to appeal to environmentalists, ecologists, eco-foresters, First Nations, community groups, and non-timber forest users concerned with community well-being and the health of BC's forest ecosystems. While the internal political economic barriers to implementing such a strategy are immense, many of the policies needed to implement it could run foul of BC/Canada's obligations under the GATT/WTO and NAFTA. Before examining the existence and severity of possible conflicts, however, we need to put some policy flesh on what is still a very schematic strategic skeleton.

Policies to Support an Eco-Conversion Strategy

Three different types of policies can be used to implement the eco-conversion strategy. Institutional policies, frequently both the most effective and the most difficult to discuss and implement, reinforce, reform, or replace existing structures of social power. Economic policies, on the other hand, seek to bring about the desired behaviour by providing incentives (usually in the form of monetary rewards), to individuals, communities, and companies. Finally, regulatory "command and control" policies set out mandatory requirements and the penalties for non-compliance.[6]

The debate on the relative merits of these policies became polarized in the 1980s and 1990s between proponents of government regulation and economic incentives. This debate focused mainly on the potential utility of either government regulations or economic incentives to achieve environmental aims.[7] Although both parties viewed institutions as vital, their preferences were shaped by differing views of the efficacy of the state and market to achieve environmental objectives. While the debate has generated a great deal of heat, it has also produced some light. Although there are few signs of a consensus appearing among the main protagonists, others have come to realize that regulatory and economic instruments are not exclusive and can, in many instances, complement each other. The critical question for many has shifted, therefore, to reforming the institutional structures that underpin our political economy.

Institutional Reform

Tenure

Institutional restructuring is an essential component of the eco-conversion strategy and must take place in at least four areas: tenure reform, competitive market establishment, community-based stewardship, and curriculum reform. Of the four, tenure reform is the most fundamental.[8] The existing tenure system is geared to meet the needs of industrial corporations. Forest Licences (FLs) and Tree Farm Licences (TFLs) are held almost exclusively by large, integrated, multinational corporations that produce commodity forest products for global markets. Several features of the FL and TFL system make it an unacceptable institutional basis for an eco-conversion strategy based on ecosystem management. Modern publicly traded corporations are global actors, planning their activities over vast distances, at large scales, with success measured by the bottom line. MNC's operate in highly competitive markets in which profitability and competitiveness are the critical ingredients of survival. Corporations have therefore little allegiance to place, and either move from forest to forest in the search for fibre supplies, or into different industries as fibre supplies are depleted. Furthermore, because corporate survival often requires large amounts of investment to finance capital-intensive production strategies, many firms are highly leveraged from borrowing on capital markets. The modern, integrated mills that constitute a major portion of this investment require large volumes of fibre throughput to generate the economies of scale necessary to make them profitable. Fibre shortages cannot be long tolerated because idle mills do not generate revenues to cover investment costs and to generate profits.[9]

Forest Valuation

Reform of the stumpage system is another important institutional innovation. The current mechanism for imputing the value of forests calculates stumpage using BC's Comparative Value Pricing System. This mechanism, which invariably undervalues the price of trees, could be replaced by competitive log markets.[10] There are currently almost no competitive log markets in BC. The Vancouver Log Market (VLM) must be excluded, because a great deal of inter- and intra-firm trading occurs among the major corporations under the close supervision by the Council of Forest Industries.[11] In contrast, the Lumby Log Market (LLM) in Vernon operates on a competitive basis. The LLM was established in 1993 under the Ministry of Forests' Small Business Forest Enterprise Program (SBFEP). The local Ministry of Forest's district manager contracts out logging operations under this program to independent loggers who cut the wood according to Vernon district forest management specifications. The district office transports the logs to Lumby, where they are sorted into bales based on species, quality, and size. Each week, an auction is held and the bales sold to the highest bidder, with average per metre log prices in the range of $105 and some logs fetching as much as $200. The establishment of a network of similar log markets around BC would constitute an important institutional innovation, encouraging better allocation of timber to end-uses.[12]

Administrative Reform

A third institutional reform involves substantial changes to BC's forest administrative structures, particularly to the operation of the Ministry of Forests. At present, the Crown, through the Ministry of Forests and the Chief Forester, is involved in every aspect of forest management ranging from reviewing tenure applications, issuing licences, setting the level of the AAC, monitoring implementation, enforcing regulations, carrying out silvicultural operations on forest land under almost all types of licences, and engaging in dispute resolution activities through a variety of local and regional processes. This centralized structure is not working well, even though the number of tenure holders is small. The structure will be completely unworkable under a new, ecoforestry tenure system, as the number and diversity of tenure holders expands and community resource boards are established. Within the new structure, the role of the Ministry of Forests would shrink to the issuing and revocation of tenure licences based on CRB recommendations. A small cadre of staff at headquarters would support a larger, community-oriented field staff that reported, in the first instance, to the CRBs in the communities where they were located.

Educational Reform

A final, urgently needed institutional reform is required in the forestry profession itself. In the past, Registered Professional Foresters (RPFs) emerging from forestry schools across Canada were narrowly trained to maximize fibre production on a given area of land. Safeguarding the health and integrity of ecosystems did not constitute an important dimension of their education or of their work. In the 1990s, significant and long-needed changes are taking place in the forestry curricula in many Canadian schools. However, the new approach is still framed within the sustained yield forest management paradigm, and on prioritizing fibre production over ecosystem health and integrity. Only fundamental reform of the forestry profession can create the New Forester to practise the New Forestry.

Economic Incentives

Institutional reforms need to be supported by the use of economic incentives to encourage the widespread practice of ecosystem-based forestry. Johnson has outlined a range of economic incentives that could be made available to individuals, communities, First Nations, and family-owned corporations to ensure that the practice of ecoforestry is remunerative.[13]

Three different types of such economic instruments are taxation, subsidization, and certification and labelling schemes.

Significant improvement to the practice of forestry in BC could be achieved simply through tax reform. Lower capital gains taxes could be applied on forest land to take account of the fact that ecosystem forestry requires a very long-term investment horizon. Capital gains as a result of the appreciation of land and timber values could be assessed at a reduced rate, protecting owners from having to sell land to meet their tax bill. Tax rebates could be introduced for certified ecoforestry operations to reward such practices. In addition, the tax and stumpage rates charged on uncertified operations could be increased to discourage unsustainable forest practices and to recoup the costs of the externalities such practices impose on society. Similarly, woodlot owners investing in ecosystem-based forestry could qualify for income tax write-offs for outfitting expenses and training courses.

Tax incentives could be complemented by subsidies to establish a competitive industry in sustainably produced forest products (SPFPs) from bona fide ecoforestry operations. The need for an eco-subsidy derives from the fact that the vast majority of timber currently sold in domestic and international markets comes from unsustainable operations based on the liquidation of natural forest ecosystems. Such UPFPs can be sold more cheaply, because the full ecological, environmental, social, and cultural costs of production are not factored into the price. Companies producing certified forest products could qualify for a subsidy from the provincial government to enable them to sell products at a competitive price. As the market for ecoforestry products takes hold, eco-subsidies can gradually be reduced and increasing reliance placed on penalizing unsustainable producers through the tax system. An ecoforestry investment fund could be set up also to encourage the establishment of woodlots, horse-logging businesses, and value-added enterprises. Such a fund could provide low-interest or even interest-free loans to qualified applicants to purchase necessary equipment, undertake additional training, and assist in covering the costs of the first few years, until the operation become established.

Another, potentially powerful, economic incentive is the adoption of bona fide eco-certification and labelling schemes. Such schemes, when properly designed and implemented, provide a guarantee to each member in the timber chain that the timber being purchased has been produced sustainably and in accordance with the principles of ecosystem management. A growing number of wholesalers, retailers, and final consumers are prepared to pay a premium over the cost of non-certified timber in order to meet the demand for eco-certified timber and because they believe it is the right thing to do.[14] In Europe, moreover, a number of large commercial retail outlets, including Sainsbury's in Great Britain, have committed themselves to buying only wood products from forests certified under the Forest Stewardship Council (FSC) process by the year 2000. This "niche" market appears to be growing rapidly and could replace the current market in uncertified wood products.[15]

The institutional and tax reforms proposed here assist value-added manu-facturers by providing access to logs at competitive prices and to credit and tax incentives to produce SPFPs. Because the value-added sector will likely consist of a large number of relatively small firms specializing in niche mar-kets, an effective and comprehensive marketing and distribution infrastruc-ture could be established to ensure that BC producers can transport, market, and distribute their products to American, European, and Japanese markets. This network would be owned and operated by the value-added sector itself, but would receive support from the BC government in the form of start-up capital and administrative and technical support.

Government Regulation

Institutional reforms and economic incentives could go a long way to restructuring the BC forest products industry to meet the requirements of an ecosystem-based approach in the forestry sector. One result of such restructuring would be a reduction, but not the elimination, of regulatory instruments. Government regulation would still be required to set and enforce provincial forest management standards, for example. However, one consequence of institutional and economic reforms proposed would be the demise of the centralized bureaucratic approach to forestry regulation, as embodied in the Forest Practices Code. The Code is the product of an attempt to balance a centralized administrative apparatus with the need to set, moni-tor, and enforce better practices in the forests. An alternative set of Ecoforestry Principles would set out the broad goals of an ecosystem-based approach. Implementation, monitoring, and enforcement would be the responsi-bility of woodlot associations and community resource boards (CRBs) which would be linked to eco-certification and labelling requirements.

The Ministry would also continue to be engaged in province-wide land use planning exercises with other relevant ministries to determine the location and boundaries of totally protected areas (TPAs) and "wholistic forest management zones" (WFMZs). This planning and land-management process would also be simplified, because WFMZs would cover the bulk of the operational forest estate, obviating the need for some of the planning currently required to designate high-intensity zones, low-intensity zones, wildlife corridors, riparian protection, and so forth. The Ministry would also continue to administer the forest tenure system, which would include tenure-holder qualifications and licensing regulations. In addition, a spe-cific unit of the Ministry would be established that would make available to the CRBs dispute resolution officers to assist them in the event of dis-putes arising at the local level.

Regulation is also required to prevent BC's efforts to develop a forest-products industry grounded in the goals and principles of ecosystem-based management from being undermined by the importation of lower cost UPFPs from other parts of Canada and the rest of the world. The danger confronting BC's eco-conversion strategy is that, notwithstanding the existence of a group of retailers and consumers prepared to pay a price premium for certified timber, the continued availability of UPFPs will simply prove too attractive to many purchasers. To create a level playing field in the forest products industry in BC, border tax adjustments (BTAs) will be required on non-certified timber products entering BC from other jurisdictions. The use of BTAs is preferred to other trade regulatory mechanisms (bans, quotas, tariffs) because they are defensible under Canada's international obligations under the GATT/WTO, because they raise revenues to fund the eco-conversion strategy required within British Columbia, and because they help create a global market for eco-certified forest products.

International Constraints on Implementing an Eco-Conversion Strategy

The feasibility of an eco-conversion strategy in the forestry sector in British Columbia depends on the development of a political consensus among certain internal social forces. That consensus is rendered ever more elusive with Canada's and BC's growing integration into an unsustainable, globalized free-trade system via the GATT/WTO and NAFTA. The provisions of these international trade agreements act as a deterrent, discouraging all levels of government from adopting policies deemed to be trade restrictive. Such agreements pave the way also for the creation of an unsustainable, single global economy, where products compete exclusively on the basis of price, quality, and availability. The policies necessary for implementing the eco-conversion strategy outlined here necessarily conflict, therefore, both directly and indirectly, with GATT/WTO and NAFTA provisions.

Direct Policy Conflicts

Direct disputes could arise between BC and its trading partners over the use of border tax adjustments, eco-subsidies, certification and labelling schemes, and government procurement programs. Because the GATT/WTO is the more comprehensive and recent of the two agreements, this discussion will focus mainly on its provisions. Given the importance of the American market for BC's forest products, however, the specific implications of NAFTA will be explored, especially where these appear to differ from those of GATT/WTO.

Border Tax Adjustments

GATT/WTO Provisions

The eco-conversion strategy includes eco-taxation measures to encourage BC consumers to purchase SPFPs in preference to USFPs. The effect of these tax measures is to remove any price disincentive to purchase SPFPs and to ensure that all locally produced goods compete on a level playing field. To ensure that foreign- and unsustainably produced forest products do not obtain a price advantage over both domestically produced UPFPs and SPFPs, it is necessary to tax imported UPFPs. The application of an eco-tax to imported UPFPs necessitates consideration of the tricky area of international trade law dealing with border tax adjustments (BTAs). While the GATT/WTO does permit the use of BTAs in certain circumstances, their legality for the purpose of environmental protection is the subject of debate among trade law experts. In reviewing the legal literature, however, it is clear that BC could make a strong case before a GATT/WTO Dispute Settlement Body (DSB), notwithstanding the GATT/WTO's past precedents and current procedures. Ultimately, the case would hinge on whether BTAs are being imposed for bona fide environmental purposes and whether UPFPs can be distinguished from SPFPs and treated as "unlike products."

Border tax adjustments have been defined as:

> any fiscal measures which put into effect, in whole or in part, the destination principle (i.e. which enable exported products to be relieved of some or all of the tax charged in the exporting country in respect of similar domestic products sold to consumers on the home market and which enable imported products sold to consumers to be charged with some or all of the tax charged in the importing country in respect of similar domestic products).[16]

The use of BTAs must be consistent with the principle of "National Treatment." GATT/WTO Article III states that:

1. The contracting parties recognize that internal taxes and other internal charges ... should not be applied to imported or domestic products so as to afford protection to domestic production.
2. The products of the territory of any contracting party imported into the territory of any other contracting party shall not be subject, directly or indirectly, to internal taxes or other internal charges of any kind in excess of those applied, directly or indirectly, to like domestic products. Moreover, no contracting party shall otherwise apply internal taxes or other internal charges to imported or domestic products in a manner contrary to the principles set forth in paragraph 1.

The import of these two paragraphs is ambiguous. While the paragraphs do not rule out the imposition of charges on goods imported into a country, such measures should not be used for discriminatory or protectionist purposes. Four critical questions arise in relation to the use of BTAs. First, is the measure a direct tax or an indirect tax? Second, is the tax applied directly to the product or is it a tax on "prior-stage" inputs? Third, does the tax measure have a discriminatory impact on "like products"? And finally, is the intention behind the tax to protect the environment or to protect domestic producers?

The first question requires us to distinguish between direct and indirect taxes. The Agreement on Subsidies and Countervailing Measures (ASCM) defines direct and indirect taxes as follows:

> The term "direct taxes" shall mean taxes on wages, profits, interests, rents, royalties, and all other forms of income, and taxes on the ownership of real property ...
>
> The term "indirect taxes" shall mean sales, excise, turnover, value added, franchise, stamp, transfer, inventory and equipment taxes, border taxes and all taxes other than direct taxes and import charges.[17]

These paragraphs draw the distinction between direct and indirect taxes on the basis of whether the tax is applied to the *producer* or the *product*. If the tax is applied to the producer (as are income taxes, corporation taxes, and payroll taxes), then it is considered to be a direct tax, ineligible for a border tax adjustment. If the tax applies to a product (such as an excise tax or a sales tax), then the tax is deemed to be indirectly applied and may be eligible for a BTA. The measure proposed for the eco-conversion strategy is evidently an indirect tax, which would apply to all domestic and imported UPFPs.

Although the eco-taxation measure proposed is an indirect tax, not all indirect taxes are BTA eligible. Demaret and Stewardson, in an extended discussion of this topic, note two further distinctions. The first is "between taxes which are part of an internal indirect tax system and those which are customs duties or other specific charges," and the second "is between those indirect taxes which, while forming part of an internal taxation system, are eligible for adjustment in respect of the import or export of the product in question, and those which are not."[18] In relation to the first distinction, the eco-taxation measure proposed here is clearly part of the internal, indirect tax system and does not fall under the category of a "customs duty or other specific charge." The second distinction, however, requires a more extended commentary in relation to the evolving GATT/WTO thinking on border tax adjustments concerning physically and non-physically incorporated components of the production process.

It is generally agreed that an indirect tax that is directly applied to a product is BTA eligible. A difficulty arises, however, in the case of taxes that do not apply directly to the product itself, but to "prior-stage" inputs. The question regarding such prior-stage taxes has, in the past, generally been resolved by distinguishing between taxes on products that are physically incorporated in the final product (which are BTA eligible) and taxes on non-physically incorporated components (which generally are ineligible). The eco-taxation measure being proposed would apply, however, specifically to UPFPs entering BC from other jurisdictions. The tax would be levied directly on the product, therefore, not on prior-stage inputs.

While it might be concluded, therefore, that the eco-taxation measure being produced is eligible for border tax adjustment, the final decision by a GATT/WTO panel would likely depend on its answers to two further questions. First, the panel would likely consider the rationale for distinguishing between UPFPs and SPFPs, querying whether these are "like" or "unlike" products. Although GATT/WTO parties have not defined the concept of "like product," panels have tended in the past to interpret the concept narrowly. In the Japanese alcohol case, for example, the panel adopted a restrictive interpretation that defined "like products" as those that were capable of serving similar end uses. If this interpretation were adopted, then a distinction between UPFPs and SPFPs could not be made, and there would be no rationale for the BTA.

Recently, however, as a result of the GATT Auto Panel ruling, a broader interpretation of the "like product" concept now appears possible. The Auto Panel ruled that "cars could be differentiated (i.e., not considered as "like products") on the basis of their fuel efficiency. A tax on fuel-inefficient cars was considered consistent with WTO rules."[19] The ruling moves beyond the narrow, "end-use" interpretation of "like products" and examines the rationale for the distinction in a broader context. Thus, the GATT Auto Panel accepted the rationale adopted by the American authorities, of an important distinction between fuel-efficient and fuel-inefficient cars, even though both served the same end-use. If a GATT/WTO panel were to apply the broader Auto Panel ruling and agree that a distinction could be made between UPFPs and SPFPs, then there is a much stronger likelihood of the BTA being deemed to be GATT/WTO compatible.

In summary, the proposed eco-taxation measure proposed constitutes an indirect, internally imposed tax that is directly applied to domestic manufactured and imported unsustainably produced forest products. If BC's trading partners tested the legality of this BTA, BC/Canada could mount a credible defence, arguing that the measure was non-discriminatory and intended to protect BC's environment. The ultimate decision of the GATT/WTO panel is impossible to predict, however, and depends on whether the panel accepts BC/Canada's claim that the measure is for the purpose of environmental protection and on whether it deems UPFPs and SPFPs to be "unlike products."

NAFTA Provisions

The NAFTA does not deal explicitly with the question of BTAs. Under Article 301, however, which sets out the "National Treatment" provisions of the agreement, the parties are bound by the GATT and subsequent agreements in this area. At a minimum, therefore, the same arguments made above in relation to GATT/WTO also apply to NAFTA. BTAs used for ecological purposes might receive more sympathetic treatment under the NAFTA, given that its Preamble commits member states to "promote sustainable development" and to "strengthen the development and enforcement of environmental laws and regulations." But NAFTA does not define the slippery concept of "sustainable development," and the strengthening of "laws and regulations" could be interpreted narrowly to refer to legislation only and not to the use of economic measures designed to achieve environmental goals.

Eco-Subsidies

GATT/WTO Provisions
The eco-conversion strategy envisages measures that would provide direct financial benefits to producers of SPFPs. The subsidization of the production of SPFPs could prompt BC/Canada's trading partners to initiate a challenge under the provisions of the Agreement on Subsidies and Countervailing Measures (ASCM). Under the ASCM, a measure is deemed to be a subsidy if:

(a) there is a financial contribution by a government or any public body within the territory of a Member ... where:
   (i) a government practice involves a direct transfer of funds ... potential direct transfers of funds or liabilities;
   (ii) government revenue that is otherwise due is foregone or not collected ... ;
   (iii) a government provides goods or services other than general infrastructure, or purchases goods;
   (iv) a government makes payments to a funding mechanism, or entrusts or directs a private body to carry out one or more of the type of functions illustrated in (i) to (iii) ... ;
or
(a) there is any form of income or price support in the sense of Article XVI of GATT 1994;
and
(b) a benefit is thereby conferred.[20]

It is clear that the proposed measures under the eco-conversion are subsidies according to the above comprehensive definition. The mere existence of a subsidy, however, does not imply BC/Canada would be liable to countervailing trade action. Our trading partners have to establish also that the measures taken are specific subsidies that are either prohibited or actionable. A specific subsidy is targeted at "specific enterprises, industries, groups of enterprises or industries, or enterprises in a specific geographical region."[21] It is distinguished from a non-specific subsidy, which is a non-targeted, generally available financial contribution by government to support transportation infrastructure, education and training, research and development, and so forth. Because the proposed eco-subsidy would grant support to a distinct sector of the BC forest industry, it constitutes a specific subsidy under the ASCM.

Not all specific subsidies can be countervailed. The ASCM classifies subsidies into three categories: non-actionable, actionable, and prohibited. Non-actionable subsidies include "all 'non-specific' subsidies, and three kinds of 'specific' subsidies granted to cover a portion of costs of research and pre-competitive development activities, to assist disadvantaged regions and to promote the adaptation of existing facilities to new environmental requirements."[22] Actionable subsidies are specific subsidies that cause "'adverse effect to the interests of other Members' by injuring their domestic industry, nullifying or impairing their benefits under the GATT, or causing them serious prejudice."[23] Prohibited subsidies are "non-agricultural subsidies which are based on export performance (export subsidies) in law or in fact and those contingent upon the use of domestic over imported goods."[24]

The proposed eco-subsidy would likely not qualify for the exemption provided for environmental subsidies under Article 8.2(c). This article allows specific subsidies for "assistance to promote adaptation of existing facilities to new environmental requirements imposed by law and/or regulations which result in greater constraints and financial burden on firms." To qualify for exemption, such environmental subsidies are subject to several restrictions, including that they be a "one-time, non-recurring measure," "limited to 20 percent of the cost of adaptation," and "available to all firms which can adopt the new equipment and/or production processes."[25] The environmental subsidies exception appears aimed exclusively at offsetting the negative impact of domestic environmental regulations, while the purpose of the proposed eco-subsidy is to encourage firms, in the absence of domestic environmental regulation, to shift production from UPFPs to SPFPs.

The subsidies proposed under the eco-conversion strategy would thus be considered either "prohibited" or "actionable" subsidies. They do not appear to qualify as prohibited subsidies, because the measures taken do not aim to give exporters of SPFPs a specific advantage over domestic producers of SPFPs. That is, the measures proposed do not meet the requirements of those prohibited in the "illustrative list of export subsidies" in ASCM, Annex 1. They would only do so if a deliberate effort was made to promote the export of SPFPs. The second category of prohibited subsidies, which refers to measures "contingent ... upon the use of domestic over imported goods," also does not apply. That is, the subsidy is being granted to producers of SPFPs, and such products could be manufactured from domestic or imported, sustainably produced, eco-certified, forest products. A company in BC that imports FSC eco-certified lumber as inputs to the production of furniture, for example, would still be entitled to receive direct financial subsidies from the government, and to benefit from attractive, government-backed loans.

Thus, it can be concluded that the proposed eco-subsidies fall within the "actionable" category of the ASCM. Article 5, which covers actionable subsidies, states:

> No Member should cause, through the use of any subsidy referred to in paragraphs 1 and 2 of Article 2, adverse effects to the interests of other Members, i.e.:
>
> (a) injury to the domestic industry of another Member;
> (b) nullification or impairment of benefits accruing directly or indirectly to other Members ... ;
> (c) serious prejudice to the interests of another Member.[26]

BC/Canada's trading partners must establish, therefore, not only the existence of a specific, actionable subsidy under the provisions of the ASCM, but also that it was injuring, nullifying, impairing, or causing "serious prejudice" to its domestic industry. Article 6 sets out the meaning of "serious prejudice" in considerable detail. A range of possible criteria are used including the level of ad valorem subsidization (in excess of 5 percent), coverage for operating losses, and debt forgiveness. Furthermore, even if the eco-subsidy measure contravenes these criteria, serious prejudice will only be deemed to have occurred if there has been market displacement, "significant" price undercutting, or an increase in world market share.[27]

In short, therefore, the countervailability of BC/Canada's eco-subsidies cannot be established a priori on the basis of their nature, because they are neither prohibited nor non-actionable subsidies under the provisions of the ASCM. Because they are actionable subsidies, any case brought against them would depend ultimately on precisely how they affected our trading partners' markets for SPFPs and UPFPs. It is extremely difficult to determine the impact in advance of the actual implementation of the eco-subsidies, because there are many unknowns. These unknowns include the actual level of subsidization; the trade impact of subsidization on the export of SPFPs from BC to other markets; and the trade impact of subsidization of SPFPs on the export of UPFPs to BC/Canada.

At the outset, the trade impact would likely be very modest. Currently, the volume of SPFPs on the market is negligible, and even a rapid increase in their production would have a limited trade impact, given the breadth and depth of the UPFP trade. While the net effect of the imposition of BTAs on UPFPs and subsidies on SPFPs will be to make the price of the latter more attractive to consumers, the fact remains that the market will, in the short term, be completely incapable of meeting consumer demand. The result will be the continued purchase of UPFPs, albeit at higher prices than before, until enough producers make the switch to manufacturing SPFPs. By the time a significant trade effect is felt, therefore, it may also be possible, as a result of economies of scale in the SPFP industry, to reduce or eliminate the eco-subsidies, and to rely exclusively on BTAs to achieve price competitiveness between the two product types.

NAFTA Provisions

As with BTAs, NAFTA does not discuss subsidies directly. The lack of subsidy provisions were a disappointment to the negotiators of the Canada-US Free Trade Agreement,[28] and no additional substantive provisions on subsidies were included in NAFTA. The procedural "solution" adopted in the FTA was extended to Mexico. Chapter 19 provides that "each Party shall replace judicial review of final antidumping and countervailing duty determinations with binational panel review."[29] The purpose of binational panels is to review, "based on the administrative record, a final antidumping or countervailing duty determination of a competent investigating authority of an importing Party to determine whether such determination was in accordance with the antidumping or countervailing duty law of the importing party."[30]

Given the lack of substantive provisions on subsidies in the NAFTA, a dispute with our major trading partner over the measures proposed would be subject to American domestic subsidy and countervailing duty law. Past experience suggests that the United States could and would launch a countervailing duty action if it felt that its trade interests were threatened by BC's eco-conversion strategy. In several high-profile cases, for example, the US Commerce Department has reviewed Canada's administrative system of timber pricing, and either threatened or imposed countervailing duties on Canadian softwood lumber imports to compensate for Canada's below-market stumpage rates.[31] It is conceivable that the US Commerce Department, under pressure from its domestic building and remanufacturing industries (as a result of a sudden increase in the cost of, and a marked decline in the availability of, softwood lumber) and its domestic timber industry (as a result of a decline in Canadian sales of its UPFPs following the imposition of eco-taxes, BTAs, and eco-subsidies), would consider launching a countervailing action against the subsidy provisions of the BC eco-conversion strategy.

The success of this countervailing action would necessarily take into consideration all of the concerns already noted in relation to the GATT/WTO provisions and would depend not only on demonstrating the existence of a subsidy but on a causal connection between the subsidy and harm caused to the American timber industry. Under the provisions of NAFTA, the findings of the Commerce Department would be subject to a

NAFTA binational review panel in which trade experts from both Canada and the United States would participate. While the existence of a binational panel provides some safeguards against the abuse of domestic US countervailing legislation, the narrow mandate, expertise, and pro-trade ethos of panel members creates a bias against trade measures taken for environmental purposes.

Government Procurement Programs

GATT/WTO Provisions

The eco-conversion strategy foresees an important role for government procurement in furthering ecosystem management in the forest products sector in British Columbia. To create a market for SPFPs, governments could issue an administrative order to departments and public agencies requiring that preference be given to SPFPs in the purchase of needed supplies and services. Suppliers, aware that there is a guaranteed market for their goods and services, would be encouraged to produce and supply a greater quantity of SPFPs.

Government procurement is big business, with the world market estimated to be over $1 trillion per annum.[32] As a result, states with well-developed, multinational corporations have been anxious to gain access to foreign markets, and governments have negotiated the multilateral Agreement on Government Procurement (AGP) to set out the international ground rules for such competition. The AGP binds only a handful of countries and regions including Canada, the United States, the European Union, and Japan. The agreement is designed to discourage states from using government purchasing power to benefit domestic industry. Article 3, for example, specifies that parties to the AGP shall ensure:

(a) that its entities shall not treat a locally-established supplier less favourably than another locally-established supplier on the basis of foreign affiliation or ownership; and

(b) that its entities shall not discriminate against locally-established suppliers on the basis of the country of production of the good or service being supplied, provided that the country of production is a Party to the Agreement.[33]

Schuman argues that a strict interpretation of this article "would forbid 'Buy America' rules" and "would have outlawed the policies of various state agencies that did not allow firms tied to South Africa to bid on public contracts."[34] The national treatment provision is apparently strengthened further by Article 16, which specifies that "entities shall not, in the qualification and selection of suppliers, products or services, or in the evaluation and award of contracts, impose, seek or consider offsets."[35] In the accompanying note, "offsets" are defined as "measures used to encourage local development or improve the balance-of-payments accounts by means of domestic content, licensing of technology, investment requirements, counter-trade or similar requirements."[36]

The bulk of the AGP sets out guidelines for the appropriate valuation of contracts, the process of bid evaluation, the product or service performance criteria required, the type of information provided to bidders, the procedures governing where and what bid-information to publish, the time limits for bid calls and submissions, the content and handling of the tender documentation, the bid challenge procedures to be used, and the institutions that deal with complaints from suppliers that a breach of the agreement has taken place. If the AGP applied in full force in BC, then it would significantly constrain the provincial government's rights to use its purchasing power in support of domestic economic development. Fortunately, however, the AGP contains numerous exclusions and loopholes. The agreement applies only to some federal government departments and Crown corporations (e.g., Canada Post Corporation, National Capital Commission) and cannot automatically be extended to cover sub-central government entities. While the Canadian government is negotiating with provincial governments to extend the agreement's scope,[37] these negotiations are proceeding slowly, and British Columbia has given no undertakings with respect to the AGP. The province is, therefore, not under any obligation to follow the tendering procedures set out in the agreement.

Even if BC were to be governed by the agreement, there is no necessity for the agreement to cover all sectors of the BC economy or to apply to all sizes of contract. The AGP as negotiated only applies to certain sectors of the federal government, and many sensitive industries are exempted. In addition, only federal contracts greater than SDR130,000 (SDR5 million for construction services) are covered by the agreement, providing scope for preferential treatment for smaller contract amounts. There are numerous clauses that permit limited tendering such as "in the absence of tenders in response to an open or selective tender," when "the products or services can be supplied only by a particular supplier and no reasonable alternative or substitute exists," "for reasons of extreme urgency brought about by events unforeseeable by the entity," "for additional deliveries by the original supplier," "when an entity procures prototypes or a first product or service," and so forth.[38]

In summary, therefore, under the current provisions of the AGP, the BC government is not restricted from adopting an SPFP procurement policy. The government could go even further and specify that preference would be given in the awarding of building contracts to wood-based buildings and renovation works. By doing so, the government could provide a significant stimulus to the SPFPs industry in BC, and position the industry to compete in the emerging global market for genuine sustainably produced, eco-certified forest products. This positioning would enable BC to gain "first-mover" advantage internationally, while simultaneously reducing the pressure on its forests to continue to produce a high volume of UPFPs.

NAFTA Provisions

The provisions for government procurement are set out NAFTA, Part Four, Chapter Ten. These provisions are substantively similar to the AGP in terms of scope, requirements for national treatment and non-discrimination, the prohibition of offsets, technical specifications and procedures for open tendering, limited tendering, exemptions, and dispute resolution. From the point of view of judging the legality of the procurement measures to be implemented under BC's eco-conversion strategy, it is important to note that the NAFTA provisions on procurement, like the AGP, do not yet apply to provincial levels of government. Article 1001.1(a), setting out the chapter's scope, states that:

> This Chapter applies to measures adopted or maintained by a Party relating to procurement: (a) by a federal government entity set out in Annex 1001.1a; a government enterprise set out in Annex 1001.1a-2; or a state or provincial government entity set out in Annex 1a-3 in accordance with Article 1024.[39] [emphasis added]

Annex 1001.1a-3 merely states that "coverage under this Annex will be the subject of consultation with the provincial governments in accordance with Article 1024." Article 1024 is thus key, and provides for further negotiation to take place between the parties according to two different time schedules depending on whether "negotiations pursuant to Article IX: 6(b) of the GATT Agreement on Government Procurement (the 'Code')" are completed or not.[40] If not, then negotiations would have commenced no later than 31 December 1998. Since negotiations on the AGP were completed in 1994, the parties are now bound under NAFTA to "immediately begin consultations with their state and provincial governments with a view to obtaining commitments, on a voluntary and reciprocal basis, to include within this Chapter procurement by state and provincial government entities and enterprises."[41] These negotiations have commenced and are proceeding slowly. At the time of writing, they have not led to the extension of the NAFTA procurement provisions to provincial governments. Currently, therefore, BC is not restricted under the NAFTA from taking the eco-conversion procurement measures proposed.

Even if the NAFTA procurement measures were to be extended to BC, however, a large number of sectors are excluded from coverage under Section B of Annex 1001.1b-1. Goods excluded include prefabricated structures, scaffolding, lumber, millwork, plywood, veneer, construction and building materials, and furniture (unless they are procured by the

Department of National Defence or the RCMP, in which case they are covered by Chapter Ten).[42] Consequently, even if the NAFTA were to be extended mutatis mutandis to BC, scope exists for procurement to take place in favour of locally produced goods.[43]

In summary, therefore, neither the NAFTA nor the AGP would restrict BC's right to adopt eco-conversion procurement measures to stimulate the supply of SPFPs. While differences exist in the technical provisions governing the two agreements relating to the size of procurement contracts, the dispute resolution procedures, the sectors included and excluded, and so forth, the substantive provisions of both agreements are similar in wording and identical in intent. It is likely that there will be increasing convergence between the two agreements, as parties continue negotiations under the provisions of NAFTA, Article 1024. While these negotiations are proceeding slowly, the existing freedom of action could be easily and quickly surrendered in subsequent rounds of federal-provincial and international trade consultations. Although the procurement window is currently wide open, therefore, the intention of trade officials is to close it and to constrain provincial government freedom of action in this area of trade policy.

Eco-Certification and Labelling Schemes

GATT/WTO Provisions

A market instrument of immense potential in the implementation of an eco-conversion strategy in the BC forest products sector is the introduction of FSC's certification and labelling scheme to enable consumers to distinguish between sustainably and unsustainably produced forest products. The FSC scheme could operate within BC and apply equally to domestically produced timber products and imports. Forest products bearing an appropriate logo shipped into BC would be deemed to be SPFPs and would compete with made-in-BC certified products. All products without a logo, or with an ineligible logo, would be deemed to be UPFPs and subject to a border tax adjustment. While the basic intention of such a certification and labelling scheme is to enable consumers to distinguish between SPFPs and UPFPs, and exercise a preference in favour of certified products, the BC government's approval and promotion of the FSC scheme could be construed as trade restrictive and discriminatory by BC's trading partners. To assess the strength of BC's case, the relevant GATT/WTO provisions, which are set out in the Agreement on Technical Barriers to Trade (ATBT), need to be examined.[44]

The ATBT makes a distinction between technical regulations and technical standards. A regulation is defined as a "document which lays down product characteristics or their related process and production methods ... with which compliance is mandatory" and a standard as a "document approved by a recognised body, that provides, for common and repeated use, rules guidelines or characteristics for products or related process and production methods, with which compliance is not mandatory."[45] The distinction between regulations and standards is important from a trade perspective, because the former prohibit the sale of products in export markets that are deemed not to meet the technical regulations, while the latter do not. The ATBT sets out the procedures to be followed by governments, requiring under Article 4 that "central government standardizing bodies accept and comply with the Code of Good Practice for the Preparation, Adoption and Application of Standards."[46] The Code of Good Practice requires that standards be developed in accordance with national treatment and most-favoured nation obligations.[47]

The compatibility of the FSC scheme with the ATBT is difficult to assess in the absence of an explicit panel ruling on the question. BC could mount a strong defence in its favour, however, by arguing that certification is voluntary, not mandatory; that the intention is environmental protection, not trade discrimination or restriction; that the provisions of the ATBT have limited applicability to sub-central governmental entities; and that eco-certified and uncertified products are not "like products" and can be distinguished. BC could begin by noting that eco-labelling schemes are either mandatory or voluntary in nature. The essential difference between the two types of schemes is that mandatory schemes prevent unlabelled products from entering the market, setting the obtaining of the label as a requirement for market entry. Voluntary schemes, on the other hand, "are not based on a *de jure* exclusion of non-environmentally friendly products from the market, but merely offer producers the possibility of applying for an eco-label."[48] Most schemes currently in operation are voluntary in this sense, and the FSC scheme is no exception. The intention is not to exclude UPFPs from market, but simply to ensure that buyers throughout the timber chain are able to distinguish SPFPs from UPFPs.[49]

While the proposal's voluntary nature means that it falls under the ATBT provisions that deal with the setting of technical standards rather than technical regulations, BC's trading partners could still deem the scheme to be incompatible with the Code of Good Practice. Article H of the Code provides that "the standardizing body within the territory of a Member shall make every effort to avoid duplication of, or overlap with, the work of other standardizing bodies in the national territory or with the work of relevant international or regional standardizing bodies." In Canada, the official standardizing body at the national level is the Canadian Standards Association (CSA), which has developed its own weak guidelines for the eco-certification and labelling of forest companies. The CSA is also seeking to have its guidelines for Sustainable Forest Management Systems accepted as the international standard for the industry through the International Organization for Standardization (ISO). BC's trade partners could thus challenge the use of FSC standards as duplicating the efforts of the CSA and of the ISO.[50]

A third issue is the degree to which the FSC scheme proves to be trade restrictive in intent and in fact. Given that the scheme applies equally to domestic and imported timber products, it is unlikely that it would be considered discriminatory in intent. The provincial government could argue that the FSC scheme is available to its trading partners and that FSC-certified products entering BC are treated similarly to domestic SPFPs. A key issue here, however, would be the degree to which SPFPs and UPFPs are regarded as "like products." If SPFPs and UPFPs are regarded as "like products" based on a narrow interpretation of the concept, then our trading partners could argue that government promotion of FSC certification and labelling is having a chilling effect on their exports, resulting in an actual decline in UPFP sales. On the other hand, if a broader interpretation of the concept of "like products" is taken, then SPFPs and UPFPs would be regarded as "unlike products." In this case it would be difficult to make the case for either a discriminatory intention or an effect.

A final key question to be determined is the degree to which the ATBT applies to sub-central governmental bodies. In regard to the preparation, adoption, and application of standards, the agreement provides that central government bodies "shall take such reasonable measures as may be available to them to ensure that local government and non-governmental standardizing bodies within their territories, as well as regional standardizing bodies of which they or one or more bodies within their territories are members, accept and comply with this Code of Good Practice." This clause would appear to permit the Canadian federal government the right to insist that the FSC-certification and labelling scheme adopted by BC conform with the rather different scheme adopted by the CSA. On the other hand, inter-provincial trade arrangements are governed by the Agreement on Internal Trade (AIT), which currently permits provinces the right to set many of their own standards. Unlike procurement agreements, however, there is no provision in the ATBT for the negotiation of sub-national coverage. There is, rather, a directive that sub-central governmental bodies be covered.

In summary, therefore, although the outcome of a GATT/WTO dispute settlement panel on BC's implementation of the FSC eco-labelling scheme is uncertain, BC could mount a strong defence in its favour, while disputing that it is under any obligation, as a sub-national entity of the Canadian federation, to be bound by the ATBT's provisions.

NAFTA Provisions
While BC could mount a good case in favour of its eco-certification and labelling scheme before a GATT/WTO panel, its case before a NAFTA panel would be even stronger. Chapter Nine of NAFTA provides more scope for provincial determination of standards, provides broader criteria for determining the legitimacy of standards adopted, and is less restrictive with regard to the standardization procedures. The first notable difference

between NAFTA and the ATBT concerns the scope of the two agreements. While NAFTA binds sub-central government bodies under Article 105, Chapter Nine weakens this requirement significantly. Article 902(2) provides only that "each Party shall seek, through appropriate measures, to ensure observance of Articles 904 through 908 by state and provincial governments and by non-governmental standardizing bodies in its territory."

More importantly, perhaps, under Chapter Nine, NAFTA permits a party to establish its own level of protection, providing it is pursuing a "legitimate objective," defined as:

(a) safety,
(b) protection of human, animal or plant life or health, the environment, or consumers, including matters relating to quality and identifiability of goods and services, and
(c) sustainable development.[51]

The inclusion of both "the environment" and "sustainable development" as legitimate objectives strengthens BC's case in defending its adoption and promotion of an FSC certification and labelling scheme. This defence is reinforced by Article 905, that a party shall use "as a basis for its standard-related measures, relevant international standards whose completion is imminent," except "where such standards would be ineffective or inappropriate to fulfil its legitimate objectives." Furthermore, in strengthening the capacity of a party to exempt itself from international standards, Article 905(3) provides that nothing in 905(1) "shall be construed to prevent a Party, in pursuing its legitimate objectives, from adopting, maintaining, or applying any standards-related measure that results in a higher level of protection than would be achieved if the measure were based on the relevant international standard."

Finally, although NAFTA, like the GATT/WTO, aims to prevent the use of standards for trade restrictive or discriminatory purposes, Article 904(4) provides considerable scope for the adoption of the FSC standards. The full provisions of Article 904(4) are:

No Party may prepare, adopt, maintain or apply any standards-related measures with a view to or with the effect of creating an unnecessary obstacle to trade between the Parties. An unnecessary obstacle to trade shall not be deemed to be created where: (a) the demonstrable purpose of the measure is to achieve a legitimate objective; and (b) the measure does not operate to exclude goods of another Party that meet that legitimate objective.

There would appear to be little doubt that BC could demonstrate that the purpose of its FSC certification scheme was to meet the legitimate objective of sustainable development, that no intention existed to discriminate against American or Mexican imports, that no such discrimination occurred in fact, and that the scheme operated exclusively as an economic instrument to further the establishment of genuine sustainable forest management.

Section Summary

This section has reviewed several of the more powerful policies required by an eco-conversion strategy to create the conditions for ecosystem management in BC's forest sector. To convert the existing unsustainable high-volume, capital-intensive, corporate-dominated, industrial forest products industry to a low-volume, labour-intensive, community-based ecoforestry industry, a range of institutional, economic, and regulatory policies are necessary, each of which strengthens and reinforces the others.

Several of these policies could conflict with BC's and Canada's commitments under the GATT/WTO and NAFTA. Although BC/Canada would be able to mount a vigorous defence of its eco-conversion strategy, the provisions of existing trade agreements, the content of past precedents, and the structure of the GATT/WTO and NAFTA dispute settlement procedures could militate against a successful outcome. On the other hand, the potential non-availability of these crucial policy options significantly weakens BC's capacity to implement a badly needed eco-conversion strategy in the forestry sector. Decisionmakers are forced to implement the eco-conversion strategy with one policy hand tied behind their backs. These potential direct constraints on the policy options available to BC decision makers are compounded by the indirect effects of international trade agreements that integrate Canada and BC ever more tightly into an unsustainable world trade and investment structure where price competitiveness and productivity are the most important measures of "success."

Eco-Conversion and International Competitiveness

Cohen defines the concept of competitiveness as follows:

> A nation's competitiveness is the degree to which it can, under free and fair market conditions produce goods and services that meet the test of international markets while simultaneously expanding the real income of its citizens. Competitiveness on the national level is based on superior productivity performance and the economy's ability to shift up to high productivity activities which in turn can generate high levels of real wages.[52]

A central feature of competitiveness is productivity, which depends on numerous components of a state's economic structure. Adler, for example, notes that "nations maintain advantaged lifestyles for their citizens by making certain that their citizens attract the most interesting and well-paying jobs in the world," which, "of course, suggests government action – not in refined border controls, but in the form of investment in superior education, research and development, and a competitive infrastructure."[53] Adler's notion of the role of the state is almost the complete opposite of its role in the early postwar period. Then, the state's role was to defend the national economic, political, and cultural realm against the designs of other states. The purpose of the new, liberal-competitive state, however, is to adjust internal economic conditions to the demands of the global marketplace and to eliminate the political, economic, and cultural barriers to successful competition in the global arena. Structures that promote increased productivity are to be supported, while those that hinder it are to be dismantled.

BC's and Canada's forest products industry is currently structured to encourage a small number of publicly quoted multinational corporations to produce a high volume of unsustainably produced, low value-added wood products for the global market. This "extroverted" strategy necessitates Canadian support of international trade and investment agreements that discourage foreign governments from using discriminatory and restrictive trade measures. For this reason, the forest industry strongly endorses free trade negotiations. One industry spokesperson has commented that "our success will still depend on moving ... Canadian forest products into world markets" and, because "the government of Canada has less trade leverage in the system to protect our export markets," we must place "greater reliance on having good trade rules and an effective trade dispute system."[54]

The eco-conversion strategy proposed in this chapter, and the institutional, economic, and regulatory policies required to implement it, conflict with the present structure and future goals of BC/Canada's UPFP industry. Consequently, the industry brands the environmental trade agenda "a witches' brew of trade restrictive government regulation, pressures for changes in international trade rules (the so-called 'greening of the GATT') and actions by our adversaries to influence the attitudes of potential buyers of Canadian export commodities."[55] Those that benefit from the industry's current structure – multinational corporations, unionized labour, and various arms of the BC and federal state – find the policies outlined here terrifying. From their perspective, the eco-conversion strategy would, if adopted, result in a massive reduction in the quantity of goods produced, labour employed, and capital invested, undermining the pursuit of the goal of increased national competitiveness.

These industry, union, and state fears sidestep two central questions that confront BC/Canada's forest sector in an era of globalization. The first concerns the capacity of BC/Canada to compete in the global, unsustainable forest products industry. The province and the country are experiencing a decline in the volume of old-growth fibre, low rates of secondary growth, and increased logging costs. The problems are compounded by technological innovation and the emergence of new, lower-cost competitors with access to fast-growing plantations (such as Chile) or the commercialization of vast tracts of previously unlogged forests (such as Russia). The consequence of these developments is that BC/Canada is increasingly unable to compete in the globalized market for commodity forest products.[56]

Opposition to the eco-conversion strategy also side-steps another central question: should BC/Canada compete in the global market for UPFPs? An affirmative answer, which downplays or even denies the degree to which current production is unsustainable, proposes further integration into a global economy through the negotiation of more comprehensive free trade and investment agreements. Proponents of this path argue that the legitimate concerns of environmentalists and workers can best be met by writing appropriately worded preambles to international agreements and negotiating side accords to protect the environment and workers' rights. Rejecting that approach and making a determined effort to develop an industry in SPFPs means a complete and fundamental restructuring of the BC forest industry.

Conclusions

In this chapter, I examined the implications of the ecosystem-based approach for the structure and operation of BC's forest industry. I argued that the significant reduction in timber volume under an ecosystem-based approach necessitates an eco-conversion strategy in the forest sector, and a shift from high-volume, low value-added production of UPFPs to low-volume, high value-added production of SPFPs. While serious internal political economic barriers need to be overcome to implement such a strategy, the focus of this chapter was on the way in which the institutional, economic, and regulatory policies required to implement the strategy could conflict with BC/Canada's increasing commitments under the GATT/WTO and NAFTA.

These international trade agreements have tended to chill the debate over policy ends, because the means to attain them are apparently prohibited. Nonetheless, I find here that BC can make good legal arguments before international dispute settlement panels in support of its eco-conversion policies. So long as these policies are based unambiguously on genuinely sustainable production, and do not lapse into protectionism for protectionism's sake, there is nothing to fear. The goals are laudable and the action is justified to preserve the health and integrity of BC's forests and to develop a genuinely sustainable forest products industry.

Acknowledgment
I would like to thank Gil Yaron for his assistance in carrying out background legal research for this paper, especially in relation to the North American Free Trade Agreement.

Notes
1 For further details on the concept and application of ecosystem management, see P. Alpert, "Incarnating ecosystem management," *Conservation Biology* 9(4) (1995): 952-5;

R.E. Grumbine, "What is ecosystem management?" *Conservation Biology* 8(1) (1994): 27-8; R.B. Keiter, "Beyond the boundary line: Constructing a law of ecosystem management," *University of Colorado Law Review* 65 (1994): 293-333; R.D. Margerum and S.M. Born, "Integrated environmental management: Moving from theory to practice," *Journal of Environmental Planning and Management* 38(3) (1995): 371-91; R.F. Noss and A.Y. Cooperrider, *Saving Nature's Legacy: Protecting and Restoring Biodiversity* (Washington, DC: Island Press 1994); and T.R. Stanley, Jr., "Ecosystem management and the arrogance of humanism," *Conservation Biology* 9(2) (1995): 255-62.

2 H. Hammond, "Forest Practices: Putting Wholistic Forest Use into Practice," in *Touch Wood: BC Forests at the Crossroads*, ed. K. Drushka, B. Nixon, and R. Travers, 96-136 (Madiera Park, BC: Harbour Publishing 1993); C. Maser, *The Redesigned Forest* (San Pedro, CA: R.E. Miles 1988); and Scientific Panel for Sustainable Forest Practices in Clayoquot Sound, *Report 5: Sustainable Ecosystem Management in Clayoquot Sound: Planning and Practices* (Victoria: Cortex Consultants Inc. 1995).

3 Herb Hammond has estimated the reduction in one stand to be in the region of 75 percent of the existing AAC, a figure which surprised even him. See M. M'Gonigle, K. Stratford, and F. Gale, eds., *The Business of Good Forestry: A Symposium of Practitioners*, Report Series 96-1 (Victoria: Eco-Research Chair of Environmental Law and Policy, University of Victoria 1996).

4 Ibid.

5 R.M. M'Gonigle and B. Parfitt, *Forestopia: A Practical Guide to the New Forest Economy* (Madeira Park: Harbour Publishing 1994).

6 There has been a tendency in the literature to divide policy instruments into two categories: government regulation and economic incentives. This dichotomy reflects a political split among policymakers, with economists favouring the use of (often theoretical) market mechanisms, and public policy analysts favouring the use of tried-and-true government regulation. Both groups do, however, advocate the use of institutional reforms to achieve policy goals, although the structure and operation of such reforms varies depending on the advocate's preferences for regulatory or market instruments. Thus, many economists advocate privatization of land ownership as a solution to resource allocation problems, while a number of policy analysts favour either the continuation of the status quo or the devolution of responsibility for forest management to the level of the local community. All three advocate the use of institutional mechanisms to achieve policy goals, albeit with rather different consequences.

7 For an account of the debate in the legal literature, see H. Latin, "Environmental deregulation and consumer decisionmaking under uncertainty," *Harvard Environmental Law Review* 6(1) (1982): 187-23; H. Latin, "Ideal versus real regulatory efficiency: Implementation of uniform standards and 'fine-tuning' regulatory reforms," *Stanford Law Review* 37 (1985): 1267-1332; and B. Ackerman and R. Steward, "Comment: Reforming environmental law" *Stanford Law Review* 37 (1985): 1333-65.

8 For full details of BC's existing tenure system and proposals for reform, see C. Burda, D. Curran, F. Gale, and M. M'Gonigle, *Forests in Trust: Reforming British Columbia's Forest Tenure, System for Ecosystem and Community Health*, Report R97-2 (Victoria: Eco-Research Chair of Environmental Law and Policy 1997).

9 For further details on the global forest industry, see P. Marchak, *Logging the Globe* (Montreal: McGill-Queen's University Press 1995).

10 Comparative studies of prices of logs sold on the open market with stumpage rates invariably reveal that stumpage prices constitute only a fraction of the market value of timber. See Forest Resources Commission, *The Future of Our Forests* (Victoria: Forest Resources Commission 1991), 65-68; and McGonigle and Parfitt, supra, note 5 at 75.

11 Thus, one consultant concluded that VLM's "transactions consist primarily of trades between the seven major tenure holders who hold 90 percent of the AAC in tree farm licences and forest licences in the Vancouver Forest Region" while a special committee of the BC legislature concluded that it had "features inconsistent with a freely competitively driven marketplace." Quoted in M'Gonigle and Parfitt, supra, note 5 at 76-77.

12 For further details on the functioning of the Lumby Log Market, see the Price Waterhouse report *Special Project on Harvesting Timber and Selling Logs: Vernon Forest District 1993/94* (Victoria: Ministry of Forests 1995).

13   K. Johnson, *Building Forest Wealth: Incentives for Biodiversity, Landowner Profitability, and Value Added Manufacturing* (Seattle, Washington: Northwest Policy Center, University of Washington 1995).

14   See F. Gale and C. Burda, supra, Ch.12.

15   As George White of Sainsbury's commented: "Certification is already a market opportunity, but ultimately it may become the means of securing continued access to all markets." Speech by George White quoted in M'Gonigle, Stratford, and Gale, supra, note 3 at 18.

16   GATT, *Analytical Index: Guide to GATT Law and Practice* (Geneva: General Agreement on Tariffs and Trade/World Trade Organization 1994), 134.

17   *Agreement on Subsidies and Countervailing Measures* (Geneva, Switzerland: World Trade Organization, 1994), Annex I, note 58, 313.

18   P. Demaret and R. Stewardson, "Border tax adjustments under GATT and EC Law and general implications for environmental taxes," *Journal of World Trade* 28(4) (1994): 5-65.

19   UNCTAD, *Newly Emerging Environmental Policies with a Possible Trade Impact: A Preliminary Discussion* (Geneva: United Nations Conference on Trade and Development 1995), 13.

20   *Agreement on Subsidies and Countervailing Measures*, in *Law and Practice of the WTO*, ed. J. Dennin (New York: Oceana Publications 1995), 271-2.

21   A. Zampetti, "The Uruguay Round Agreement on Subsidies: A forward-looking assessment," *Journal of World Trade* 29(6) (1995): 11-12.

22   Ibid., 14.

23   Ibid., 13.

24   Ibid., 12.

25   Supra, note 20 at 283-4.

26   Supra, note 20 at 276.

27   Supra, note 20 at 276-9.

28   S. Arndt, W. Kaempfer and T. Willett, "Subsidy and Countervailing Duty Issues in the Context of North American Economic Intergration," in *Negotiating and Implementing a North American Free Trade Agreement,* ed. L. Waverman (Toronto: The Fraser Institute/Centre for International Studies 1992), 114.

29   *North American Free Trade Agreement* (Ottawa: Ministry of Supply and Services 1993), Article 1904(1).

30   Ibid., Article 1901 and 1902.

31   See Scarfe, infra Ch. 9. For a good review of all three challenges to Canada's softwood lumber pricing practices, see S. Sinclair, "The Use of Trade Remedy Laws: Case Study of the Softwood Lumber Disputes," in *The Environmental Implications of Trade Agreements* (Toronto: Queen's Printer 1993), 195-221.

32   G. de Graaf and M. King, "Towards a more global government procurement market: The expansion of the GATT government procurement agreement in the context of the Uruguay Round," *International Lawyer* 29(2) (1995): 435-52.

33   *Agreement on Government Procurement, 1994,* in *International Trade Regulation,* ed. E. McGovern (Exeter: Globefield Press 1995), 36.19.

34   M.H. Schuman, "GATTzilla v. communities," *Cornell International Law Journal* 27(3) (1994): 545.

35   Supra, note 33 at 36.19.

36   Ibid.

37   C. Nicholls, "Government procurement after the Uruguay Round," *C.D. Howe Institute Commentary* 74 (1995): 3-5.

38   Supra, note 33 at 36.18-.20.

39   Supra, note 29 at Part Four, Chapter Ten, Article 1001.1(a), 10-1.

40   Supra, note 29 at Part Four, Chapter Ten, Article 1024, 10-26.

41   Ibid.

42   Supra, note 29 at Part Four, Chapter Ten, Annex 1001.1b-1, 10-40, 10-41.

43   The question arises about the extent to which BC could exercise discrimination in favour of BC products over products produced elsewhere in Canada. The answer to this question is found in a consideration of the Agreement on Internal Trade (AIT), which has been signed between the provincial and federal governments and which is under constant renegotiation.

44 *Agreement on Technical Barriers to Trade 1994,* in *International Trade Regulation,* ed. E. McGovern (Exeter: Globefield Press 1995), 18-1 to 18-18-22.

45 Ibid.

46 Supra, note 44 at Article 4.

47 Supra, note 44 at Annex 3, paras. D and E.

48 C. Tietje, "Voluntary eco-labeling programmes and questions of state responsibility in the WTO/GATT legal system," *Journal of World Trade* 29(5) (1995): 123-4.

49 The voluntary nature of the scheme distinguishes it from the Austrian tropical timber eco-label. See B. Chase, "Tropical forests and trade policy: The legality of unilateral attempts to promote sustainable development under the GATT," *Hastings International and Comparative Law Review* 17(2) (1994): 349-88.

50 For details of the CSA scheme, see C. Upton and S. Bass, *The Forest Certification Handbook* (Delray Beach, FL: St. Lucie Press 1996); C. Elliott and A. Hackman, *Current Issues in Forest Certification in Canada* (Toronto: WWF-Canada 1996); and F. Gale and C. Burda, supra Ch.12.

51 Supra, note 29 at 9-14 to 9-15.

52 S.S. Cohen, "Discussion dossier on competitiveness: Section 1: Krugman revisited," papers presented at the Trinational Institute on Innovation, Competitiveness and Sustainability, Whistler, British Columbia (Vancouver: SFU 1994), 5. For other definitions of the concept of "competitiveness," see F. Ezeala-Harrison, "Canada's global competitiveness challenge: Trade performance versus total factor productivity measures," *American Journal of Economics and Sociology* 54(1) (1995): 57-78; and N.J. Adler, *Globalization, Government and Competitiveness* (Ottawa: Canadian Centre for Management Development, Ministry of Supply and Services 1994).

53 Adler, supra, note 52 at 5.

54 G. Elliot, "Forest products trade: The challenge of environmental trade barriers," *Forestry Chronicle* 71(1) (1995): 41-44.

55 Ibid., 41.

56 For details of the growing challenges confronting the BC and Canadian forest industries, see Industry Canada, Forest Industries and Building Products Branch, *Sector Competitiveness Frameworks: Forest Products: Part 1-Overview and Prospects* (Ottawa: Ministry of Supply and Services 1996); and C. Burda and F. Gale, *Trading in the Future: An Examination of British Columbia's Commodity Export Strategy in Forest Products,* Report D-96-8 (Victoria: Eco-Research Chair of Environmental Law and Policy, University of Victoria 1996).

Part 5:
Conclusion

15
Conclusion
Chris Tollefson

This book bridges two distinct, yet related, public policy debates. The first concerns the policy implications of sustainable forestry. As we have documented, there is a broad consensus that industrial forestry, as we have come to know it in North America, has and must continue to change – in short, that our pattern of resource exploitation to date has been unsustainable. What it would take, in specific policy terms, to support a transition to "sustainable forestry," recognizing the amorphous and contested nature of this concept, is a central concern of many of the contributions to *The Wealth of Forests*.

In contemplating policy alternatives for achieving sustainable forestry, the book intersects with another highly controversial debate: the relative merits of market as opposed to regulatory instruments to promote public policy. Frequently, these two debates are strategically linked by business and other interests that claim sustainable forestry can be promoted through the market, without the need for prescriptive regulation.

Many politicians, indeed even regulators, find the concept of market-based solutions appealing. A case in point is provided by recent experience in British Columbia, the province that has legislatively gone further than any other Canadian jurisdiction to promote sustainable forestry. But less than two years after bringing into force Canada's first legally binding Forest Practices Code, BC's NDP government – now led by Glen Clark – has decided to backtrack, introducing significant amendments that streamline Code procedures and require the weighing of economic factors in environmental protection decision-making.

The Clark government's desire to harness market forces for what it has characterized as "sustainable forestry" is even more evident in the high profile Jobs and Timber Accord (JTA) announced in summer 1997.[1] In entering into the JTA with the province's largest forest companies, the government has sought a *modus vivendi* with forest sector employers and unions that claim the industry is suffering as a result of recent stumpage increases, cut reductions and delays, and forest practices requirements. The JTA aims to promote forest sector employment and industry stability through a variety of incentive measures, many of which involve increasing industry access to timber supply. For example, the JTA promises to bring actual cut levels up to the allowable cut levels set by the Chief Forester by, among other things, expediting the harvest permitting process. Companies that meet JTA job targets will also be able to increase their AAC by qualifying for a waiver of the until now legislatively mandated 5 percent cut "clawback" (which normally applies on licence transfer or expiry).[2] The JTA also seeks to create new jobs by increasing the proportion of FRBC funding available to forest companies to cover new operational planning, inventory, and road construction costs related to Forest Practices Code compliance.

But while the government has championed the JTA as another important step towards ensuring forest sustainability, environmentalists have condemned it as sustaining only an unsustainable status quo.[3] In entering into the JTA, they contend, the government has ignored the need for an immediate, significant cut reduction to mitigate the "falldown" effect expected to occur when the province's supply of old growth is depleted. Moreover, they argue, the JTA squanders the government's last opportunity to manage the transition to a new sustainable forest economy – founded on a restructured community-based tenure system and dramatically expanded value-added production – by consolidating the existing volume-based, industrial tenure system.

In considering the future of Canadian forest policy in the context of the debate over policy instruments, several important themes deserve to be revisited in this concluding chapter. A first theme concerns the limits of market-oriented methods for promoting sustainable forestry. While varying in their enthusiasm for market instruments, none of the authors in this volume would suggest that the market, on its own, offers a pathway to sustainable forestry. Inevitably, a future sustainable forestry would, it would seem, involve a mix of regulatory and market instruments. As Kuttner has recently expressed it, "Everything cannot be regulated. But neither can everything be deregulated."[4] This perspective is captured in Pearse's opening essay which contends that while state regulation will always be necessary to protect particularly vulnerable features of forest ecosystems, other forest resources (including fish, wildlife, water, and recreational amenities) would be more effectively conserved if they were made subject to a pricing, as opposed to a more traditional regulatory, regime.

This argument leads into a second theme: the potential of various policy instruments – both market and regulatory – to promote a sustainable forestry strategy. While sceptical about the potential for market-based reforms to promote this goal, many contributors to this book are, for the most part, equally dubious of the likelihood that sustainable forestry can be achieved through regulatory tools currently in use. What is needed, many contributors contend, are new approaches, that depend neither on command and control edicts nor on individuated property rights, as exemplified in proposals to create new community- or group-based forms of forest tenure, eco-certification of forest products, and zoning initiatives, among other options.

In some important respects, however, whether we can successfully move from unsustainable to sustainable forestry depends less on choosing the most appropriate policy instrument(s) than it does on other legal and political factors. The third major theme explored in this book, therefore, is how various legal and political arrangements – including international trade agreements, the law of "takings," and, indeed, even the Western liberal democratic system of governance – challenge, and potentially constrain, movement towards sustainable forestry.

The Failings of Governments and of Markets

In the introduction to this book, I contend that, in several key respects, contemporary Canadian forest policy has failed.[5] The dimensions of this failure are, I argue, threefold. Perhaps the most prominent dimension has been the failure of government policy to protect the multitude of non-timber values of our forests. But it is equally apparent that governments have failed to promote efficient use, conservation, and regeneration of the timber resource, and to ensure that the resource is utilized in a way that maximizes forest sector employment and community stability.

This book has grappled, from differing disciplinary and philosophical perspectives, with the challenge of identifying sustainable policy alternatives that would correct the "government failures" in these three areas. In part because these failures have come to be so closely associated with government, as notional protector of the public interest in our Crown forests, some have argued that the time has come for governments to step aside in favour of markets. Stronger, longer, and more clearly defined property rights in timber and non-timber values would, it is contended, greatly enhance incentives for prudent use and conservation of timber and non-timber forest resources. In a similar vein, it has been argued that reducing or eliminating the regulation of timber harvesting and increasing private access to timber supplies would have an immediate positive effect on local economies.

Of course, the perception that governments have failed where markets can (and should be allowed to) succeed is by no means unique to the realm of forest policy. For over a generation, political discourse in North America has been characterized by a profound scepticism about the wisdom and efficacy of state regulation and an abiding faith in, even a reverence for, markets and market instruments. This simultaneous denigration of government and celebration of markets are closely connected phenomena.[6]

Despite their differing disciplinary and philosophical perspectives, among the contributors to this book there is little optimism about the potential for markets alone to provide a vehicle for achieving sustainable forestry. To be effective, markets require price signals. However, many of the benefits associated with functioning, sustainable forest ecosystems – including storing and recycling essential nutrients, maintaining hydrological cycles, supporting biodiversity, providing fish and wildlife habitat, sequestering carbon, and so on – are either not priced, or are seriously underpriced by the market. Scarfe points out that two important implications flow from the failure of the market to price these and other non-timber forest values effectively.[7] The first is that the private cost of timber harvesting will be less than the social cost; the second is that the private benefit of responsible forest stewardship (including reforestation and silviculture) will be less than the social benefit of such practices.[8]

In addition, markets tend to create a variety of socially undesirable distributional outcomes in the forestry context. As Haley and Luckert point out, insofar as sustainable forestry implies a concern for community stability, job creation, and the promotion of value-added processing, markets – given their dependence on price signals – will frequently yield a less than optimal social result.[9] Another distributional failing of markets is inter-temporal in nature. Future human and non-human species that benefit from various non-timber values associated with fully functioning forest ecosystems cannot participate in markets. As such, the forest ecosystem functions and attributes that benefit these future interests go largely unvalued in the market.

Markets are also prone to delivering unsustainable outcomes where consumers lack the information necessary to make informed consumption decisions. In environmental markets, this information is particularly scarce as a result of the inherent complexity, differentiation, interdependence, local variability, and non-linearity of human impacts on the environment. As Cook puts it, "Whatever their preferences, typical consumers simply do not know what the environmental and social consequences of their discrete market decisions might be."[10]

To the extent that these failings arise because the market does not price, or vastly underprices, timber and non-timber values, in theory the potential exists for governments to correct the problem by intervening to establish appropriate prices for these values. In the United States, the federal government has adopted this market-based approach to control air pollution through a complex, transferable emission-permitting system. Under this system, companies purchase quantified rights to pollute that can be used, sold, or banked under the auspices of a regulated air pollution credits market. The potential for deploying such a system as a vehicle for promoting sustainable forestry, however, seems limited. As Stanbury and Vertinsky note, there is little likelihood that such an approach could be adapted to the forestry context. Even if there were political support for such an approach, they believe that a variety of technical problems, related to the complex, interactive environmental impacts of tree harvesting, would fatally thwart such an initiative.[11] Many of these problems are a function of the fact, as they put it, that "promoting sustainable forestry" is a qualitatively different policy problem than "reducing pollutant X below level Y." Pollution reduction entails reducing a very specific, easily targeted negative externality. In contrast, achieving sustainable forestry is, as they put it, a policy problem "framed in the positive – i.e., we wish to induce behaviour which will contribute to the goal. But many behaviours appear to be consistent with the goal. Behaviours adverse to the goal cannot simply be defined as the opposite of those consistent with it."[12] Presumably, therefore, this is why the Forest Practices Code requires certain kinds of behaviours *and* prohibits others."[12]

Finally, it is important to recognize that, even in the absence of market failure, there remains a strong rationale for structured public involvement in forest decision-making. While both ecological and economic analyses offer useful insights into the various costs and benefits associated with alternative approaches to "sustainable forestry," ultimately the process of balancing these costs and benefits involves calculations that cannot and should not be left to the market. In short, decisions about the future of our forest resources are necessarily political in nature. As Rayner emphasizes, in the context of priority-use zoning, the challenge that thus presents itself is to design democratic decision-making processes to ensure the decisions flowing from political processes reflect social preferences while giving due weight to credible scientific and technical input.[13]

Towards Sustainable Forestry: The Path Ahead
If we take as our point of departure a broad notion of sustainable forestry – a vision of forestry that sustains the timber *and* non-timber values of our forests as well as the human interests that depend upon these values – what lessons emerge from the contributions to this book?

While markets and market instruments do not appear to offer ready answers to this question, it appears equally apparent that, if the challenge of sustainability is to be met, existing approaches to forest policy must be dramatically overhauled.

In British Columbia, a number of legal and policy initiatives that have been undertaken – including land zonation, forest practices legislation, the Forest Renewal program, and, most recently, the Jobs and Timber Accord – reflect a perception that prevailing approaches must change. Yet, as many contributors vigorously argue, it is important to neither overstate the magnitude of the change that has occurred nor underestimate the magnitude of the challenge that lies ahead.

Perhaps the most critical obstacle to sustainability is forest tenure reform. The forest tenure system in most Canadian provinces has remained virtually unchanged for most of this century. Designed to promote rapid expansion and development of timber resources, these tenures vest virtually exclusive harvesting rights to the forest resource in a small handful of large, vertically and laterally integrated forest companies. To meet the challenge of sustainability, current tenure arrangements, described by Haley and Luckert as "anachronistic," must give way to new forms of tenure that, as they put it "serve today's varied and frequently conflicting public objectives."[14] To this end, both Haley and Luckert, and M'Gonigle, advocate experimentation with a variety of new forms of tenure that vest individuals, communities, and First Nations with rights to use, benefit from, and steward the timber resource.[15] Despite the fundamental importance of tenure, both as an obstacle to *and* a vehicle for sustainable forestry, Canadian governments have stubbornly resisted the idea of tenure reform. This situation is true even in reform-oriented British Columbia, where – despite early indications that tenure reform might be on the agenda of a second-term NDP government – the Clark government's Jobs and Timber Accord portends consolidation, as opposed to dismantling, of the industrial forestry tenure system.

Nor does the JTA augur well for reforming the process by which the province's annual allowable cut is determined. Dellert pronounces the current "sustained yield" system of cut regulation – in place in British Columbia, and many other Canadian jurisdictions, for over fifty years – to have been an abject regulatory failure. In her view, the sustained yield approach has yielded neither a sustainable timber supply nor sustainable local forest economies.[16] This failure, she concludes, results from a technocratic model of cut determination that has erroneously equated timber supply with social and economic sustainability, and has resisted rather than adapted to fundamental changes in social values and scientific knowledge.[17] However, despite uniformly dire predictions about the economic and ecological consequences of maintaining current cut levels, little progress towards the goal of reforming cut regulation is being made. Indeed, the JTA, which contemplates *increasing* the rate of cut to stimulate job creation, suggests that current approach is firmly entrenched.

In other areas of forest policy arguably greater progress towards sustainability has been made. While Cook defends British Columbia's embattled Forest Practices Code as a necessary corrective for the negative externalities associated with timber harvesting, she contends that its effectiveness, as a policy instrument, has been undermined by lax enforcement, inefficient administration, and its failure to reward good forest practices as opposed to simply punishing poor ones.[18] While Code monitoring and enforcement remains scandalously weak, the government has taken steps to expedite the harvesting approvals process as part of the JTA. As well, under legislation introduced in 1996, the government is promoting enhanced silviculture, above Code requirements, by offering companies that enter into so-called innovative forest practice agreements the incentive of additional cutting rights.[19] Meanwhile, in response to continuing industry pressure for so-called Code rollbacks, the government has watered down Code environmental protections as part of a package of changes introduced in 1997.[20]

Cautious approval is also given to British Columbia's timber pricing policies, which since 1994 have financed the Forest Renewal Plan. Scarfe supports maintaining stumpage levels at their new higher levels not only to pre-empt American trade retaliation, but also, as he puts it, "to reflect the economic value of timber resources to society."[21] Scarfe also approves of stumpage charges being channelled to and administered by FRBC but only on the condition that the funds are dedicated to the long-neglected task of replenishing our forest inventories.[22] Recent events – including the BC government's well-publicized raid of "surplus" FRBC funds to help pay off its budget deficit, and the JTA's earmarking of FRBC monies to subsidize industry compliance with Code harvesting requirements – suggest that Scarfe's concerns in this regard are well founded.

Rayner offers an even more restrained assessment of the potential of priority-use zoning as a means of achieving sustainability. He contends that underlying the growing popularity of zoning as a "sustainable solution" are two diametrically opposed visions of what zoning is meant to achieve: a conservationist ecological conception and a productionist economic conception.[23] Ultimately, he argues, zoning will not satisfy proponents of either of these visions, given the inevitability of political compromises in actual zoning decisions. As a means of supporting a transition to sustainable forestry, at best, in his view, "zoning has a modest role to play in making tenure reform more attractive and promoting economically rational investment in forest resources. It can also help in determining priority areas for protection."[24]

Another policy instrument that is frequently touted as being capable of supporting a transition to forest sustainability is eco-certification. In their examination of forest eco-certification, however, Gale and Burda sound a note of caution.[25] In their view, certification has significant potential to promote sustainable forestry through the marketplace by enhancing the ability of consumers to make informed purchasing decisions. Whether or not this potential is realized, however, depends on the design and operational characteristics of the certification scheme that ultimately becomes the prevailing standard. According to Gale and Burda, the Canadian Standards Association (CSA) certification scheme currently favoured by large segments of the North American forest industry will, at best, only have a marginal positive impact on forest practices. They contend that the CSA's main competitor scheme, run by the Forest Stewardship Council (FSC), offers significantly more promise as a vehicle for promoting sustainable forestry. In their view, however, it is highly uncertain whether FSC certification has any realistic prospect for widespread adoption, given the current nature and breadth of support for CSA.

Elsewhere in this volume, as part of his "eco-conversion" blueprint for the BC forest industry, Gale identifies two further non-regulatory, "institutional" reforms he regards as necessary steps towards the elusive goal of sustainable forestry.[26] A pressing priority is to reform the cumbersome, centralized forest management administrations currently in place in most Canadian jurisdictions. The need to reform forest bureaucracies has, in BC, been recognized for some time. To date, however, the main thrust of reform has been decentralization; little progress has been made towards the goal of improving administrative productivity.[27] According to Gale, the need for (and the problems associated with) our current highly centralized control structures will largely disappear as the centrepiece of his eco-conversion strategy – community-based forest tenures – come on stream under watchful locally elected "community resource boards." Another "urgently needed" reform, according to Gale, concerns the profession of forestry itself. Only in recent years has the protection of ecosystems been a recognized component of the curricula in Canadian forestry schools. In Gale's view, forestry schools can and should be doing much more than they are to prepare their graduates for the rapidly evolving demands of "New Forestry."

Barriers to Sustainable Forestry

As Gale points out, it would be a mistake to conclude that "Sustainable forest management is not occurring in BC for want of policy."[28] On the contrary, as virtually all of the contributions to this collection confirm, there are a multitude of reforms capable of being implemented by a variety of market-based, regulatory, and institutional "instruments" that could bring us closer to the goal of a forestry that sustains humans, trees, and other forest resources.

In the face of virtually unanimous agreement that contemporary industrial forestry is unsustainable and must change, what explains the reluctance of Canadian governments to meet this challenge by embracing sustainable policy alternatives?

Some have argued that our governments have no choice but to carry on business as usual; that legal constraints foreclose them from doing otherwise. In the domestic legal arena, a principal constraint is said to be the obligation of government to compensate private parties when it substantially interferes with (in legal terminology, "takes") private property interests. According to this line of argument, by undertaking tenure reform (and even, arguably, other less far-reaching sustainability-oriented policies), government would be required to pay multi-million dollar "takings" compensation to existing major industrial tenure-holders. Cohen and Radnoff are sceptical that, under existing law, environmental regulation aimed at promoting sustainability would indeed legally trigger a compensation requirement.[29] In any event, they would support, on normative and efficiency grounds, a presumption against compensation for regulatory takings in the environmental context, a presumption that, in the interests of business certainty, could be legislated at the provincial level.[30]

International trade obligations under GATT/WTO and NAFTA are also often regarded as limiting the potential for implementing sustainable forestry policies at the provincial level. That such policies may be challenged on trade grounds, Gale readily acknowledges. But, in his view, the prospect of defending these policies in international dispute resolution arenas should not deter us from adopting them in the first place. As long as the challenged policies are "based unambiguously on genuinely sustainable production, and do not lapse into protectionism for protectionism's sake," he remains optimistic that such challenges would fail.[31]

Many readers will likely concur with these authors that the domestic and international legal constraints on adopting a sustainable forestry strategy have been overstated. At the same time, some will undoubtedly contend that the legal status of such reforms is far less salient than the manner in which they are politically perceived. In short, the mere perception that sustainability-related reforms are in potential conflict with domestic or international norms or obligations may be more than enough to discourage governments from embarking on the path of sustainable forestry reform, particularly when confronted with the prospect of dealing with the social upheavals, corporate resistance, and disinvestment implications that might flow from such a strategy.

As recent experience in BC culminating in the JTA amply demonstrates, business and labour resistance to sustainable forestry reform has and will continue to be a potent political constraint militating against the adoption of eco-transition policies of this kind. Making this transition even more politically problematic, argues M'Gonigle, is the very structure of the modern liberal democratic state.[32] Centralized, hierarchical, and sustained by distant resources that flow out of local environments and communities, our state institutions are simply incapable of delivering urgently needed ecosystem-based solutions to the challenges faced by the forestry sector.[33] What is needed, he posits, are local decision-making institutions that facilitate the *recirculation* – as opposed to the *linear extraction* – of natural and human capital. To this one might add, as Meidinger has argued elsewhere, that these local decision-making institutions must be designed to identify and respond to surprises in ecological and social processes, build ongoing social dialogue around ecological issues, and adapt to changes in knowledge and social values.[34] In short, before we can hope for ecosystem-based solutions to our forestry crisis, we must first have ecosystem-based governance. Thus we should not, M'Gonigle warns, be lulled into believing that achieving sustainable forestry is a matter of selecting the right mix of state and market instruments. Fundamentally, he contends, the challenge is to transform the character of industrial forestry and, with it, the state itself."[35]

Never before have the demands on our public forest resources been so great. These demands not only serve to test the capacity of our public decision-making and dispute resolution processes, they also force us to re-evaluate the policy instruments we rely on to promote and protect the public interest in our forests. It is unlikely that the ongoing debate in the political realm over the meaning and implications of sustainable forestry, triggered by these demands, will soon subside. Given this political fluidity, as well as the scientific and social uncertainties that characterize forest policy-making, a key challenge is to design public institutions and policy instruments that can adapt to, indeed even flourish in, an environment of change. The nature of the governance and related reforms necessary to meet this challenge are only now being explored. This collection is a contribution to that process of exploration, offered in the hope that we can come closer to realizing the true wealth of our forests.

Notes

1 Office of the Premier, "*Jobs and Timber Accord* to create thousands of new jobs," news release, Prince George, BC (19 June 1997).
2 Companies will also receive priority consideration as candidates for AAC increases contemplated under "innovative forest practice" provisions of the Forest Act made possible by Bill 7, an amendment to the Forest Act enacted in 1996 (see *Forest Act*, R.S.B.C. 1996, c. 157, s. 59.1, as amended).
3 See, for example, the critique of the JTA prepared by M. Haddock (25 June 1997) of BC Policy Watch; on file with the author.
4 Kuttner, *Everything for Sale: The Virtues and Limits and Markets* (New York: Knopf 1997), 329.
5 See the editor's introduction at 7-8.
6 Kuttner, supra, note 4 at 361.
7 B. Scarfe, supra, Ch.8 at 198.
8 Ibid.
9 D. Haley and M. Luckert, supra, Ch.6 at 133-4.
10 T.L. Cook, supra, Ch.9 at 218.
11 W. Stanbury and I. Vertinsky, supra, Ch.3 at 67-8.
12 Ibid.
13 J. Rayner, supra, Ch.10 at 251-2.
14 Haley and Luckert, supra, note 9 at 136.
15 Ibid. See also M'Gonigle, supra, Ch.7.
16 L. Dellert, supra, Ch.11 at 272-4.
17 Ibid.
18 Cook, supra, note 10 at 213-7.
19 *Bill 7*, supra, note 2.
20 Reference here to Cook's postscript, supra, Ch.9.
21 Scarfe, supra note 7 at 201.
22 Ibid.
23 J. Rayner, supra, note 13 at 233-4.
24 Ibid.
25 F. Gale and C. Burda, supra, Ch.12 at 285 et seq.
26 F. Gale, supra, Ch.14 at 345-6.
27 Improving the efficiency with which the Ministry of Forests processes harvesting applications is one of the goals of the JTA (supra, note 1).
28 F. Gale, supra, note 26 at 342.
29 D. Cohen and B. Radnoff, supra, Ch.13.
30 During the early years of the Harcourt administration, the BC government commissioned an inquiry into the issue of compensation for the taking of resource interests. The recommendations of the Schwindt Commission, which included establishing a compensation policy and dispute resolution procedures, were never acted on. See the *Report of the Commission of Inquiry into Compensation for the Taking of Resource Interests* (Vancouver: Queen's Printer 1992).

31 Gale, supra, Ch.14 at 366.
32 M. M'Gonigle, supra, Ch.5 at 108 et seq.
33 Ibid., 109.
34 See E. Meidinger, "Organizational and Legal Challenges to Ecosystem Management," in *Creating a Forestry for the 21st Century: The Science of Ecosystem Management*, ed. K. Kohm and J. Franklin (Washington, DC: Island Press 1997), 361. See also discussion in Dellert, supra, Ch. 11 at 273-4.
35 M'Gonigle, supra, Ch.5 at 113.

Contributors

Cheri Burda is the Senior Researcher with the Eco-Research Chair of Environmental Law and Policy at the University of Victoria, and the principal author of *Forests in Trust: Reforming British Columbia's Forest Tenure System for Ecosystem and Community Health* (1997). She writes, speaks, and organizes extensively on community forestry, forest tenure reform, and alternatives to industrial forestry.

David Cohen is a Professor in the Faculty of Law, University of Victoria, where he has also been Dean since 1994. He teaches and researches in the areas of law and regulatory policy, commercial law and planning, contract law, law and economics, governmental liability, product safety regulation, dispute resolution, and environmental policy and regulation. He is a member of the Strategic Steering Committee on the Environment of the Canadian Standards Association. He was a member of the federal Environmental Choice Advisory Board from 1988-93, and is currently on the board of directors of the West Coast Environmental Law Association. In addition to being the author of numerous articles, he is also a co-editor of *Managing Natural Resources in British Columbia* (1995).

Tracey L. Cook received her LL.B. from the University of Victoria in 1996. She is interested in environmental law and policy reform.

Lois Dellert received a B.Sc. degree in forestry from the University of Alberta in 1979. She was employed by the BC Forest Service and spent ten years conducting and managing the determination of the province's sustainable supply of timber. In 1992, she was appointed BC's first Deputy Chief Forester. In 1994, she completed her master's degree in Environmental Studies at York University, where she explored the complex pattern of forces influencing sustained yield forest policy. She is currently studying sculpture and landscape design in Toronto, where she communicates her passion for the forest through environmental art.

Rod Dobell is Professor of Public Policy at the University of Victoria. Following faculty appointments at Harvard (Economics) and Toronto (Political Economy), he served as Deputy Secretary (Planning) in the Treasury Board Secretariat of the Government of Canada (1973-6), President of the Institute for Research on Public Policy (1984-91), and Director of the School of Public

Administration, University of Victoria (1978-84), where he was later appointed to the Francis G. Winspear Chair for Research in Public Policy (1991-97).

Fred Gale is a Visiting Scholar in the Department of Political Science, University of Victoria. He is the author of *The Tropical Timber Trade Regime* (1998), which details the struggles of environmental civil society organizations to influence the policies and practices of the International Tropical Timber Organization. Dr. Gale has also published on such themes as power in ecological economics, sustainable trade, and international regime theory. He is currently working on an edited volume that develops an ecological political economy approach to the interrelations of nature, production, and power.

David Haley is Professor of Forest Economics and Forest Policy in the Department of Forest Resources Management, University of British Columbia, and an associate of UBC's Forest Economics and Policy Analysis Research Unit (FEPA). For many years, Professor Haley's main interests have been the institutional and public policy environments in which forestry is practised, with particular emphasis on property rights and fiscal arrangements. He has a long-term interest in smaller scale forestry and has been an outspoken advocate for community forestry in British Columbia. In addition to his academic pursuits, Professor Haley has worked extensively as a consultant to governments and the private sector, both domestically and internationally.

Martin K. Luckert is a Professor in the Department of Rural Economy at the University of Alberta. Professor Luckert teaches and researches in forest and natural resource economics and policy in Canadian and international contexts. He has published extensively on the economic implications of forest tenures.

Michael M'Gonigle is Professor of Law and currently holds the Eco-Research Chair of Environmental Law and Policy in the Faculty of Law and School of Environmental Studies at the University of Victoria. A lawyer, political economist, and long-time environmental activist, he is co-author of the award-winning *Stein: The Way of the River* (1989), *Forestopia: A Practical Guide to the New Forest Economy* (1994), and *Forests in Trust: Reforming British Columbia's Forest Tenure System for Ecosystem and Community Health* (1997). He has participated extensively in the wilderness/forestry debate in British Columbia, and is a co-founder of Greenpeace International, a founding director of the Sierra Legal Defence Fund, and a former Chair of Greenpeace Canada.

Peter H. Pearse is a specialist in natural resource management and policy, having spent much of his career as Professor of Economics and Forestry at the University of British Columbia. He has conducted two Royal Commissions of Inquiry on resources policy, and has chaired two other public inquiries on Canada's natural resources, one on water and one on freshwater fisheries. He has been a frequent advisor to foreign governments and international organizations on natural resources policy and has published numerous articles, books, and reports dealing mainly with government policy of forests, fisheries, and water resources. Dr. Pearse has received a number of academic awards and professional distinctions, including the Forestry Achievement Award and the Distinguished Forester Award. In 1988, he was appointed a member of the

Order of Canada. He currently is the president of the private investment and consulting firm Pearse Ventures Ltd. and is also Professor Emeritus at UBC.

Brian Radnoff graduated in 1997 from the University of Victoria Faculty of Law after completing a B.Comm and an M.A. in economics from the University of Toronto. He is an associate at Davis & Co. in Vancouver.

Jeremy Rayner teaches political science at Malaspina University-College, Nanaimo, BC, and is an adjunct Associate Professor in the Department of Political Science, University of Victoria. The author of many papers on forest policy, he is currently researching the impact of contested science on forest policy and is co-authoring a manuscript on efforts to promote biodiversity conservation in British Columbia and the Pacific Northwest states.

Brian L. Scarfe is president of BriMar Consultants Ltd. Dr. Scarfe has held academic and/or senior administrative positions at universities in all four western provinces, and has published widely in the areas of macroeconomics and natural resource economics. BriMar Consultants Ltd. has recently completed several major reports for various BC government departments. Reports for the BC Ministry of Forests include: *An Independent Review of Timber Royalty Rates in BC* (1995), *The Allocation of Softwood Lumber Export Quotas* (1996), *The BC Forests Products Industry: Current Status and Future Prospects* (1997), and *Forest Resource Rents and the BC Timber Pricing System* (1998).

W.T. Stanbury is UPS Foundation Professor of Regulation and Competition Policy in the Faculty of Commerce and Business Administration at the University of British Columbia. He has over 300 publications, and those that have focussed on forest policy have addressed such issues as the stumpage system, the impact of environmental groups on both firms and public policy, water pollution regulations applied to pulp and paper mills, and the wood fibre market in British Columbia. Professor Stanbury has also been a consultant to a number of BC forest companies.

Chris Tollefson is an Associate Professor in the Faculty of Law, University of Victoria. He has written extensively on a variety of environmental and resource law topics. As executive director of the Victoria-based Environmental Law Centre, he oversees Canada's only clinical program in public interest environmental law. He is a consultant to private, public, and NGO-sector organizations on environmental and resource issues, in Canada and abroad. Currently, he chairs the Sierra Legal Defence Fund. In 1997, he was appointed as a Canadian associate to the Rockefeller Foundation's LEAD (Leadership through Environment and Development) program.

Ilan Vertinsky is the director of the Forest Economics and Policy Analysis Research Unit (FEPA), Vinod Sood Professor of International Business Studies, and a Professor in the Institute of Resources and the Environment at the University of British Columbia. His many published papers on forest policy deal with a variety of aspects related to questions of sustainable forest management, market access, and trade-offs between timber and other values of the forest. He is a co-editor of *Forest Policy: International Comparisons* (forthcoming).

Index

Note: MUSY means multiple use-sustained yield; (t) denotes a policy.

AAC. *See* Allowable annual cut

Aboriginals. *See entries under* First Nations; Land claims, First Nations

Accelerated development thesis (with takings of property), 317, 337n112

Administrative monetary penalties (AMPs), 228n31

Adverse selection (in insurance), 313-4, 334n78

Advisory processes, 57-8

Agreement on Government Procurement (AGP), 357-9

Agreement on Subsidies and Counter-vailing Measures (ASCM), 353-6

Agreement on Technical Barriers to Trade (ATBT), 360-2

Allocation of resources, market failures, 123-4

Allowable annual cut (AAC): calculation factor, operability, 258-9; calculation factor, rotation age, 259-60; calculation factor, second-growth yield, 260-2; changing the limits, 261, 272-3; Chief Forester's role, 55-6, 236-7, 256-7, 261; economic values, influence of, 105, 257-8; failure to achieve sustainability, 255-6; impact of AAC reductions, 257; and integrated resource management guidelines, 271-2; limits too optimistic, 105, 256-8; scientific expertise v. economics and politics, 256-62; silviculture, 131-2, 260-1, 274nn4-5; in sustained yield policy, 55-6, 72n34, 105, 200, 263, 264-5

Anthropocentrism, 226n4

Association of BC Professional Foresters, 173

BandQ and eco-certification, 288

BC Forest Resources Commission (1991), 135, 144, 145

BC Forest Service, proposed management of forest lands, 144

B.C.M.A. v. *British Columbia* (1985), 304

Bill 47 (Forest Statutes Amendment Act, 1997), 225

Biodiversity and conservation biology, 241-4

British Columbia, as focus of book, 12

British Columbia Forest Renewal Act (1994), 132

British Columbia Jobs and Timber Accord (1997), 10, 134, 266, 373-4, 378

British Columbia v. *Tener* (1984), 302-3

Bureaucratic covenants, 92

Canadian Standards Association (CSA), 282, 283-4, 285, 379

Casamiro Resources Corp. v. *British Columbia* (Attorney-General) (1991), 303, 326n18

'Chain of custody' (in eco-certification), 287-90

Chief Forester: allowable annual cut, competing values, 256-7; Chief Forester's Standards of Forest Practices Code, 208; Forest Act and allowable annual cut, 55-6; integrated resource management and allowable annual cut, 236-7; reducing allowable annual cut, 261

'chain of custody' and consumer demand, 287-90; constraints by GATT/ WTO, 360-2; constraints by NAFTA, 362-4; in eco-conversion strategy, 347-8; ECSOs (environmental civil society organizations), 280-1; first-, second-, third-party schemes, 281-2; Forest Stewardship Council (*see* Forest Stewardship Council); future in Canada, 290-2, 379; green forestry, 280-1; industry attitude towards, 289, 290; life-cycle analysis (LCA), 280; management-systems process, 283-4, 285; market-based incentive, 223, 279-80, 287; objectives, 284-5; performance-based process, 283-4, 285, 290-1; pitfalls, 285-6; as step towards sustainable forestry, 170, 171-2, 379

Eco-conversion strategy: economic incentives, 346-8; GATT/WTO constraints on border tax adjustments, 349-52; GATT/ WTO constraints on eco-certification and labelling, 360-2; GATT/WTO constraints on eco-subsidies, 353-6; GATT/WTO constraints on government procurement, 357-9; institutional reforms, 345-6, 379-80; international competitiveness, 364-6; NAFTA constraints on border tax adjustments, 352-3; NAFTA constraints on eco-certification and labelling, 362-4; NAFTA constraints on eco-subsidies, 356-7; NAFTA constraints on government procurement, 359-60; overview, 344-5

Ecoforestry: goal and principles, 343; strategy (*see* Eco-conversion strategy)

Eco-labelling. *See* Eco-certification

Ecological centralism, 108-10, 113-4, 153

Ecology, political. *See* Political ecology

Economic institutions, definition, 21

Economic (market-based) instruments: to achieve social goals, 86, 96-7; advantages, 22, 50-1; compared with political ecology, 103-4; for a competitive forest industry, 10-1; consumer demand for non-timber values, 52-3; eco-certification (*see* Eco-certification); ecological context, 81-6; feedback mechanisms required, 88, 92; formal economy, 83; free market environmentalism, 86, 100n21, 205-6, 226n4; growing support for, 4, 19-20; limitations, 51, 52-3, 71n23, 374-5, 375-7; log markets, 170-2, 345-6; marketable pollution

permit regimes, 4; narrow and broad incentives, 51-2, 57; need for ethical framework, 89-90, 97; pricing (*see* Timber pricing policies); property rights (*see* Property rights); regulation v. incentives, 10, 19-20, 21-2, 28; 'sunshine and scrutiny' (reporting and audit), 79-80, 98, 99n3; taxes and subsidies (*see* Taxes and subsidies); tenures (*see* Forest tenures); types, 22, 27; voluntary compliance mechanisms (VCMs), 79, 90-3, 96, 101n32

Ecosphere: ecosystem and private ownership, 94-6; ethical framework needed for sustainability, 89-90, 97; externalities, 82-3, 91; 'full world,' 78; impact of humans on natural world, 81-3, 84; linking sustainability objectives to individual behaviour, 87-90; natural capital, 78, 83, 84, 94-6; need for discipline and self-restraint, 84-6, 95; role of forests, 43, 82; social capital and equity, 78

Ecosystem-based management. *See* Community-based forest management; Eco-conversion strategy; Stewardship; Sustainable forestry

Ecosystems: and conservation biology, 241-4; forest (*see* Ecosystems, forest); political ecology, 103-4, 108-10, 111-3

Ecosystems, forest: goal of sustainability, 3, 6, 272-3; part of 'full world,' 78, 80; stewardship, 78, 80-1. *See also* Eco-conversion strategy

ECSOs (environmental civil society organizations), 280-1. *See also* Forest Stewardship Council

Employment: Jobs and Timber Accord (1997), 10, 134, 266, 373-4, 378; and sustained yield policy, 269-70

Environmental Advantage consulting group, 170, 183n76

Environmental Choice Program (ECP), 280

Environmental civil society organizations (ECSOs), 280-1

Environmental Greenprint Program, 288

Environmental movement: conservation biology, 241-4; environmental awareness, 9, 236; and Forest Practices Code, 64-5; goals in policy making, 45(t); and integrated resource management, 238; and Jobs and Timber Accord 266-7, 373-4. *See also* Eco-certification; Eco-conversion strategy; Stewardship

Environmental regulation: acting as

Set in Stone by Val Speidel

Printed and bound in Canada by Friesens

Copy editor: Maureen Nicholson

Indexer: Pat Buchanan